ACS SYMPOSIUM SERIES **733**

# Polymer Membranes for Gas and Vapor Separation

## Chemistry and Materials Science

**B. D. Freeman,** EDITOR
*North Carolina State University*

**I. Pinnau,** EDITOR
*Membrane Technology and Research, Inc.*

American Chemical Society, Washington, DC

**Library of Congress Cataloging-in-Publication Data**

Polymer membranes for gas and vapor separation : chemistry and materials science / B. D. Freeman, editor, I. Pinnau, editor.

p. cm.—(ACS symposium series ; 733)

Developed from a symposium sponsored by the Division of Polymer Materials: Science and Engineering at the 214th National Meeting of the American Chemical Society, Las Vegas, Nev. Sept. 7–11, 1997.

Includes bibliographical references and index.

ISBN 0–8412–3605–4 (cloth)

1. Gases—Separation—Congresses. 2. Membranes (Technology)—Congresses. 3. Polymers—Congresses.

I. Freeman, B. D. (Benny D.) II. Pinnau, I. (Ingo) III. American Chemical Society. Division of Polymeric Materials: Science and Engineering. IV. American Chemical Society. Meeting (214th : 1997 : Las Vegas, Nev.). V. Series.

TP242.P655 1999
660′.2842—dc21

99–25293
CIP



The paper used in this publication meets the minimum requirements of American National Standard for Information Sciences—Permanence of Paper for Printed Library Materials, ANSI Z39.48–1984.



Distributed by Oxford University Press



PRINTED IN THE UNITED STATES OF AMERICA

# Advisory Board

**ACS Symposium Series**

# Foreword

THE ACS SYMPOSIUM SERIES was first published in 1974 to provide a mechanism for publishing symposia quickly in book form. The purpose of the series is to publish timely, comprehensive books developed from ACS sponsored symposia based on current scientific research. Occasionally, books are developed from symposia sponsored by other organizations when the topic is of keen interest to the chemistry audience.

Before agreeing to publish a book, the proposed table of contents is reviewed for appropriate and comprehensive coverage and for interest to the audience. Some papers may be excluded in order to better focus the book; others may be added to provide comprehensiveness. When appropriate, overview or introductory chapters are added. Drafts of chapters are peer-reviewed prior to final acceptance or rejection, and manuscripts are prepared in camera-ready format.

As a rule, only original research papers and original review papers are included in the volumes. Verbatim reproductions of previously published papers are not accepted.

ACS BOOKS DEPARTMENT

# Contents

## INDEXES

# Preface

Gas separation using polymer membranes has emerged as a rapidly growing, commercially viable alternative to traditional methods of gas separation such as adsorption, absorption, and cryogenic distillation. In the 1980s, the selective permeation properties of polymers were harnessed on an industrial scale for gas separations applications. Polymer membranes were commercialized to selectively remove $H_2$ from mixtures with larger molecules ($N_2$, CO, $CH_4$, etc.), $O_2$ from air, and $CO_2$ from natural gas. These membrane separations are now practiced at installations throughout the world. Continued research and development of polymeric membranes for these applications contribute to a firm foothold for this technology in the chemical, petrochemical, and gas processing industries. In the 1990s, vapor separation using membranes was commercialized on a large scale. For vapor separation, polymer characteristics are quite different from those used in the gas separations applications just described. Vapor separation polymers must selectively remove large organic vapors (vinyl chloride monomer, propylene, ethylene, gasoline, Freons, etc.) from mixtures with permanent gases ($N_2$, air) or selectively remove higher hydrocarbons (propane, butane, etc.) from mixtures with natural gas or hydrogen. For both gas and vapor separations, fundamental research in chemistry and materials science of polymer membranes contributes to the commercial success of this technology by providing polymers with competitive separation characteristics and stability in industrial processing environments. This book provides a snapshot of state-of-the-art research in this field from leading academic, industrial, and government laboratories around the world. As such, we believe it provides a glimpse into the future of new materials and materials design strategies for the next generation of polymeric gas separation membranes.

The genesis of this book was the Chemistry and Materials Science of Synthetic Membranes Symposium held at the Fall National Meeting of the American Chemical Society (ACS) in Las Vegas, Nevada, on September 8–11, 1997. This symposium, sponsored by the Polymeric Materials: Science and Engineering Division of ACS was the largest symposium at the meeting, comprising 10 oral presentation sessions plus a poster session. Approximately 100 scientists and engineers from academic, industrial, and government laboratories participated as speakers or poster presenters at this symposium. The authors and coauthors of contributions to this symposium were from 18 countries. Although various professional organizations sponsor several symposia each year dedicated to membranes, these meetings typically explore a wide variety of topics ranging from fundamental scientific studies to highly applied engineering analyses. We used this symposium as an opportunity to provide a worldwide forum focused specifically on theoretical as well as experimental aspects of the chemistry and materials science of synthetic membranes.

Based on the overwhelming enthusiasm of the membrane research community toward this symposium, it seemed appropriate to organize a more permanent record of some of the excellent, leading-edge contributions presented in Las Vegas. Therefore, we decided, in concert with the symposium participants, to organize a brace of ACS Symposium Series Books. This book and its companion, *Membrane Formation and Modification*, focus on tailoring separation properties and control of porous membrane formation. These topics represent the two major areas of membrane science in which polymer chemistry, polymer physics, and materials science play key roles in determining the ultimate property sets of membranes, which, in turn, define the ability of membranes to participate competitively in the commercial separations arena. In this regard, these two books are unusual. Although recently several books have been published related to gas separation using polymer membranes, these are the only two books we know of that focus exclusively on structure–property relations in polymers for membrane-based separations.

This book is organized into several sections, beginning with the overview chapter, which provides a short introduction to the fundamentals of gas separations using polymers and a rapid overview of the two classes of polymers (i.e., vapor selective and gas selective) that are used commercially. Then, several contributions regarding the chemistry and materials science of each class of polymers are presented. The contributions related to vapor-selective polymers focus primarily on substituted polyacetylenes, which have the most attractive permeability–selectivity property profile for many vapor separation applications. These chapters describe synthesis and gas transport characterization of new and existing disubstituted polyacetylenes. The impact of processing history and blending different polyacetylenes on the physical aging properties of these glassy polymers is described in three chapters. As the unusual transport properties of this class of polymers are linked, partly to very high free volume, one chapter describes state-of-the-art characterization of free volume using positron annihilation lifetime spectroscopy. The next two chapters focus on facilitated transport membranes, based on blends of silver salts with polymers, for vapor phase olefin–paraffin and liquid phase aromatic–aliphatic separations. The final chapter of this section describes the use of plasma treatment to enhance the gas separation properties of poly(dimethylsiloxane), which is used commercially in large-scale vapor separation applications. Several chapters then focus on polymers for gas separation. Two chapters provide new theoretical approaches to predicting sorption and transport properties of gases in glassy polymers. The remaining contributions complement the theoretical studies with a focus on synthesis and characterization of gas sorption and transport properties of highly aromatic, glassy polyimides and related structures. Afterward, two chapters describe the effect of inorganic additives and thermal processing history to alter polymer chain packing in the solid state to systematically control gas sorption and transport properties. Because air separation is one of the most important gas separation applications, it is appropriate that the final chapter in the book describes measurement methodology for detection of oxygen in gas mixtures.

We hope that this book will be of interest to specialists in gas separation research or technology as well as novices and students of this field. The overview chapter provides some introductory material to equip a nonspecialist with the information to further appreciate the more focused chapters that comprise the bulk of the book. As the focus of the book is on polymeric membranes for gas and vapor separation and, more specifically, on structure–property relationships in such materials, we hope that readers from the polymer chemistry, materials science, and polymer physics communities will also find the book of interest. We believe the reader can develop an appreciation for the direction of development of new polymeric materials for gas separations based on this book.

We express our sincere gratitude to the authors of the contributions in this book. Their time, energy, and commitment to excellence are evident in each and every chapter. We are also very grateful to the session organizers at the Las Vegas ACS meeting. Their contributions were instrumental in the great success of the symposium. On behalf of the participants in the symposium, we warmly thank Air Products and Chemicals, Inc., Daicel Chemical Industries, Ltd., the ACS Petroleum Research Fund, and the PMSE Division of ACS for their generous financial support of the symposium. The staff at ACS Books, particularly Anne Wilson, provided excellent guidance and encouragement during the entire manuscript preparation process. Finally, this symposium series book would not have been possible without the very important contributions from our support staff, Rosemary Robinson, Melba Garcia, and Marie Brizzard.

The editors dedicate this book to Richard Baker, Donald Paul, and Vivian Stannett. These pioneers in gas and vapor separation with polymer membranes have set a standard of excellence in materials science research that is both a model for and an inspiration to us.

B. D. Freeman
Chemical Engineering Department
North Carolina State University
113 Riddick Hall, Campus Box 7905
Raleigh, NC 27695–7905

I. Pinnau
Membrane Technology and Research, Inc.
1360 Willow Road, Suite 103
Menlo Park, CA 94025

Chapter 1

# Polymeric Materials for Gas Separations

**B. D. Freeman[1] and I. Pinnau[2]**

**[1]Chemical Engineering Department, North Carolina State University, 113 Riddick Hall, Campus Box 7905, Raleigh, NC 27695–7905**
**[2]Membrane Technology and Research, Inc., 1360 Willow Road, Suite 103, Menlo Park, CA 94025**

Industrial applications of polymers as gas separations membrane materials fall broadly into two categories: (1) separations using polymers that are more permeable to the smaller components in a gas mixture and (2) separations requiring polymers that are more permeable to larger components in a gas mixture. Optimum polymeric membrane materials for the first type of separations are stiff chain, glassy, moderately high free volume polymers that sieve penetrant molecules rather strictly according to penetrant size. Optimum polymers for the second type of separations are either highly flexible, rubbery polymers or certain ultrahigh free volume, glassy substituted polyacetylenes. Such polymers sieve gas molecules weakly based on size. Therefore, larger, more soluble penetrants are more permeable than smaller, less soluble ones. This overview provides an introduction to the fundamentals of gas separations with polymers, provides examples of polymers for both categories of separations and provides current theoretical and practical design rules for preparing polymeric membranes with optimum performance characteristics.

Separation of gases using polymer membranes is an important unit operation that competes effectively with well-established processes such as cryogenic distillation, absorption and pressure-swing adsorption (*1*). Commercially, the most widely practiced gas separations using membranes are the production of high purity nitrogen from air, recovery of hydrogen from mixtures with larger components such as nitrogen, methane, and carbon monoxide, and purification of natural gas by removal of carbon dioxide (*1*). In these separations, membranes with adequately high fluxes of the more permeable components (oxygen, hydrogen, and carbon dioxide, respectively) and sufficient selectivity have been developed for membranes to be competitive with other gas separation technologies. The membrane materials used in these separations are glassy polymers, which derive high selectivity in large measure from their ability to separate gases based on subtle differences in penetrant size (*2*). Such polymers are most permeable to the smallest components in a mixture and least permeable to the largest components.

The separation of organic vapors from supercritical gases is also an application of considerable industrial importance and is performed economically using membranes (*3-10*). The removal of volatile organic compounds (*e.g.* vinyl chloride monomer (*11*), propylene, ethylene, gasoline, Freons (*4*)) from mixtures with air or nitrogen is practiced commercially. The removal of higher hydrocarbons from refinery hydrogen purge streams or from methane in natural gas represents promising future applications (*3, 5*). Because organic vapors are typically the minor components in these streams and because it is often desirable to keep the supercritical gas components at high pressure (*e.g.* methane in natural gas), membranes with high organic vapor/supercritical gas selectivity and high organic vapor flux are required for economical processes (*3-5*). Such membranes sieve penetrant molecules based primarily on relative penetrant solubility in the polymer since larger penetrants are usually more soluble than smaller penetrants.

The remainder of this overview chapter provides fundamental background information related to transport of small molecules in polymers and then describes materials design strategies to prepare polymers with excellent permeability and selectivity properties for both supercritical gas separations and vapor separations. In addition to high permeability and selectivity, membranes must also be stable in industrial process environments, which may be chemically and/or thermally challenging. For example, due to chemical stability and thermal transition temperatures of polymers used in gas separations, these materials are typically used at or near ambient temperatures. The chapter by Bayer *et al.* in this book describes the use of selective crosslinking of polyimides to prepare high performance membrane materials that are stable to 300°C.

## Background

The permeability of a polymer film of thickness $\ell$ to penetrant A, $P_A$, is (*2, 12, 13*):

$$P_A = \frac{N_A \, \ell}{p_{2_A} - p_{1_A}} \tag{1}$$

where $N_A$ is the steady state gas flux through the film, and $p_{2_A}$ and $p_{1_A}$ are the upstream and downstream partial pressures of A, respectively ($p_{2_A} > p_{1_A}$). Methods to measure permeability and other gas transport properties are described by Felder and Huvard (*14*). The chapter by Stevens and Rezac in this book describes a novel electrochemical probe to measure gas permeability.

When the downstream pressure is much less than the upstream pressure and fickian diffusion is the rate limiting step in penetrant transport, permeability is written as (*2, 13*):

$$P_A = S_A \times D_A \tag{2}$$

where $S_A$, the apparent solubility coefficient, is the ratio of the dissolved penetrant concentration in the upstream face of the polymer to the upstream penetrant partial pressure in the contiguous gas or vapor phase, and $D_A$ is the concentration averaged apparent penetrant diffusion coefficient (*13*).

Equation 2 is an analytical statement of the solution-diffusion model of penetrant transport in polymers, which is the most widely accepted explanation of the mechanism of gas permeation in nonporous polymers (*15*). According to this model, penetrants first dissolve into the upstream (*i.e.* high pressure) face of the film, diffuse through the film, and desorb at the downstream (*i.e.* low pressure) face of the film. Diffusion, the second step, is the rate limiting process in penetrant permeation. As a result, much of the fundamental research related to the development of polymers with improved gas separation properties focuses on manipulation of penetrant diffusion coefficients via systematic modification of polymer chemical structure or superstructure and either chemical or thermal post-treatment of polymer membranes. Many of the fundamental studies recorded in this book describe the results of research projects to explore the linkage between polymer structure, processing history, and small molecule transport properties.

Polymers separate gas mixtures because they are more permeable to some components in the mixture than to others. The ability of a polymer to be more permeable to component A in a mixture of A and B is characterized by the ideal selectivity, $\alpha_{A/B}$, which is the ratio of permeabilities of the two components (*2*):

$$\alpha_{A/B} \equiv \frac{P_A}{P_B} = \left[\frac{D_A}{D_B}\right] \times \left[\frac{S_A}{S_B}\right] \tag{3}$$

In equation 3, $D_A/D_B$ is the diffusivity or mobility selectivity, which is the ratio of the concentration-averaged effective diffusion coefficients of components A and B. $S_A/S_B$ is the ratio of solubility coefficients and is called the solubility selectivity. Ideal membranes have high permeability coefficients and high selectivity values.

Permeation properties vary dramatically from one polymer to another. Table I provides permeability and selectivity data for several membrane polymers ranging from glassy, relatively low permeability polysulfone, PSF, to rubbery crosslinked poly(dimethylsiloxane), PDMS, to glassy poly(1-trimethylsilyl-1-propyne), PTMSP, the most permeable polymer known. Oxygen permeability values vary by nearly four orders of magnitude across this series of polymers, but $O_2/N_2$ selectivity varies by less than a factor of four. These polymers and, in fact, all known polymers, are more permeable to oxygen than to nitrogen because oxygen is both more mobile than nitrogen (*i.e.* $D_{O_2} > D_{N_2}$) owing to the smaller effective size of oxygen and more soluble than nitrogen (*i.e.* $S_{O_2} > S_{N_2}$) due to the more condensable nature of $O_2$. In striking contrast, *n*-butane permeability coefficients span more than seven orders of magnitude, and *n*-butane/methane selectivity values vary by more than a factor of 1000. Moreover, some polymers, such as PSF, are more permeable to methane than to *n*-butane while others, such as PDMS and PTMSP, are much more permeable to *n*-butane, which is less mobile but more soluble than methane.

Polymers such as PSF are used for separation of gas mixtures where the more permeable component is the smaller component in the mixture. Examples include air separation, removal of $H_2$ from mixtures with $N_2$, and removal of $CO_2$ from natural gas mixtures (*2, 16*). Polymers such as PDMS and PTMSP are useful for the selective removal of higher hydrocarbons (*i.e.* propane, *n*-butane, *etc.*) from natural gas or hydrogen and for the removal of organic compounds from air or nitrogen. These polymers are more permeable to larger components in a mixture. Figure 1 presents

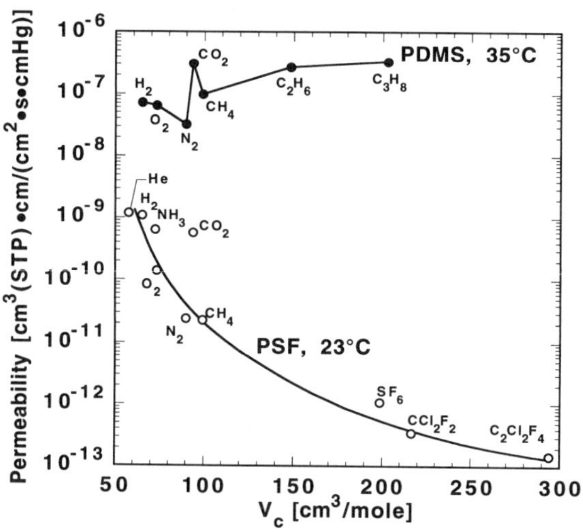


Figure 1. Effect of penetrant size on permeability in glassy polysulfone at 23°C (*64*) and in rubbery poly(dimethylsiloxane) at 35°C and infinite dilution (*76*). $V_c$ is penetrant critical volume.

the permeability of PDMS and PSF to a series of penetrants as a function of critical volume, a convenient measure of penetrant size. In both polymers, permeability coefficients of the light supercritical gases $H_2$, $N_2$, and $O_2$ decrease with increasing

Table I. Pure Gas Permeability and Selectivity Properties of Selected Polymers

| | PSF[a] | PDMS[b] | PTMSP[c] |
|---|---|---|---|
| Polymer Type | Low free volume glassy | Rubbery | High free volume glassy |
| Oxygen Permeability × $10^{10}$ [$cm^3$(STP)•cm/($cm^2$•s•cmHg)] | 1.4 | 770 | 9,700 |
| $O_2/N_2$ selectivity | 5.6 | 2.1 | 1.5 |
| *n*-Butane Permeability × $10^{10}$ [$cm^3$(STP)•cm/($cm^2$•s•cmHg)] | 0.007 | 48,500 | 78,800 |
| *n*-Butane/Methane selectivity | 0.026 | 40 | 5 |
| Glass transition temperature [°C] | 186 (*17*) | -127 (*18*) | >250 (*19*) |
| Fractional free volume, FFV | 0.14 (*17*) | 0.18[d] | 0.29[e] |

PSF=polysulfone 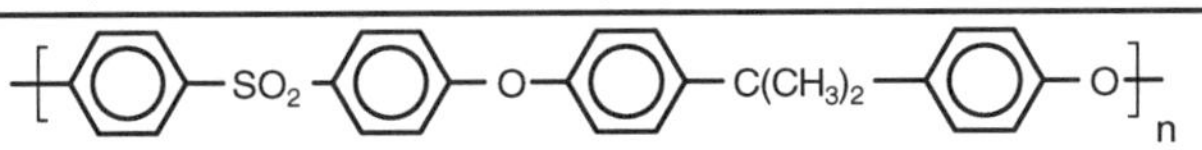

PTMSP=poly(1-trimethylsilyl-1-propyne) 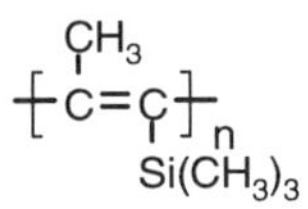

PDMS=poly(dimethylsiloxane) 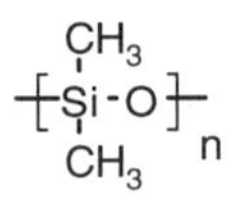

[a] Pure gas permeability and selectivity data at 35°C, $p_{feed}$ = 5 atm, and $p_{permeate}$ = 0 atm for all penetrants except *n*-$C_4H_{10}$ (*17*). *n*-$C_4H_{10}$ pure gas permeability reported at 35°C, $p_{feed}$ = 0.68 atm, $p_{permeate}$ = 0 atm.

Pure gas permeability and selectivity data at 25°C for a 225 μm thick, unfilled, PDMS film. $p_{feed}$ = 4.4 atm, $p_{permeate}$ = 1 atm for all penetrants except *n*-$C_4H_{10}$. *n*-$C_4H_{10}$ permeability reported at 25°C, $p_{feed}$ = 1.34 atm, $p_{permeate}$ = 1 atm.

[c] Pure gas permeability coefficients at 23°C, $p_{feed}$ = 4.4 atm, $p_{permeate}$ = 1 atm for all penetrants except *n*-butane, where $p_{feed}$=1.34 atm, $p_{permeate}$ = 1 atm (*8*).

[d] based on a density of 0.98 g/$cm^3$(*10*) and Bondi's method of FFV estimation (*20*).

[e] based on a density of 0.75 g/$cm^3$(*19*) and Bondi's method of FFV estimation (*20*).

penetrant size. However, in rigid, glassy PSF, permeability coefficients generally decrease strongly with increasing penetrant size (in the absence of strong plasticization effects). For example, gas permeability coefficients in PSF decrease by approximately three orders of magnitude as penetrant critical volume increases from 65 $cm^3$/mole ($H_2$) to approximately 200 $cm^3$/mole (similar in size to propane). In contrast, in soft, rubbery PDMS, the permeability of the hydrocarbon penetrants increases with increasing size, and all of the hydrocarbons are more permeable than even $H_2$, the smallest penetrant for which data are presented. The permeability of PDMS to propane is five times *higher* than the permeability to hydrogen. To rationalize the wide differences in permeation properties in Figure 1, the individual contributions of solubility and diffusivity to permeability must be considered.

**Solubility of Gases in Polymers.** Figure 2 presents the solubility of a series of gases and hydrocarbon vapors at infinite dilution in glassy PSF, a conventional polymer used for supercritical gas separations, and rubbery PDMS, which is used commercially for removal of organic vapors from mixtures with permanent gases. Gas solubility increases with increasing critical temperature in both polymers. Like gas dissolution in liquids, penetrant dissolution in polymers is regarded as a two step thermodynamic process: (1) condensation of the gaseous penetrant to a liquid-like density and (2) mixing the pure compressed penetrant with the polymer segments (*12*). The first step is governed by penetrant condensability, and the second depends on polymer-penetrant interactions. For penetrants that do not undergo specific interactions with the polymer, the first effect is often dominant, and penetrant solubility scales with measures of penetrant condensability such as penetrant boiling point, critical temperature, or the force constant in the Lennard-Jones intermolecular potential (*12, 20, 21*). For such cases, there is a simple relation between the logarithm of solubility and one of these measures of penetrant condensability. In this regard, the following relation between penetrant critical temperature and infinite dilution gas solubility at 35°C was derived using a classical thermodynamics model (*20, 22, 23*):

$$\ln S_A = M + 0.016 T_{C_A} \tag{4}$$

In this expression, M is a parameter that depends primarily on polymer-penetrant interactions, polymer free volume, and $T_{C_A}$ is the critical temperature of penetrant A. While M varies from polymer to polymer, Van Krevelen recommends average values of -9.7 and -8.7 for rubbery and glassy polymers at 35°C, respectively, when solubility is expressed in $cm^3(STP)/(cm^3 \bullet cmHg)$) (*20*). These constants represent averages from several polymers and should only be used for approximate estimates of penetrant solubility in polymers. Although equation 4 is strictly valid for equilibrium rubbery materials such as PDMS, the solubility data for PSF are also well-described by it. The values of M determined from Figure 2 for PDMS and PSF are -8.7 and -7.7, respectively, which are somewhat lower than the average values from Van Krevelen, and provide an indication of the range of M for various materials. A more complete list of values of M is provided by Toi *et al.* (*24*).

Penetrant solubility in glassy PSF is higher than that of rubbery PDMS for all penetrants. This result is ascribed to the existence of nonequilibrium excess volume in all glassy polymers (*25, 26*). This nonequilibrium excess volume provides excess sorption sites relative to an analogous equilibrium rubbery material and, consequently, glassy polymers have higher solubility coefficients than rubbery polymers of similar structure (*27*). Since glassy polymers are nonequilibrium materials, their properties are history dependent (*28*). In this regard, the chapter by Tsutsui *et al.* in this book describes the influence of thermal annealing on sorption properties of $CO_2$ in glassy, syndiotactic polystyrene, and the chapters by Morisato *et al.*, Nakagawa *et al.*, Costa *et al.*, and Nagai *et al.* present physical aging properties of substituted polyacetylenes.

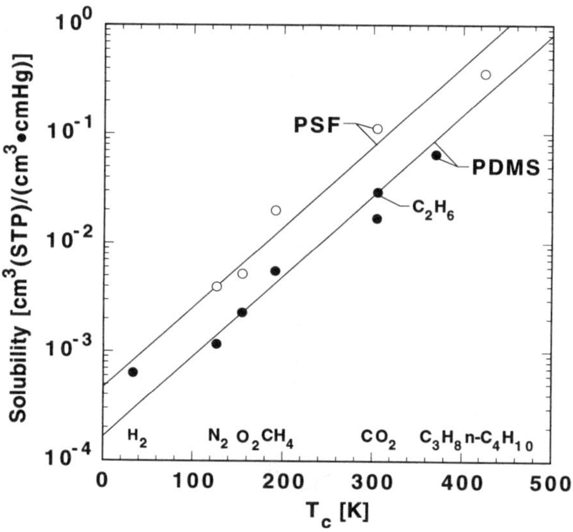


Figure 2. Penetrant solubility in polysulfone and poly(dimethylsiloxane) at 35°C and infinite dilution. The lines through the data are the best fits of the data to Equation 4. The values of M for PDMS and PSF are $1.7\times 10^{-4}$ and $4.7\times10^{-4}$ $cm^3(STP)/(cm^3\bullet cmHg)$, respectively. The polysulfone data are from Ghosal *et al.* (*77*) and the authors' laboratories. The poly(dimethylsiloxane) data are from Merkel *et al.* (*31*).

Gas condensability is often the most important effect in determining the relative solubility of different gases in a polymer. Consequently, equation 4 usually provides good estimates of solubility selectivity. Moreover, equation 4 indicates that while penetrant solubility will change from polymer to polymer, solubility selectivity changes little from one polymer to another, which contributes to the earlier observation that $O_2$ and $n$-$C_4H_{10}$ permeability, for example, varies much more from polymer to polymer than $O_2/N_2$ and $n$-$C_4H_{10}/CH_4$ selectivity, respectively. Of course, other effects also impact gas solubility in polymers. If polymer-penetrant interactions are important, then the chemical structure of the polymer can have a significant impact on penetrant solubility. For example, polar or quadrupolar penetrants such as $CO_2$ are more soluble in polar polymers than would be predicted based solely on penetrant critical temperature as a result of favorable polymer-penetrant interactions (*29, 30*). Similarly, perfluorinated vapors (*i.e.* $CF_4$, $C_2F_6$, $SF_6$, *etc.*) are less soluble in hydrocarbon-based polymers than their critical temperature indicates due to unfavorable interactions between fluorinated compounds and chemically-dissimilar hydrocarbon polymers (*e.g.* PDMS, polyethylene) (*31, 32*). Facilitated transport membranes, such as those described in the chapters by Koval *et al.* and Eriksen *et al.* in this book rely upon specific interactions between penetrants and polymers to provide high solubility selectivity and, in turn, high overall selectivity. As polymer chain packing becomes more ordered (higher levels of crystallinty, liquid crystalline order or chain orientation), gas solubility decreases (*33*). However, the effect of such improvements in the efficiency of chain packing (and the resulting loss of free volume) is rather nonspecific (*33*). That is, changes in the solid state packing of chains in a polymer matrix influence the sorption properties of all penetrants in the polymer to approximately the same degree. In general, the more disordered the polymer matrix or the higher the free volume in the polymer, the higher the gas solubility. Gas solubility in polymers also depends on gas concentration, though this effect is usually negligible for light gases (*e.g.* $H_2$, He, $N_2$, $O_2$) (*12*). For more condensable gases (*e.g.* $CO_2$, hydrocarbon vapors), solubility is more sensitive to gas pressure, and models of the effect of gas pressure on gas solubility in rubbery and glassy polymers are available (*12*). The chapter by Doghieri, *et al.* in this book describes a new theory to predict the solubility of gas mixtures in glassy polymers based on a nonequilibrium equation of state approach.

Gas solubility also depends on temperature, and solubility coefficients obey a van't Hoff relation in temperature ranges away from thermal transitions (*i.e.* glass/rubber transition, melting point) of the polymer. The heat of sorption reflects the balance between the enthalpy change on condensation and the enthalpy change associated with mixing penetrant with polymer segments and can be, therefore, either positive or negative depending on the relative magnitude of these two factors. In general, the larger the penetrant, the lower (or more negative) the heat of sorption (*20*). For a given penetrant, the heat of sorption is typically lower in glassy polymers than in rubbery polymers of similar structure (*20*).

**Diffusivity of Gases in Polymers.** *Influence of Free Volume on Diffusivity.* The rate limiting step in penetrant diffusion in polymers is the creation, adjacent to a penetrant molecule, of a molecular scale gap or opening of sufficient size to accommodate a penetrant molecule, thereby permitting a diffusion step (*12*). These gaps or free volume elements are created and continuously redistributed throughout the polymer by the ceaseless, thermally-stimulated molecular motion of the polymer chain segments. The size, concentration, and mobility of such free volume elements are critical to the diffusion of gases in polymers. Among amorphous polymers, which are used in virtually all commercial gas separation membranes, within a given polymer type (*i.e.* rubbery or glassy) and class (*e.g.* polyimides, polysulfones, or polycarbonates), much of the polymer-to-polymer variation in diffusion coefficients can be correlated with the average free volume accessible to the penetrant molecules. The accessible free volume depends, in turn, on the efficiency of chain packing and polymer segmental mobility (*34*).

Cohen and Turnbull developed a statistical mechanics model of diffusion in a liquid of hard spheres. This model has been widely adapted to describe the diffusion of small molecules in polymers (*35*). It provides the following expression for the diffusion coefficient (*36*):

$$D_A = A \exp\left(-\frac{\gamma}{\langle v_f \rangle} v_A^*\right) \tag{5}$$

where A is a pre-exponential factor which depends weakly on temperature. Penetrant size is strongly correlated with $v_A^*$, the minimum free volume element size that can accommodate the penetrant. $\gamma$ is an overlap parameter introduced to avoid double counting of shared free volume elements, and $\langle v_f \rangle$ is the average free volume in the matrix. From equation 5, diffusion coefficients decrease very strongly with increasing minimum free volume element size, $v_A^*$. This model, based on diffusion of spheres, cannot provide insight into the effect of penetrant shape on diffusion coefficients. While this model does not capture the complex cooperative segmental chain dynamics that govern penetrant transport in polymers, it provides an intuitively useful qualitative rationale for the effect of free volume and penetrant size on transport properties of polymers. In many of the contributions in this book, the effect of polymer structure variations on gas diffusivity and permeability properties is rationalized based on the notion that changes in polymer structure impact solid state chain packing (*i.e.* free volume and free volume distribution) and, in turn, transport properties.

Equation 5 predicts that in two polymer matrices with different average free volumes, the effect of penetrant size on diffusion coefficients is weaker (*i.e.* diffusivity selectivity is lower) in the higher free volume polymer. For example, PTMSP has by far the largest amount of free volume among the polymers presented in Table 1 and, even though it is a stiff, rigid, glassy polymer (a characteristic commonly associated with polymers having high diffusivity selectivity), it has the lowest $O_2/N_2$ selectivity of all polymers, indicating weak size-sieving ability.

Diffusion coefficients are very sensitive to free volume and free volume distribution. However, a direct, unambiguous measurement of free volume and free volume distribution in polymers is not available. The most common characterization of solid state chain packing is average fractional free volume, FFV, which is defined as follows (*13*):

$$FFV = \frac{V - V_o}{V}. \tag{6}$$

In this expression, V is the polymer specific volume, and $V_O$ is the so-called occupied volume of the polymer, which is commonly estimated by group contribution methods (*20*). For correlations of transport properties with free volume, the FFV defined in equation 6 is used in place of $\langle v_f \rangle$ in equation 5, and the parameters A and $\gamma v_A^*$ are treated as empirical adjustable constants. The chapter by Laciak *et al.* in this book describes a new methodology for estimating permeability coefficients *a priori*. Their approach does not rely explicitly on correlations between transport properties and free volume. Rather, their group contribution method allows a direct calculation of permeability coefficients based only on the chemical structure of the polymer.

Figure 3a presents an example of a typical correlation of gas diffusivity in a systematic series of amorphous, glassy, aromatic polyimides with reciprocal free volume. Permeability data, rather than diffusivity data, are often presented as a function of reciprocal free volume since the effect of free volume on solubility is usually much weaker than the effect of free volume on diffusivity (*37*). Therefore, both permeability and diffusivity usually obey the functional dependence on free volume indicated in equation 5. This notion is demonstrated in Figure 3b, which shows the strong correlation between methane permeability and methane diffusivity in this family of materials. Therefore, it is the variation in diffusivity (not solubility) from one polymer to another that is responsible for most of the variation in permeability. Polyimide membranes are used commercially in separation of supercritical gas mixtures because of their outstanding combinations of permeability, selectivity, and processability. More detailed information about the gas separation properties of polyimides and their correlation with polymer backbone structure, fractional free volume and mechanical properties is provided in the chapter by Hirayama *et al.* in this book. Other chapters presenting structure/property relations in polymers containing imide moieties are those by Langsam, Fritsch, Maier *et al.*, and Mahajan *et al.*

Positron annihilation lifetime spectroscopy (PALS) is a more recent tool used to probe free volume and free volume distribution in polymers (*38, 39*). PALS uses orthoPositronium (oPs) as a probe of free volume in the polymer matrix. oPs resides in regions of reduced electron density, such as free volume elements between and along chains and at chain ends (*38*). The lifetime of oPs in a polymer matrix reflects the mean size of free volume elements accessible to oPs. The intensity of oPs annihilations in a polymer sample reflects the concentration of accessible free volume elements. The oPs lifetime in a polymer sample is finite (on the order of several nanoseconds), so PALS probes the availability of free volume elements on nanosecond timescales (*40*). The minimum free volume cavity diameter required by oPs for localization is 3.8Å (*41*), which is equal to the kinetic diameter of methane (*42*). Thus, PALS probes the dynamic availability of free volume elements similar in size to those important for gas separations applications. Several recent studies demonstrate the strong correlation of PALS parameters and transport properties in polymers (*34, 38, 43-45*). The chapter by Yampol'skii and Shantarovich in this book describes the use of PALS to characterize free volume distribution in membrane polymers.

*Effect of Penetrant Size on Diffusion Coefficients.* For a given polymer, diffusion coefficients decrease with increasing penetrant size. Figure 4 presents examples of the effect of penetrant size on diffusion coefficients in low free volume, glassy PSF and rubbery PDMS. Whereas the gas solubility in PDMS is within a factor of 2-4 of the gas solubility in PSF for all penetrants in Figure 2 except $CO_2$, the differences between penetrant diffusion coefficients in PDMS and PSF are enormously larger. As penetrant size increases from $H_2$ to perfluoropropane ($V_c$=299.8 $cm^3$/mol), diffusion coefficients in PDMS decrease by a factor of 40. In contrast, in the glassy PSF matrix, diffusion coefficients decrease by a factor of almost 10,000 over the same range in critical volume. Thus, a fundamental difference between liquid-like, rubbery polymers such as PDMS and stiff, low-mobility, glassy polymers such as PSF is a very strong dependence of penetrant diffusion coefficient on penetrant size in glassy PSF and a weak dependence of penetrant diffusion coefficient on penetrant size in rubbery PDMS. This difference is due mostly to the much higher chain flexibility (as characterized by glass transition temperature) of PDMS. From the data in Figure 4, the diffusion coefficients in these

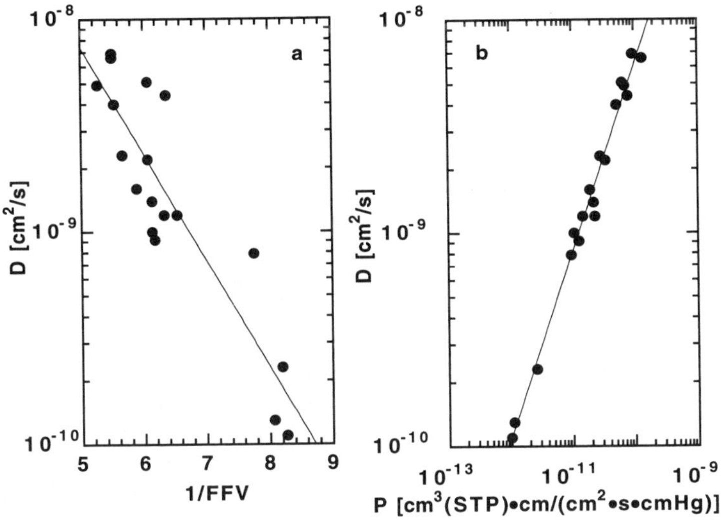


Figure 3. Methane diffusivity and permeability in a series of glassy, aromatic polyimides at 35°C (*78*). (a) Methane diffusion coefficients as a function of reciprocal free volume. (b) Correlation between methane diffusivity and methane permeability.

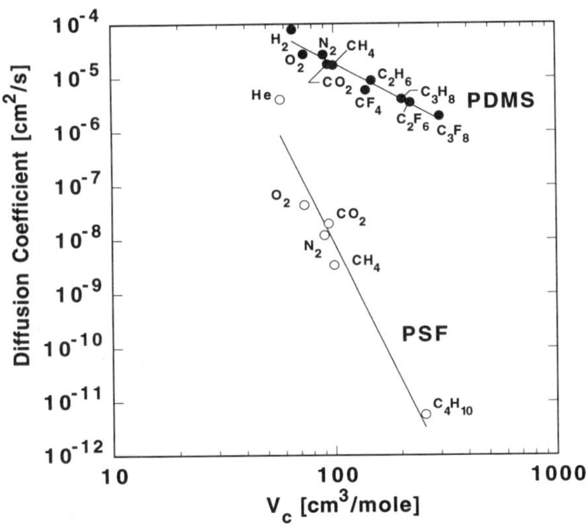


Figure 4. Diffusion coefficients in rubbery poly(dimethylsiloxane) and glassy polysulfone at 35°C as a function of penetrant critical volume. PDMS data are extrapolated to infinite dilution (*76*). The butane diffusivity in polysulfone is extrapolated to infinite dilution. All other polysulfone diffusion coefficients determined at 10 atm (*25, 77*). The lines through the data are best fits to equation 5. The parameter values are $\tau$=0.43 and $\eta$=2.2 for PDMS and $\tau=4.8\times10^{8}$ and $\eta$ =8.4 for PSF when diffusivity is in $cm^2/s$ and critical volume is in $cm^3/mol$.

polymers obey, to good approximation, a simple power law relation with penetrant critical volume:

$$D_A = \frac{\tau}{V_{C_A}^{\eta}} \tag{7}$$

where $\tau$ and $\eta$ are adjustable constants. While equation 7 is empirical, its form is based on analogy with correlations of diffusion coefficients of small molecules in liquids (*46-48*). This result is consistent with equation 5 if critical volume is correlated with the minimum free volume element size required by for a diffusion step. A correlation of diffusion coefficients with penetrant critical volume should only be applicable for penetrants small enough that the entire penetrant molecule participates in the diffusion step. For larger penetrants (*e.g.* long chain hydrocarbons), diffusion steps may occur via segmental steps of small sections of the molecule, so critical volume would not be expected to capture the effective size of a penetrant unit participating in a diffusion step. In this case, diffusion coefficients should become less sensitive to penetrant size than indicated in equation 7. Kinetic diameter, which characterizes the smallest zeolite window through which a penetrant can fit (*42*), is often used as a measure of penetrant size appropriate for correlation of penetrant diffusion coefficients in polymers. This parameter is particularly appropriate for small penetrants ranging in size from He up to approximately $CH_4$. However, for larger penetrants, kinetic diameter does not give a good measure of penetrant size for transport property correlations, (*49*) and kinetic diameter values are not available for many larger penetrants. Therefore, we use critical volume to characterize penetrant size when penetrants of widely varying sizes are considered. Of course, critical volume, an equilibrium property, cannot capture the influence of penetrant shape on diffusion coefficients, whereas kinetic diameter is sensitive to penetrant shape. When only such light gases are considered (*i.e.* He, $H_2$, $N_2$, $O_2$, $CO_2$, and $CH_4$), we use kinetic diameter to characterize penetrant size. A useful contribution to the field would be the development of consistent measures of penetrant size from first principles to provide accurate correlating parameters for transport studies.

The exponent in equation 7, $\eta$, is a crude measure of the size-sieving ability of the polymer. Polymers with larger values of $\eta$ will have higher diffusivity selectivity than polymers with lower values of $\eta$. Based on a fit of the experimental data in Figure 4 to equation 7, the values of $\eta$ for PDMS and PSF are 2.2 and 8.4, respectively. The large difference in these exponents is indicative of the much stronger size-sieving ability of low free volume, glassy PSF relative to rubbery PDMS. For comparison, low molar mass organic liquids, such as hexane, heptane, and benzene, have even weaker size-sieving ability than PDMS. The values of $\eta$ for these liquids is 0.45 based on the Tyn and Calus correlation of gas diffusion in liquids with gas critical volume (*46-48*). At the other extreme, glassy polymers such as poly(vinyl chloride) have even more stringent size-sieving ability ($\eta$=10.5) than PSF (*50*).

Based on the results in Figures 2 and 4, as penetrant size increases, penetrant solubility increases and diffusivity decreases. The effect of penetrant size on permeability reflects the interplay between these two factors. In soft, flexible, rubbery PDMS, diffusion coefficients decrease much more slowly with increasing penetrant size than in glassy, rigid PSF. In contrast, while the absolute solubility levels are higher in glassy PSF than in rubbery PDMS, the change in solubility from penetrant to penetrant is similar for both polymers. Therefore, the higher permeability of hydrocarbons (such as propane) than small penetrants (such as hydrogen) and the increase in permeability with increasing hydrocarbon size reflects the weak size-sieving ability of PDMS. However, in PSF, the much stronger decrease in penetrant diffusion coefficients with increasing penetrant size more than offsets the increase in solubility with increasing penetrant size, and permeability coefficients decrease with increasing penetrant size.

Factors other than free volume, polymer type (*i.e.* rubbery or glassy), and penetrant size also influence diffusion coefficients. Diffusivity can depend on penetrant concentration. For light, highly supercritical gases, this dependence is practically negligible in both glassy and rubbery polymers. For more soluble penetrants such as $CO_2$ and organic vapors, however, diffusion coefficients can change significantly with penetrant concentration. Models for such effects have been published for rubbery (*51, 52*) and glassy polymers (*53*). Usually, diffusion coefficients increase with increasing penetrant concentration (*53*), but examples of decreasing diffusion coefficients with increasing penetrant concentration in both rubbery (*10*) and glassy polymers (*54*) are also reported. Polymer morphology influences diffusion coefficients. For example, amorphous polymers are used in most gas separations membranes because polymer crystallites act as low free volume, impermeable barriers in a polymer matrix, increasing the path length for diffusion, thereby reducing effective diffusion coefficients and, in turn, permeability of all penetrants (*33*). Diffusion is an activated rate process and, consequently, diffusion coefficients obey an Arrhenius dependence on temperature with a positive activation energy of diffusion, $E_{D_A}$, (*12*):

$$D_A = D_{o_A} \exp\left[-E_{D_A}/RT\right] \tag{8}$$

where $D_{o_A}$ is the front factor in the Arrhenius equation. Therefore, diffusion coefficients always increase with increasing temperature (*12*). In general, the larger the penetrant, the larger the activation energy of diffusion (*20*). For a given penetrant, activation energy of diffusion is a complex function of polymer chain stiffness, cohesive energy density, and free volume. In general, in the absence of specific polymer/penetrant interactions, activation energy of diffusion increases with increasing chain stiffness and cohesive energy density (*12, 55, 56*) and decreases with increasing accessible free volume (*20*).

In the remaining sections of this overview, polymer materials design strategies to achieve high performance in supercritical gas separation based on diffusivity selectivity and in vapor separation based on high solubility selectivity will be discussed.

## Membrane Materials Based on Diffusivity Selectivity

In the separation of supercritical gases such as air separation, $CO_2$ removal from natural gas, and hydrogen recovery from a variety of chemical and petrochemical processes, the smaller penetrant molecules ($O_2$, $CO_2$ and $H_2$, respectively) are the more permeable penetrants, and membrane materials research focuses on developing materials with high diffusivity and high diffusivity selectivity (*57*). For such polymers, a rather general tradeoff relation has been recognized between permeability and selectivity for such materials: polymers which are more permeable are generally less selective and *vice versa* (*57, 58*). An example of this relation is given in Figure 5, which presents hydrogen permeability coefficients and $H_2/N_2$ separation factors for many polymers. The most desirable materials would have properties in the upper right corner (*i.e.* high permeability *and* high selectivity). However, materials with permeability/selectivity combinations above and to the right of the line drawn in this figure are rare. This line defines the "upper bound" combinations of permeability and selectivity of known polymers for this particular gas pair. Lines such as the one shown in Figure 5 were constructed empirically for many gas pairs using available permeability and selectivity data (*57, 58*). The properties of polymers defining the upper bound were fit to the following equation (*57*):

$$\alpha_{A/B} = \frac{\beta_{A/B}}{P_A^{\lambda_{A/B}}} \tag{9}$$

where $\lambda_{A/B}$ and $\beta_{A/B}$ are parameters which depend on the gas pair. Robeson reported values of these parameters for many common gas pairs (*57, 58*).

A recent theory provides the following analytical expressions for the coefficients in equation 9 (*59*):

$$\lambda_{A/B} = \left(\frac{d_B}{d_A}\right)^2 - 1 \tag{10}$$

and

$$\beta_{A/B} = \frac{S_A}{S_B} S_A^{\lambda_{A/B}} \; e^{-\lambda_{A/B}\left[b - f\left(\frac{1-a}{RT}\right)\right]} \tag{11}$$

In equations 9 and 10, $d_B$ and $d_A$ are the kinetic diameters of the larger and smaller penetrants, respectively. The parameters a and b in equation 11 are from the linear free energy relation between activation energy of diffusion, $E_{D_A}$, and $D_{o_A}$, the front factor in the Arrhenius expression for the diffusion coefficient (*60, 61*):

$$\ln D_{o_A} = a\frac{E_{D_A}}{RT} - b \tag{12}$$

The parameter a is 0.64 (*62*). The parameter b has values of 9.2 and 11.5 in rubbery and glassy polymers, respectively, when $D_{o_A}$ has units of $cm^2/s$ (*20*). The polymers that define the upper bound relation in Figure 5 and related plots for other light gas pairs are stiff chain, amorphous glassy polymers that achieve high selectivity based on their strong size-sieving tendency, so the value of b is set to 11.5 in the following calculations. The parameter f is from the following empirical relation between activation energy of diffusion and penetrant size (*59*):

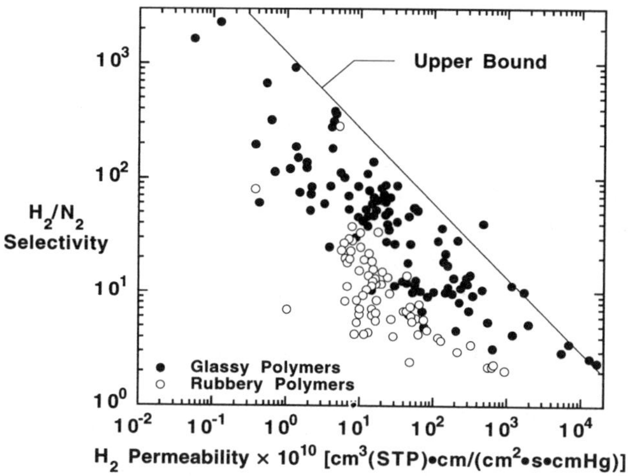


Figure 5. Hydrogen permeability and hydrogen/nitrogen selectivity of glassy and rubbery polymers. Data from Robeson (*57*).

$$E_{D_A} = cd_A^2 - f \quad (13)$$

Within the context of this model, c and f are adjustable constants that vary from polymer to polymer. The ratio $\sqrt{f/c}$ is a crude measure of the average interchain separation. As polymer chain stiffness increases and free volume decreases, c should increase (*20*), and as fractional free volume increases, $\sqrt{f/c}$ should increase. This theory is based on four hypotheses: (1) the solution-diffusion mechanism (equation 2) governs gas transport; (2) diffusion obeys an Arrhenius expression (equation 8); (3) the linear free energy relation (equation 12) is valid; and (4) the effect of penetrant size on activation energy is given by equation 13.

Equation 4 was used to calculate $S_A$ and $S_B$ in equation 11 for several gas pairs, and $\lambda_{A/B}$ and $\beta_{A/B}$ values were calculated using equations 10 and 11. The results, presented in Figures 6 and 7, are compared with the empirically-determined values of $\lambda_{A/B}$ and $\beta_{A/B}$. To complete the calculation of $\beta_{A/B}$ using equation 11, a value of f is required. For simplicity, the value of f which gave the least deviation between the calculated and observed values of $\beta_{A/B}$ Figure 7 was estimated, and its value is 12,600 cal/mol. In general, the agreement between the calculated and observed slopes is strikingly good given the approximate nature of the slopes and the fact that the predictions are based on a theory with only a single adjustable parameter, f. Based on the results presented in Figures 6 and 7, $\lambda_{A/B}$ increases and $\beta_{A/B}$ decreases with increasing difference in the size of the penetrant molecules.

From this theory, the slope of the upper bound or tradeoff curves, $\lambda_{A/B}$, depends only on relative gas diameter. Consequently, $\lambda_{A/B}$ is not expected to change significantly with further polymer development efforts. In contrast, $\beta_{A/B}$ contains variables that can be systematically tuned to simultaneously improve permeability and selectivity characteristics. Higher performance polymeric membranes for supercritical gas separations which rely on high diffusivity selectivity to achieve high overall selectivity will result from materials development strategies which systematically increase $\beta_{A/B}$, either through solubility selectivity enhancement and/or increases in chain stiffness (*i.e.* increasing c), while simultaneously increasing interchain spacing (*i.e.* fractional free volume and f), to increase selectivity while maintaining or increasing permeability.

Increasing interchain separation to increase permeability without sacrificing selectivity should only be effective as long as the interchain separation is not so large that penetrant diffusion coefficients are no longer governed by thermally stimulated polymer segmental motions. Materials that are near or beyond this limit may already exist. PTMSP, the most permeable polymer known and the polymer which has the highest free volume of all hydrocarbon-based polymers, also has the lowest selectivities for permanent gases (*e.g.* $O_2/N_2$) of all polymers and exhibits many permeation characteristics similar to those of microporous materials (*8, 9*).

The chapters in this book by Langsam, Xu *et al.*, Hirayama *et al.*, Fritsch, and Maier *et al.* focus on polymer structure modification to improve the performance of gas separation membranes relative to the upper bound tradeoff relations. Mahajan *et al.* describe characteristics of hybrid inorganic/organic membranes as a route to break the simple rules that result in equations 8 and 9, possibly resulting in materials with properties which are above and beyond the upper bound lines. Koval *et al.* and Eriksen *et al.* describe facilitated transport membranes. They seek to strongly enhance solubility selectivity for penetrant pairs (*i.e.* olefin/paraffin) where

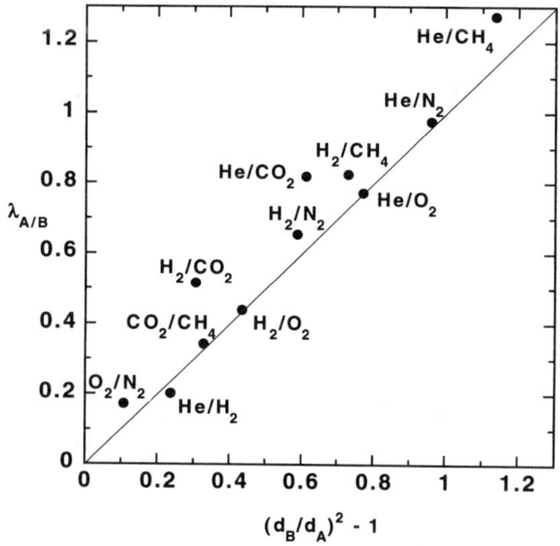


Figure 6. Comparison of slopes of ln $\alpha_{A/B}$ vs. ln$P_A$ plots, $\lambda_{A/B}$, reported by Robeson (*57, 58*) with theoretical prediction (solid line). Gas pairs are listed as A/B (*i.e.* $O_2/N_2$ implies that $O_2$=A and $N_2$=B).

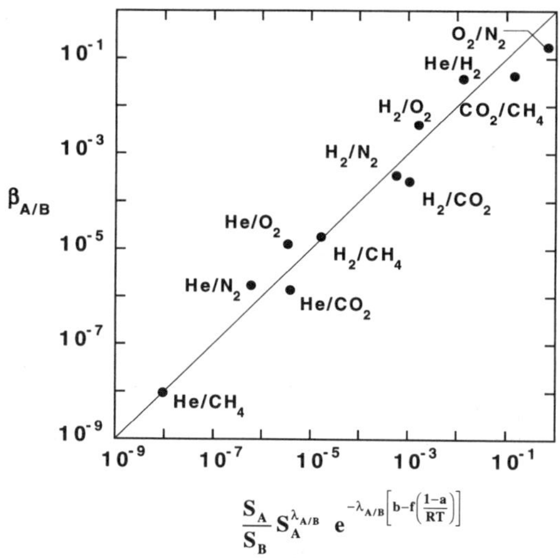


Figure 7. Comparison of front factors of ln $\alpha_{A/B}$ vs. ln$P_A$ plots, $\beta_{A/B}$, reported by Robeson (*57, 58*) with theoretical prediction (solid line). The best fit of the data to the theory is obtained with the parameter f set to 12,600 cal/mole. $\beta_{A/B}$ has units of $(\mathrm{cm^3(STP)\bullet cm/(cm^2\bullet s\bullet cmHg)})^{\lambda_{A/B}}$.

diffusivity selectivity alone is not sufficiently high to prepare commercially useful membranes. An alternative approach to improving gas separation properties of polymers is described in the chapter by Maki *et al.* They describe the use of low temperature plasma treatment to modify the surface of PDMS membranes in order to improve $CO_2/CH_4$ gas separation properties.

## Membrane Materials Based on Solubility Selectivity

Polymers that are more permeable to larger penetrants than to smaller supercritical gas components are useful for selective removal of volatile organic compounds from air or nitrogen and selective removal of higher hydrocarbons (*e.g.* propane, butane, *etc.*) from natural gas or hydrogen streams in petrochemical processes. Such materials must derive high organic vapor/supercritical gas selectivity from high solubility selectivity. For example, to remove higher hydrocarbons from natural gas, ideal membranes should be extremely permeable to the higher hydrocarbon components of the mixture and relatively impermeable to methane. Conventional glassy polymers, such as polysulfone, are not suitable because they are more permeable to methane than to higher hydrocarbons (*63, 64*). If a conventional glassy polymer membrane was used, the bulk of the natural gas would have to permeate through the membrane. Not only would this require a very large membrane area, but the methane-rich stream would be produced at low pressure. Recompressing the entire treated stream to pipeline pressure would be economically impractical. Thus, optimized materials for this separation should have high organic vapor permeability and high organic vapor/supercritical gas selectivity. As organic vapors in these separations are larger than the supercritical gas components, membrane materials that sieve molecules strictly based on size are not useful.

As shown in Figure 4, diffusion coefficients decrease with increasing penetrant size. Therefore, for an organic vapor/supercritical gas separation, diffusivity selectivity is always less than one:

$$\frac{D_{\text{organic vapor}}}{D_{\text{supercritical gas}}} < 1 \qquad (14)$$

For overall organic vapor/supercritical gas selectivity to be greater than one, solubility selectivity must be sufficiently greater than one to offset the unfavorable diffusivity selectivity. As indicated in Figure 2, larger penetrant molecules are, in general, more soluble than smaller penetrants. Thus, for organic vapor/supercritical gas separations, solubility selectivity generally favors the larger organic vapor components:

$$\frac{S_{\text{organic vapor}}}{S_{\text{supercritical gas}}} > 1 \qquad (15)$$

At present, there are two classes of polymers that have sufficiently low diffusivity selectivity to permit:

$$\frac{P_{\text{organic vapor}}}{P_{\text{supercritical gas}}} > 1 \qquad (16)$$

They are rubbery polymers and ultrahigh free volume glassy materials (*5*). PDMS is an important member of the first class, and PTMSP is an example of the second class. PDMS has the lowest diffusivity selectivity of any rubbery polymer. It has a very flexible polymer backbone as indicated by its extremely low glass transition temperature. As a result, PDMS has a very weak ability to sieve penetrant molecules based on size (*cf.* Figure 4) and can, therefore, achieve the criterion of equation 15. Because of its good balance of permeability and selectivity for a wide range of organic vapor/supercritical gas separations, PDMS is used commercially for the separation of organic compounds from air (*3, 4*).

Recent studies have demonstrated that some disubstituted polyacetylenes have high vapor permeability and high vapor selectivity for gas/vapor separations (*5, 8, 9*). These rigid, amorphous polymers contain alternating double bonds along the chain backbone and bulky side groups that hinder chain segmental motion. They are characterized by very high glass transition temperatures (> 200°C), high fractional free volumes, and high gas permeabilities. Attracting the most attention within this family of polymers is PTMSP. For hydrocarbon vapor separations, PTMSP possesses the highest mixture $C_{3+}$/methane and $C_{3+}$/hydrogen selectivities combined with the highest $C_{3+}$ permeabilities of all known polymers (*9*).

An example of the extraordinary separation properties of PTMSP for hydrocarbon/hydrogen mixtures is given in Figures 8a and 8b, which present propane and hydrogen permeability coefficients and mixture separation factors as a function of propane relative pressure in the feed gas. As the propane relative pressure increases, propane permeability increases by less than a factor of two, from $26{,}000\times10^{-10}$ to $41{,}000\times10^{-10}$ $cm^3(STP)\cdot cm/(cm^2\cdot s\cdot cmHg)$. In striking contrast, the permeability of PTMSP to hydrogen *decreases* by more than an order of magnitude, from $22{,}000\times10^{-10}$ $cm^3(STP)\cdot cm/(cm^2\cdot s\cdot cmHg)$ for pure hydrogen to about $1{,}000\times10^{-10}$ $cm^3(STP)\cdot cm/(cm^2\cdot s\cdot cmHg)$ when the propane relative pressure is approximately 0.8. As a result, propane/hydrogen selectivity increases by more than an order of magnitude, from approximately one to more than 25, in this relative pressure range. These data illustrate the strong dependence of hydrocarbon/supercritical gas selectivity on the relative pressure of the hydrocarbon component for PTMSP. Similar results are also reported for butane/hydrogen mixture permeation in PTMSP (*65*).

A structurally similar glassy polyacetylene, poly(4-methyl-2-pentyne), PMP, is somewhat less permeable than PTMSP but is also more permeable to large organic vapors than to small supercritical gas components (*66, 67*). PMP is a hydrocarbon-based, disubstituted polyacetylene synthesized more than a decade ago by Masuda *et al.* (*68, 69*). It has a low density of 0.78 $g/cm^3$ and is the most permeable, hydrocarbon-based polymer known (*66, 67*). In hydrocarbon/supercritical gas mixtures, the permeability of the supercritical gas component in PMP is also markedly reduced as the relative pressure of the hydrocarbon increases (*66, 67*).

The strong reduction in light gas permeability in the presence of higher hydrocarbon copermeants is reminiscent of blocking of light gases by heavier

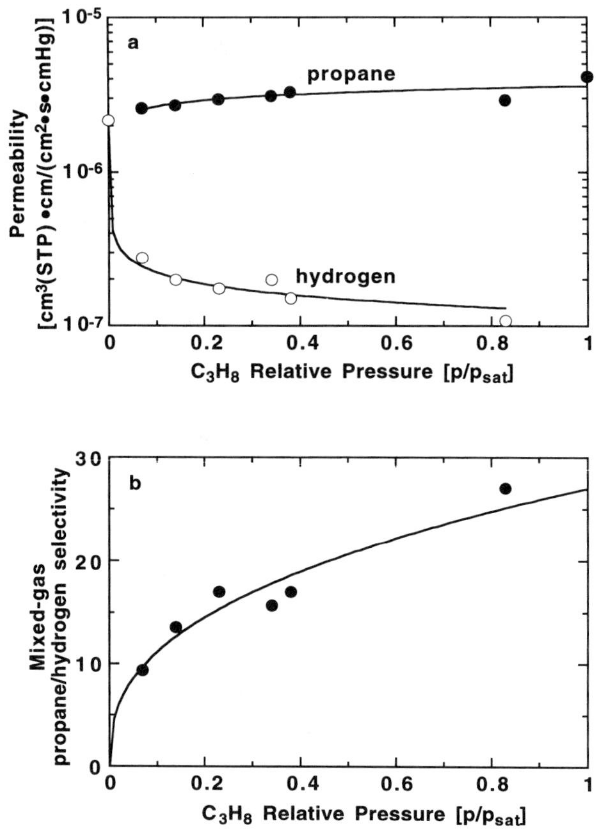


Figure 8. Hydrocarbon/light gas separation properties of PTMSP. (a) Effect of propane relative pressure on propane and hydrogen permeability coefficients in PTMSP at 25°C. (b) Effect of propane relative pressure on $C_3H_8/H_2$ mixture selectivity. The feed pressure of the propane/hydrogen binary mixture was 200 psig and the permeate pressure was atmospheric (0 psig) (*9*). Propane relative pressure is $p/p_{sat}$, where p is the partial pressure of propane in the feed and $p_{sat}$ is the saturation vapor pressure of propane.

hydrocarbon components in microporous carbon (*70*). In fact, transport properties in PTMSP have been compared with those of microporous carbon (*8, 9, 71*). Furthermore, PALS results suggest that PTMSP has a free volume distribution consistent with the existence of interconnected free volume elements that span the sample (*44*). These composite data suggest that PTMSP and some other disubstituted polyacetylenes may represent a new class of glassy polymers, so-called superglassy polymers, that have sufficient free volume so that some free volume elements in the polymer matrix are interconnected and span the sample, acting as very "fast" permeation pathways for penetrant molecules. When heavy hydrocarbons copermeate with light, supercritical gases, sorption of the hydrocarbons into the interconnected free volume elements effectively blocks most of the permeation of light gas components. If so, such materials represent a new class of materials for vapor separation applications. In this regard, the chapters by Masuda *et al.*, Costa *et al.*, Nakagawa *et al.*, Nagai *et al.*, and Doghieri *et al.* in this book describe in-depth studies of the sorption, diffusion, and permeation of gases and vapors in PTMSP and related high-permeability disubstituted polyacetylenes.

More generally, good materials for organic vapor-gas separations will have diffusivity selectivity values as near unity as possible and high solubility selectivity. To illustrate this point, pure gas permeability coefficients and separation factors for propane/methane for a number of rubbery and glassy polymers as well as liquids are presented in Figure 9. The permeability coefficients of the organic liquids included in Figure 9 were estimated as the product of the known solubility of the penetrant (propane or methane) in the liquid (*72*) and the diffusion coefficient of the penetrant in the liquid estimated using the Tyn and Calus correlation (*46, 47*). Two features are of interest. First, the permeability and selectivity values span an extremely broad range. Propane permeability coefficients vary by more than eight orders of magnitude. Propane/methane selectivity values vary by nearly four orders of magnitude. Propane/methane solubility selectivity values are between 3 and 30 for this gas pair, and solubility coefficients for either component do not vary by more than a factor of 40. Therefore, most of the variation in propane permeability and selectivity results from the large range of diffusion coefficients and diffusivity selectivity values. For example, propane diffusion coefficients vary by seven orders of magnitude, and propane/methane diffusivity selectivity varies by almost three orders of magnitude. Second, for materials that are more permeable to propane than to methane, as propane permeability increases, propane/methane selectivity increases. That is, more permeable materials are more selective. This trend is in stark contrast to the tradeoff curves for conventional supercritical gas separations where increases in permeability are usually accompanied by decreases in selectivity (*cf.* Figure 5).

The most permeable materials presented in Figure 9 are organic liquids. These compounds have diffusivity selectivity values closest to one ($D_{propane}/D_{methane} \approx 0.7$) among the materials in this figure and the highest diffusion coefficients of the materials in Figure 9. The fact that low free volume, glassy, strongly size-sieving polymers (*i.e.* polyimide, polycarbonate, and polystyrene in Figure 9) have the lowest propane permeabilities and propane/methane selectivities and weakly size-sieving materials such as rubbery polymers, organic liquids, and ultrahigh free volume, glassy, disubstituted polyacetylenes have the highest combinations of propane permeability and propane/methane selectivity suggests design rules for materials for this class of separations. Diffusivity selectivity values should be as close to one as possible. Increasing the number of mobile linkages in the polymer backbone and increasing the free volume of the polymer matrix are two strategies to achieve this goal when rubbery polymers are used. For superglassy polymers, a principal design

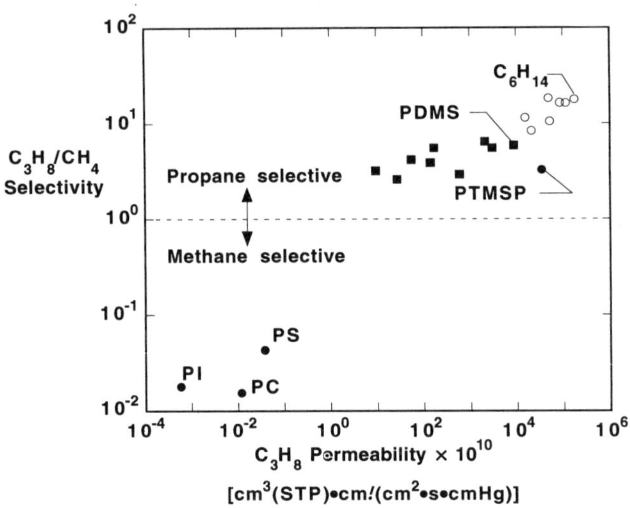


Figure 9. Pure gas propane permeability and propane/methane selectivity for a series of selected organic liquids (○), rubbery siloxane-based polymers (■), and glassy polymers (●). The glassy polymers include PI, a polyimide (*79*), PC, polycarbonate (*80*), PS, polystyrene (*81*), and PTMSP (*82*). Data for the siloxane-based rubbery polymers are from Stern *et al.* (*83*). The solubility of propane and methane in selected organic liquids (hexane, heptane, octane, acetone, benzene, methanol, and ethanol) is from the compilation by Fogg and Gerrard (*72*). Diffusion coefficients of propane and methane in these liquids were estimated using the Tyn and Calus correlation (*46-48*).

rule should be to increase polymer free volume through chain backbone modifications that frustrate packing efficiency or reduce cohesive energy density.

Additionally, the permeability/selectivity properties of membrane polymers which are more permeable to larger components in a gas mixture can be enhanced via increased solubility selectivity by preparing polymers which have a favorable interaction with one component in a mixture, thereby increasing its solubility. If such polymers are rubbery, diffusivity selectivity could be maintained at low values. This design strategy has been demonstrated for $CO_2/H_2$ separations where rubbery polyethers are observed to have the highest combinations of $CO_2$ permeability and $CO_2/H_2$ selectivities of all non-facilitated transport polymers explored to date (*29*). The ether linkages exhibit favorable interactions with quadrupolar $CO_2$, thereby increasing solubility selectivity, and the rubbery matrix has weak size-sieving ability (*73*). These polymers can be approximately 7 times more permeable to $CO_2$ than to $H_2$ (*29*). As another example, solid polymer electrolytes, composed of a rubbery polymer matrix laced with silver salts, which strongly enhance the solubility of olefins such as propylene relative to paraffins such as propane (*74*), may be used to prepare polymers with very high fluxes of propylene or ethylene and propylene/propane or ethylene/ethane selectivity values greater than 1,000 (*75*).

## Conclusions

Gas separations applications of interest for polymer membranes may be organized broadly into two categories, those for which membranes which are more permeable to smaller components in a gas mixture are desirable and those for which membranes which are more permeable to larger components in a gas mixture are useful. The materials design rules to prepare optimal polymeric membrane materials for these categories of separations are quite different. For separation of mixtures of small, supercritical gases, where polymers that sieve penetrant molecules as strictly as possible based on effective penetrant size, moderate free volume, glassy polymers are most efficient at the present time. However, as a result of the strong size-sieving nature of these polymers, such materials follow immutable tradeoff relations, and future improvements in polymer properties are likely to come from increases in solubility selectivity while maintaining high diffusivity selectivity or through a careful combination of polymer backbone modifications which increase the average free volume in the polymer matrix while maintaining backbone stiffness. In contrast, for removal of organic vapors from mixtures with light gases, removal of higher hydrocarbons from mixtures with hydrogen or methane, or for the other examples mentioned in the previous section, polymers which sieve penetrant molecules based primarily on relative penetrant solubility are required. Such materials may be prepared by minimizing diffusivity selectivity. Two routes to achieve weak size-sieving polymers appear to be available: (1) increase backbone flexibility (*i.e.* use rubbery polymers) or (2) increase free volume in glassy polymers sufficiently to prepare materials with interconnected free volume elements which span the sample and have separation properties similar to those of microporous carbon. Solubility selectivity may also be enhanced by modifying polymer structure to increase the solubility of one component in a mixture or adding agents which can complex with a desired penetrant in a mixture. In these ways, polymer membrane materials may be prepared which are suitable for a wide variety of separations applications in the chemical, petrochemical, and other industries and are competitive with more traditional gas separations technologies.

## Acknowledgments

The authors would like to acknowledge partial support of this work by the National Science Foundation (Young Investigator Award CTS-9257911-BDF).


## Literature Cited


1. Prasad, R.; Shaner, R. L.; Doshi, K. J. In *Polymeric Gas Separation Membranes*; Paul, D. R. Yampol'skii, Y. P., Eds.; CRC Press: Boca Raton, FL, 1994; pp 513-614.
2. Stern, S. A. *J. Membrane Sci.* **1994**, *94*, 1-65.
3. Baker, R. W.; Yoshioka, N.; Mohr, J. M.; Khan, A. J. *J. Membrane Sci.* **1987**, *31*, 259-271.
4. Baker, R. W.; Wijmans, J. G. In *Polymeric Gas Separation Membranes*; Paul, D. R. Yampol'skii, Y. P., Eds.; CRC Press: Boca Raton, FL, 1994; pp 353-397.
5. Freeman, B. D.; Pinnau, I. *Trends in Polymer Science* **1997**, *5*, 167-173.
6. Morisato, A.; Freeman, B. D.; Pinnau, I.; Casillas, C. G. *J. Polym. Sci.: Polym. Phys. Ed.* **1996**, *34*, 1925-1934.
7. Morisato, A.; Shen, H. C.; Sankar, S. S.; Freeman, B. D.; Pinnau, I.; Casillas, C. G. *J. Polym. Sci.: Polym. Phys. Ed.* **1996**, *34*, 2209-2222.
8. Pinnau, I.; Toy, L. G. *J. Membrane Sci.* **1996**, *116*, 199-209.
9. Pinnau, I.; Casillas, C. G.; Morisato, A.; Freeman, B. D. *J. Polym. Sci.: Polym. Phys. Ed.* **1996**, *34*, 2613-2621.
10. Singh, A.; Freeman, B. D.; Pinnau, I. *J. Polym. Sci.: Polym. Phys. Ed.* **1998**, *36*, 289-301.
11. Lahiere, R. J.; Hellums, M. W.; Wijmans, J. G.; Kaschemekat, J. *Ind. Eng. Chem. Res.* **1993**, *32*, 2236-2241.
12. Petropoulos, J. H. In *Polymeric Gas Separation Membranes*; Paul, D. R. Yampol'skii, Y. P., Eds.; CRC Press: Boca Raton, 1994; pp 17-81.
13. Ghosal, K.; Freeman, B. D. *Polymers for Advanced Technologies* **1994**, *5*, 673-697.
14. Felder, R. M.; Huvard, G. S. In *Methods of Experimental Physics, Vol. 16C*; Fava, R., Ed.; Academic Press: New York, 1980; pp 315-377.
15. Wijmans, J. G.; Baker, R. W. *J. Membrane Sci.* **1995**, *107*, 1-21.
16. Koros, W. J.; Fleming, G. K. *J. Membrane Sci.* **1993**, *83*, 1-80.
17. Ghosal, K.; Chern, R. T.; Freeman, B. D. *J. Polym. Sci.: Polym. Phys. Ed.* **1993**, *31*, 891-893.
18. Peyser, P. In *Polymer Handbook, Third Edition*; Brandrup, J. Immergut, E. H., Eds.; Wiley: New York, 1989; pp VI/209-VI/277.
19. Ichiraku, Y.; Stern, S. A.; Nakagawa, T. *J. Membrane Sci.* **1987**, *34*, 5-18.
20. Van Krevelen, D. W., *Properties of Polymers: Their Correlation with Chemical Structure; Their Numerical Estimation and Prediction from Additive Group Contributions*; Elsevier: Amsterdam, 1990; pp 875.
21. Stannett, V. T. In *Diffusion in Polymers*; Crank, J. Park, G. S., Eds.; Academic Press: New York, 1968; pp 41-73.
22. Gee, G. *Quart. Revs.* **1947**, *1*, 265-298.

23. Barrer, R. M.; Skirrow, G. *J. Polym. Sci.* **1948**, *3*, 564-575.
24. Toi, K.; Morel, G.; Paul, D. R. *J. Appl. Polym. Sci.* **1982**, *27*, 2997.
25. Erb, A. J.; Paul, D. R. *J. Membrane Sci.* **1981**, *8*, 11-22.
26. Paul, D. R. *Ber. Bunsenges Phys. Chem.* **1979**, *83*, 294-302.
27. Lipscomb, G. G. *AIChE J.* **1990**, *36*, 1505-1516.
28. Struik, L. C. E., *Physical Aging in Amorphous Polymers and Other Materials*; Elsevier: Amsterdam, 1978.
29. Bondar, V. I.; Freeman, B. D.; Pinnau, I. *Proceedings of the American Chemical Society Division of Polymeric Materials: Science and Engineering* **1997**, *77*, 311-312.
30. van Amerongen, G. J. *Rubber Chem. Tech.* **1964**, *37*, 1065-1152.
31. Merkel, T. C.; Bondar, V.; Nagai, K.; Freeman, B. D. *Macromolecules* **1999**, *32*, 370-374.
32. Michaels, A. S.; Bixler, H. J. *J. Polym. Sci.* **1961**, *50*, 393-412.
33. Weinkauf, D. H.; Paul, D. R. In *Barrier Polymers and Barrier Structures*; Koros, W. J., Ed.; American Chemical Society: Washington, D.C., 1990; pp 60-91.
34. Freeman, B. D.; Hill, A. J. In *Structure and Properties of Glassy Polymers*; Tant, M. R. Hill, A. J., Eds.; ACS: Washington, DC, 1999; pp 306-325.
35. Vrentas, J. S.; Duda, J. L. *J. Polym. Sci.: Polym. Phys. Ed.* **1977**, *15*, 403-416.
36. Cohen, M. H.; Turnbull, D. *J. Chem. Phys.* **1959**, *31*, 1164-1169.
37. Pixton, M. R.; Paul, D. R. In *Polymeric Gas Separation Membranes*; Paul, D. R. Yampol'skii, Y. P., Eds.; CRC Press: Boca Raton, Florida, 1994; pp 83-153.
38. Hill, A. J.; Weinhold, S.; Stack, G. M.; Tant, M. R. *Eur. Polym. J.* **1996**, *32*, 843-849.
39. Kobayashi, Y.; Haraya, K.; Hattori, S.; Sasuga, T. *Polymer* **1994**, *35*, 925-928.
40. Hill, A. J. In *High Temperature Properties and Applications of Polymeric Materials*; Tant, M. R.; Connell, J. W. McManus, H. L. N., Eds.; ACS Books: Washington, DC, 1995; pp 65-79.
41. Bartenev, G. M.; Varisov, A. V.; Goldanskii, V. I.; Mokrushin, A. D.; Tsyganov, A. D. *Soviet Physics-Solid State* **1971**, *12*, 2806.
42. Breck, D. W., *Zeolite Molecular Sieves*; Wiley-Interscience: New York City, 1974.
43. Singh, A.; Bondar, V.; Dixon, S.; Freeman, B. D.; Hill, A. J. *Proceedings of the American Chemical Society Division of Polymeric Materials: Science and Engineering* **1997**, *77*, 316-317.
44. Consolati, G.; Genco, I.; Pegoraro, M.; Zanderighi, L. *J. Polym. Sci.: Polym. Phys. Ed.* **1996**, *34*, 357-367.
45. Yampol'skii, Y. P.; Shantorovich, V. P.; Chernyakovskii, F. P.; Zanderleigh, L. *J. Appl. Polym. Sci.* **1993**, *47*, 85-92.
46. Tyn, M. T.; Calus, W. F. *Processing* **1975**, *21*, 16-17.
47. Tyn, M. T.; Calus, W. F. *J. Chem. Eng. Data* **1975**, *20*, 106-109.
48. Reid, R. C.; Prausnitz, J. M.; Poling, B. E., *The Properties of Gases and Liquids*; McGraw-Hill: New York, 1987; pp 741.

49. Tanaka, K.; Taguchi, A.; Hao, J.; Kita, H.; Okamoto, K. *J. Membrane Sci.* **1996**, *121*, 197-207.
50. Berens, A. R.; Hopfenberg, H. B. *J. Membrane Sci.* **1982**, *10*, 283-303.
51. Fujita, H. In *Diffusion in Polymers*; Crank, J. Park, G. S., Eds.; Academic Press: London, 1968; pp 75-105.
52. Fujita, H. *Fortschr. Hoch-polym.-Forsch* **1961**, *3*, 1-47.
53. Stern, S. A.; Saxena, V. *J. Membrane Sci.* **1980**, *7*, 47-59.
54. Doghieri, F.; Biavati, D.; Sarti, G. C. *Ind. Eng. Chem. Res.* **1996**, *35*, 2420-2430.
55. Brandt, W. W. *J. Phys. Chem.* **1959**, *63*, 1080-1084.
56. Meares, P. *J. Am. Chem. Soc.* **1954**, *76*, 3415-3422.
57. Robeson, L. *J. Membrane Sci.* **1991**, *62*, 165-185.
58. Robeson, L. M.; Burgoyne, W. F.; Langsam, M.; Savoca, A. C.; Tien, C. F. *Polymer* **1994**, *35*, 4970-4978.
59. Freeman, B. D. *Macromolecules* **1999**, *32*, 375-380.
60. Barrer, R. M. *Trans. Faraday Soc.* **1942**, *38*, 322-331.
61. van Amerongen, G. J. *J. Appl. Phys.* **1946**, *17*, 972-985.
62. Barrer, R. M.; Skirrow, G. *J. Polym. Sci.* **1948**, *3*, 549-563.
63. Pauly, S. In *Polymer Handbook, 3rd Edition*; Brandrup, J. Immergut, E., Eds.; John Wiley and Sons: New York, 1989; pp VI/435-VI/449.
64. *Permeability and Other Film Properties*; Plastics Design Library: New York, 1995.
65. Pinnau, I.; Casillas, C. G.; Morisato, A.; Freeman, B. D. *J. Polym. Phys.: Polym. Phys. Ed.* **1997**, *35*, 1483-1490.
66. Morisato, A.; Pinnau, I. *J. Membrane Sci.* **1996**, *121*, 243-250.
67. Morisato, A.; He, Z.; Pinnau, I. In *Proceedings of the American Chemical Society Division of Polymeric Materials: Science and Engineering*; Cocuzzi, D. A., Ed.; American Chemical Society: Washington, DC, 1997; pp 254-255.
68. Masuda, T.; Kawasaki, M.; Okano, Y.; Higashimura, T. *Polym. J.* **1982**, *14*, 371-377.
69. Takahashi, T.; Masuda, T.; Higashimura, T. *Polym. Prepr. Jpn.* **1982**, *31*, 1193-1196.
70. Ash, R.; Barrer, R. M.; Sharma, P. *J. Membrane Sci.* **1976**, *1*, 17-32.
71. Srinivasan, R.; Auvil, S. R.; Burban, P. M. *J. Membrane Sci.* **1994**, *86*, 67-86.
72. Fogg, P. G. T.; Gerrard, W., *Solubility of gases in liquids*; John Wiley & Sons: New York, 1991; pp 332.
73. Bondar, V. I.; Freeman, B. D.; Pinnau, I. *J. Polym. Sci., Polym. Phys. Ed.* submitted.
74. Sunderrajan, S.; Freeman, B. D.; Pinnau, I. In *Proceedings of the American Chemical Society Divsion of Polymeric Materials: Science and Engineering*; Cocuzzi, D. A., Ed.; ACS: Washington, DC, 1997; pp 267-268.
75. Pinnau, I.; Toy, L. G.; Sunderrajan, S.; Freeman, B. D. In *Proceedings of the American Chemical Society Divsion of Polymeric Materials: Science and Engineering*; Cocuzzi, D. A., Ed.; ACS: Washington, DC, 1997; pp 269-270.
76. Merkel, T. C.; Bondar, V. I.; Nagai, K.; Freeman, B. D. *J. Polym. Sci.: Polym. Phys. Ed.* in press.

77. Ghosal, K.; Chern, R. T.; Freeman, B. D.; Daly, W. H.; Negulesco, I. I. *Macromolecules* **1996**, *29*, 4360-4369.
78. Tanaka, K.; Kita, H.; Okano, M.; Okamoto, K. *Polymer* **1992**, *33*, 585-592.
79. Haraya, K.; Obata, K.; Hakuta, T.; Yoshitome, H. *Maku (Membrane)* **1986**, *11*, 48-52.
80. Chen, S. P.; Edin, J. A. D. *Polym. Eng. Sci.* **1980**, *20*, 40-50.
81. Barrie, J.; Williams, M. J. L.; Munday, K. *Polym. Eng. Sci.* **1980**, *20*, 20-29.
82. Nagai, K.; Higuchi, A.; Nakagawa, T. *J. Appl. Polym. Sci.* **1994**, *54*, 1207-1217.
83. Stern, S.; Shah, V.; Hardy, B. *J. Polym. Sci.: Polym. Phys. Ed.* **1987**, *25*, 1263-1298.

# Chapter 2

# Synthesis and Gas Permeability of Poly(diphenylacetylenes) with Substituents

**Toshio Masuda, Masahiro Teraguchi, and Ryoji Nomura**

**Department of Polymer Chemistry, Kyoto University, Kyoto 606–01, Japan**

Polymerization of diphenylacetylenes with a variety of substituents is presented, and the properties of the polymers are surveyed. The $TaCl_5$/cocatalyst ($n$-$Bu_4Sn$, $Et_3SiH$, 9-BBN, BuLi, etc.) systems have proven to be quite effective for the polymerization of diphenylacetylenes, providing high molecular weight polymers in good yields. Most of the produced polymers are completely soluble in organic solvents. High thermal stability and film-forming ability of the polymers make it possible to prepare thin tough membranes with high gas permeability. Interestingly, quite large $P_{O2}$ values (1100 barrers) are observed with poly(diphenylacetylenes) bearing bulky round-shaped *para*-substituents such as *tert*-butyl, trimethylsilyl, and trimethylgermyl groups.

Two essential factors for gas-separation membranes are permeability and permselectivity, and polymers with high permeability have gathered much attention due to both fundamental interest and practical importance (*1–3*). There are several groups of highly permeable polymers; one of them is substituted polyacetylenes whose large free volume is considered to contribute to their high gas permeability. Poly[(1-trimethylsilyl)-1-propyne] [poly(TMSP)], a silicon-containing disubstituted acetylene polymer, has been recognized as the most permeable polymer known (*4–9*). Its oxygen permeability ($P_{O2}$) is about ten times larger than that of poly(dimethylsiloxane) [poly(DMS)]. This polymer is glassy at room temperature and gives a thin tough film, which is in contrast to poly(dimethylsiloxane). These findings prompted many membrane scientists into intensive research on the gas permeability of this polymer (*10–15*). However, despite its potential utility as a gas separation membrane, there are several problems to be solved, among which is its insufficient thermal stability (*6*). Thus, the design and synthesis of polyacetylenes with both high permeability and excellent thermal stability are desirable for practical applications.

Poly(diphenylacetylene) [poly(DPA)] exhibits the highest thermal stability among substituted polyacetylenes. In this polymer, weight loss is observed only above 500 °C. However, insolubility of the polymer in any solvent has prevented detailed studies of its gas permeability (*16–17*). On the other hand, the introduction of substituents into DPA allows the formation of polyacetylenes with inflated molecular weight as well as high solubility in organic solvents (*16, 18–20*). Our recent efforts demonstrated that poly(DPAs) with suitable substituents, especially bulky ones, exhibit high solubility and high gas permeability. The present article intends to focus on the synthesis and properties, in particular gas permeability, of poly(DPAs) with a variety of substituents.

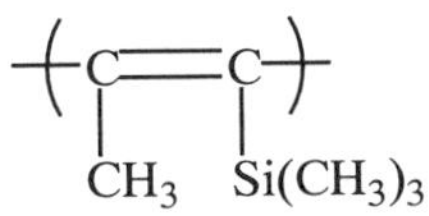

poly[(1-trimethylsilyl-1-propyne)]
[poly(TMSP)]

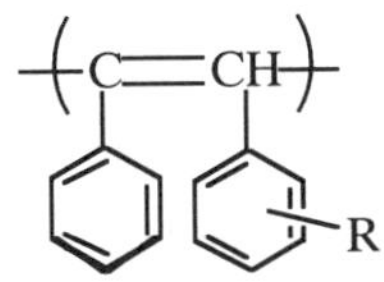

poly(DPAs) with substituents

## Synthesis of Poly(diphenylacetylenes)

**Polymerization of DPAs with Hydrocarbon Substituents.** The hydrocarbon DPAs (**1a–1e**; Table I) employed bear bulky substituents such as *tert*-butyl, *n*-butyl, adamantyl, and benzyl groups (*21–22*). The monomers are usually prepared in satisfactory yields by the Pd-coupling reaction of phenylacetylene with the respective *p*-iodobenzenes.

**Table I. Polymerization of 1 by the $TaCl_5$/*n*-$Bu_4Sn$ system**

| Monomers | Yield[a] (%) | $M_w/10^4$[b] | $T_0$[c] (°C) | $P_{O_2}$[d] (barrer) |
|---|---|---|---|---|
| PhC≡C$C_6H_4$-*p*-*t*Bu (**1a**) | 84 | 360 | 380 | 1100 |
| PhC≡C$C_6H_4$-*p*-*n*Bu (**1b**) | 82 | 130 | 320 | 100 |
| PhC≡C$C_6H_4$-*p*-Adm (**1c**) | 73 | 430 | 420 | 55 |
| PhC≡C$C_6H_4$-*p*-Bz (**1d**) | 74 | 87 | 330 | 18 |
| PhC≡C$C_6H_4$-*p*-Ph (**1e**) | 82 | insol. | 450 | – |

[a] Methanol-insoluble part. [b] Estimated by GPC (PSt. standards). [c] Measured in air.
[d] 1 barrer = 1 x $10^{-10}$ $cm^3$(STP)•cm/($cm^2$•s•cmHg).

The most effective catalysts for the polymerization of **1** involve combinations of $TaCl_5$ with cocatalysts such as *n*-$Bu_4Sn$, 9-borabicyclo[3.3.1]nonane (9-BBN), and $Et_3SiH$. These catalyst systems are so active that they achieve good yields of the polymers. No polymers are obtained without the cocatalysts. Group-6 metal chlorides such as $MoCl_5$ and $WCl_6$, which are quite active for the polymerization of phenylacetylenes, are not effective for **1** at all. NMR and IR spectroscopic analyses support the formation of polymers with alternating double bonds in the main chains; information on the regiospecificity is, however, not available.

Poly(**1**)s except poly(**1e**) are completely soluble in organic solvents and possess quite high molecular weights. For example, poly(**1a**) dissolves in a variety of solvents such as toluene, cyclohexane, $CCl_4$, $CHCl_3$ and THF, and GPC analysis of poly(**1c**) showed a weight-averaged molecular weight of 4.3 X $10^6$ (*23*). These polymers are thermally fairly stable; e.g., the onset temperature of weight loss ($T_0$) of poly(**1e**) in air is observed at 450 °C, which is comparable to that of poly(DPA). The polymers from **1a–1d** give free-standing films when cast from toluene solution.

**Polymerization of DPAs with Silyl or Germyl Groups.** As mentioned above, silicon-containing polyacetylenes such as poly(TMSP) and poly(dimethylsiloxane) show high gas permeability. It is, therefore, expected that silicon-containing poly(DPA)s will exhibit high gas permeability along with high solubility and thermal stability. A variety of DPAs with silyl or germyl groups were synthesized and polymerized by the $TaCl_5$-cocatalyst systems (*24–27*). The monomers investigated by the authors, **2a–2o**, are listed in Table II.

**Table II. Polymerization of 2 by the $TaCl_5$/*n*-$Bu_4Sn$ system**

| Monomers | Yield[a] (%) | $M_w/10^4$[b] | $T_0$[c] (°C) | $P_{O_2}$[d] (barrer) |
|---|---|---|---|---|
| *para*-substituted | | | | |
| PhC≡$CC_6H_4$-*p*-$SiMe_3$ (**2a**) | 85 | 220 | 420 | 1100 |
| PhC≡$CC_6H_4$-*p*-$SiEt_3$ (**2b**) | 84 | 160 | 350 | 95 |
| PhC≡$CC_6H_4$-*p*-Si*i*$Pr_3$ (**2c**) | 83 | 110 | 370 | 20 |
| PhC≡$CC_6H_4$-*p*-$SiPh_3$ (**2d**) | 84 | 220 | 430 | 3.8 |
| PhC≡$CC_6H_4$-*p*-$SiMe_2$*i*Pr (**2e**) | 74 | 160 | 350 | 200 |
| PhC≡$CC_6H_4$-*p*-$SiMe_2$*t*Bu (**2f**) | 61 | insol. | 350 | – |
| PhC≡$CC_6H_4$-*p*-$SiMe_2$Ph (**2g**) | 63 | insol. | 350 | – |
| PhC≡$CC_6H_4$-*p*-$GeMe_3$ (**2h**) | 52 | 180 | 350 | – |
| *meta*-substituted | | | | |
| PhC≡$CC_6H_4$-*m*-$SiMe_3$ (**2i**) | 87 | 140 | 400 | 1200 |
| PhC≡$CC_6H_4$-*m*-$SiEt_3$ (**2j**) | 63 | 54 | 350 | 92 |
| PhC≡$CC_6H_4$-*m*-$Me_2$*t*Pr (**2k**) | 71 | 105 | 350 | 3.9 |
| PhC≡$CC_6H_4$-*m*-$SiMe_2$*t*Bu (**2l**) | 51 | 50 | 350 | 110 |
| PhC≡$CC_6H_4$-*m*-$SiMe_2$Ph (**2m**) | 42 | 45 | 350 | 26 |
| PhC≡$CC_6H_4$-*m*-$GeMe_3$ (**2n**) | 60 | 180 | 390 | 1100 |
| disubstituted | | | | |
| PhC≡$CC_6H_3$-*m,m*-$(SiMe_3)_2$ (**2o**) | 0 | – | – | – |

[a] Methanol-insoluble part. [b] Estimated by GPC (PSt. standards). [c] Measured in air.
[d] 1 barrer = 1 x $10^{-10}cm^3$(STP)•cm/($cm^2$•s•cmHg).

The monomers are prepared in high yields similarly to hydrocarbon DPAs. A similar tendency was recognized with the polymerization catalysts; i.e., combinations of $TaCl_5$ with proper cocatalysts (*n*-$Bu_4Sn$, $Et_3SiH$, 9-BBN, *n*-BuLi, etc.) exhibit high catalytic activity. No polymers are obtained with $NbCl_5$, $MoCl_5$ and $WCl_6$-based catalysts. Yields of the polymers are usually satisfactory, and most of the produced polymers dissolve in common solvents such as toluene and chloroform. The molecular weights of the polymers are fairly high. For instance, polymerization of the DPA with *para*-trimethylsilyl group (**2a**) gives a polymer of very high molecular weight ($M_w = 220 \times 10^4$) in good yield.

There are no significant differences in the yields between *para*- and *meta*-substituted polymers, but the molecular weights for *meta*-substituted polymers are generally lower than those for *para*-substituted counterparts. The molecular weights of the polymers decrease when bulky substituents are incorporated onto the silicon atom, especially in the case of *meta*-substituted ones. Such a tendency was also found in the polymerization of TMSP (*28–29*). Poly(**2**)s are highly thermally stable; most of the polymers begin to lose weight in air only around 400 °C and above.

The high polymerizability of phenylacetylenes having two ring substituents [e.g., *o,p*-$(Me_3Si)_2C_6H_3C{\equiv}CH$] provides in good yields polymers whose gas permeability values are generally superior to those of the monosubstituted ones (*30*). On the contrary, **2o**, a disubstituted DPA with two trimethylsilyl groups shows poor polymerizability. For example, an attempt to polymerize **2o** by the $TaCl_5$/*n*-$Bu_4Sn$ system resulted in low conversion (34%) and no polymer formation. However, copolymers of **2o** with a few DPAs are obtained even in low yields as shown in Table III. It is noteworthy that completely soluble copolymers with high molecular weights are obtained when the feed ratio of **2o** to DPA was above 1:1.

**Table III. Copolymerization of 2o ($M_1$) with DPAs ($M_2$) by the $TaCl_5$/*n*-$Bu_4Sn$**

| Comonomer ($M_2$) | Feed Ratio | Convn. (%) $M_1$ | Convn. (%) $M_2$ | Yield[a] (%) | $M_w/10^4$ [b] | $T_0$ [c] (°C) | $P_{O_2}$ [d] (barrer) |
|---|---|---|---|---|---|---|---|
| none | – | 34 | – | 0 | – | – | – |
| DPA | 1:2 | 71 | 71 | insoluble | | 420 | 1100 |
| DPA | 1:1 | 38 | 46 | 33 | 37 | – | 95 |
| DPA | 2:1 | 37 | 30 | 14 | 8.7 | 370 | 20 |
| **2a** | 1:1 | 34 | 42 | 21 | 20 | 430 | 3.8 |
| **2i** | 1:1 | 18 | 17 | 7.4 | 11 | – | 200 |

[a] Methanol-insoluble part. [b] Estimated by GPC (PSt. standards). [c] Measured in air.
[d] 1 barrer = $1 \times 10^{-10} cm^3(STP){\bullet}cm/(cm^2{\bullet}s{\bullet}cmHg)$.

**Polymerization of DPAs with Other Heteroatoms.** Polymerization of other heteroatom-containing DPAs (**3a–3d**) was investigated with the motivation of evaluating the effect of polar groups on the polymerization behavior and polymer properties. The monomers employed involve alkoxy-, phenoxy-, and carbazolyl-substituted DPAs as shown in Table IV (*31–32*).

These monomers are synthesized in good yields by the Pd-catalyzed coupling reaction of the corresponding 4-iodophenyl ethers and *N*-(4-iodophenyl)carbazole with phenylacetylene. *O*- and *N*-containing polymers are obtained in good yields by using the $TaCl_5$/*n*-$Bu_4Sn$ system. Addition of cocatalysts such as *n*-$Bu_4Sn$, $Et_3SiH$ and 9-BBN is quite effective for the formation of high polymers, whereas no polymers are obtained without them, which is consistent with results of the other DPAs. Of particular note is that polymerization of DPAs in the presence of ether solvents does not lead to the formation of polymers (*20*), whereas rapid conversion of **3a–3d** is accomplished despite the presence of heteroatoms. For example, although the polymerization of **2i** with $TaCl_5$/*n*-$Bu_4Sn$ in anisole is completely suppressed, **3a–3c** are consumed within 15 min by using the same catalyst system to give the corresponding polymers. High molecular weight polymer with film-forming ability is attainable in good yield from **3c**. The nitrogen atom in **3d** does not inhibit its polymerization, and the respective polymer with high molecular weight is accessible in good yield. The most part of this polymer dissolves in toluene, chloroform, THF, anisole, 1,4-dioxane, DMF, and methyl benzoate. The polymers from **3c** and **3d** exhibit high thermal stability; the onset temperatures of weight loss for poly(**3c**) and poly(**3d**) are 420 and 470 °C, respectively, which are comparable to that of poly(DPA).

**Table IV. Polymerization of 3 by the $TaCl_5$/*n*-$Bu_4Sn$ system**

| Monomers | Yield[a] (%) | $M_w/10^4$[b] | $T_0$[c] (°C) | $P_{O2}$[d] (barrer) |
|---|---|---|---|---|
| PhC≡C$C_6H_4$-*p*-OMe (**3a**) | 58 | 54 | 310 | – |
| PhC≡C$C_6H_4$-*p*-O-*n*-Bu (**3b**) | 58 | 54 | 220 | – |
| PhC≡C$C_6H_4$-*p*-OPh (**3c**) | 69 | 170 | 420 | 37 |
| PhC≡C$C_6H_4$-*p*-*N*-Cz[e] (**3d**) | 67 | 49 | 470 | 2 |

[a] Methanol-insoluble part. [b] Estimated by GPC (PSt. standards). [c] Measured in air. [d] 1 barrer = 1 x $10^{-10}$cm$^3$(STP)•cm/(cm$^2$•s•cmHg). [e] Cz = Carbazolyl

## Gas Permeability of the Polymers

Figures 1, 2 and 3 plot the oxygen permeability coefficients ($P_{O2}$) versus the separation factor against nitrogen ($P_{O2}/P_{N2}$) of the present polymers along with those of poly(DMS) and poly(TMSP).

As shown in Figure 1, the substituents on the phenyl ring remarkably affect the permeability. It is notable that the polymers having *tert*-butyl (**1a**), trimethylsilyl (**2a, 2i**), and trimethylgermyl (**2n**) groups exhibit very large $P_{O2}$ values, up to 1100–1200 barrers, which is about a quarter as large as that of poly(TMSP) and approximately twice as large as that of poly(dimethylsiloxane). Along with poly(4-methyl-1-pentyne) (*33*), poly(**1a**) is one of the hydrocarbon polymers whose $P_{O2}$ values are larger than that of poly(dimethylsiloxane). In contrast, the $P_{O2}$ values of poly(**1b**) and poly(**1d**) are much smaller than that of poly(**1a**). From these results, one can conclude that the $P_{O2}$ value is greatly dependent on the shape of ring substituents and that introduction of bulky round-shaped substituents generally favors high oxygen permeability. However, even though round-shaped and bulkier substituents such as adamantyl (**1c**), triisopropylsilyl (**2c**), and triphenylsilyl (**2d**) groups are introduced, the permeability does not increase, but rather becomes much smaller (Figure 2). It is noted that there are no differences in the $P_{O2}$ values between poly(**2a**) and poly(**2i**). It is an indication that the position of the substituents does not play an important role in the permeability.

Copolymers of **2o** with either DPA or **2a** show appreciably small $P_{O2}$ values (Figure 3), which reveals that the presence of more than two trimethylsilyl groups is

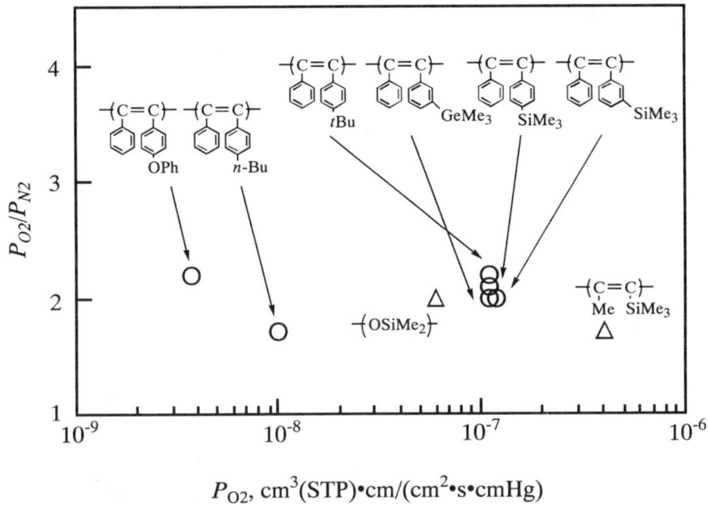


Figure 1. The $P_{O2}$ and $P_{O2}/P_{N2}$ of poly(DPA)s with variously shaped ring-substitutents (25 °C).

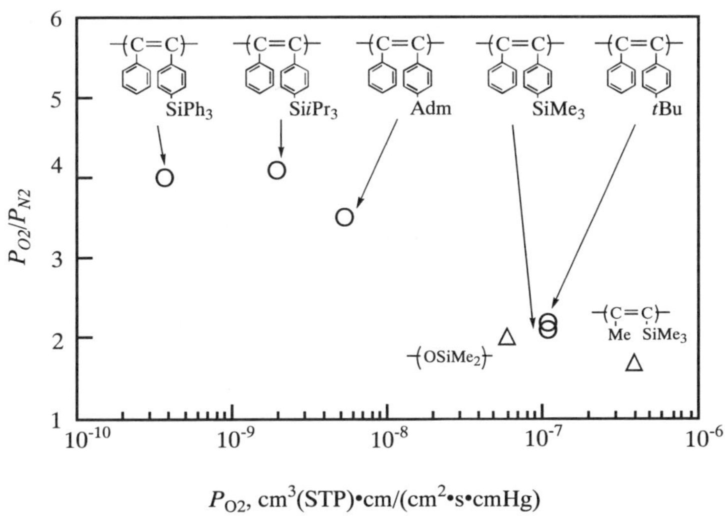


Figure 2. The $P_{O2}$ and $P_{O2}/P_{N2}$ of poly(DPA)s with round-shaped ring-substitutents (25 °C).

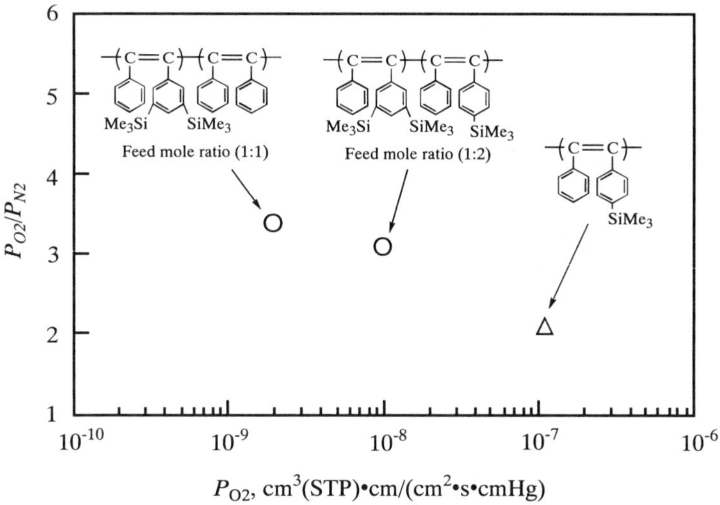


Figure 3. The $P_{O2}$ and $P_{O2}/P_{N2}$ of copolymers of **2o** with either DPA or **2a** (25 °C).

not indispensable for the formation of highly permeable membranes. The $P_{O2}$ values of the polymers with oxygen and nitrogen atoms are rather low; e.g., the oxygen permeability values of poly(**3c**) and poly(**3d**) are 37 and 2.0 barrers, respectively, which are similar to that of natural rubber (18 barrers).

Figure 4 illustrates the permeability coefficients of poly(**2a**), poly(**2c**) and poly(TMSP) for several gases at 25 °C. These polymers show polygonal lines with very similar patterns, although the permeability varies to a large extent depending on the polymer structure. These results suggest that the permeation mechanism is identical for these polymers.

## Summary

This paper has provided an overview of the polymerization of DPAs bearing a variety of substituents and the polymer properties including thermal stability and gas permeability. The high ability of the $TaCl_5$/cocatalyst system to polymerize DPAs permits the formation of polymers with high molecular weight in satisfactory yields. Poly(DPA)s are, in general, thermally stable, and many of them are comparable to the most thermally stable poly(DPA). Several poly(DPA)s with roundshaped substituents exhibited large $P_{O2}$ values which are about twice as large as that of poly(dimethylsiloxane). Although a systematic discussion of the relationship between the structure of substituents and gas permeability is not feasible yet, some tendencies of the effect of substituents on gas permeability were demonstrated. The high permeability as well as good thermal stability of the series of polymers offers the possibility to find practical applications for these interesting materials in the future.

## References

1. Kesting, R. E.; Fritzsche, A. K. *Polymeric Gas Separation Membranes*; Wiley: New York, 1993.
2. Stern, S. A. *J. Membr. Sci*. **1994**, *94*, 1.
3. Cabasso, I. *Encyclopedia of Polymer Science and Engineering 1*; Kroschwitz, J. I., Ed.; 2nd Ed.; Wiley: New York, 1987, Vol. 9; pp 509–579. I
4. Masuda, T.; Iguchi, Y.; Tang, B-Z.; Higashimura, T. *Polymer* **1988**, *29*, 2041.
5. Masuda, T.; Isobe, E.; Higashimura, T.; Takada, K. *J. Am. Chem. Soc*. **1983**, *105*, 7473.
6. Takada, K.; Matsuya, J.; Masuda, T.; Higashimura, T. *J. Appl. Polym. Sci.* **1985**, *30*, 1605.
7. Ichiraku, Y.; Stern, S. A.; Nakagawa, T. *J. Membr. Sci.* **1987**, *34*, 5.
8. Witchey-Lakshmanan, L. C.; Hopfenberg, H. B.; Chern, R. T. *J. Membr. Sci.* **1990**, *48*, 321.
9. Plate, N. A.; Bokarev, A. K.; Kalieuzhnyi, N. E.; Litvinova, E. G.; Khotimskii, V. S.; Volkov, V. V.; Yampolskii, Y. P. *J. Membr. Sci.* **1991**, *60*, 13.
10. Savoca, A. C.; Surnamer, A. D.; Tien, C-F. *Macromolecules* **1993**, *26*, 6211.
11. Chen, G.; Griesser, H. J.; Mau, A. W. H. *J. Membr. Sci.* **1993**, *82*. 99.
12. Yampolskii, Y. P.; Shantorovich, V. P.; Chernyakovskii, F. P.; Kornilov, A. I.; Plate, N. A. *J. Appl. Polym. Sci.* **1993**, *47*, 85.
13. Pope, D. S.; Koros, W. J.; Hopfenberg, H. B. *Macromolecules* **1994**, *27*, 5839.
14. Consolati, G.; Genco, I.; Pegoraro, M.; Zanderighi, L. *J. Polym. Sci., Part B: Polym. Phys.* **1996**, *34*, 357.
15. Toy, L. G.; Freeman, B. D.; Spontak, R. J.; Morisato, A.; Pinnau, I. *Macromolecules* **1997**, *30*, 4766.

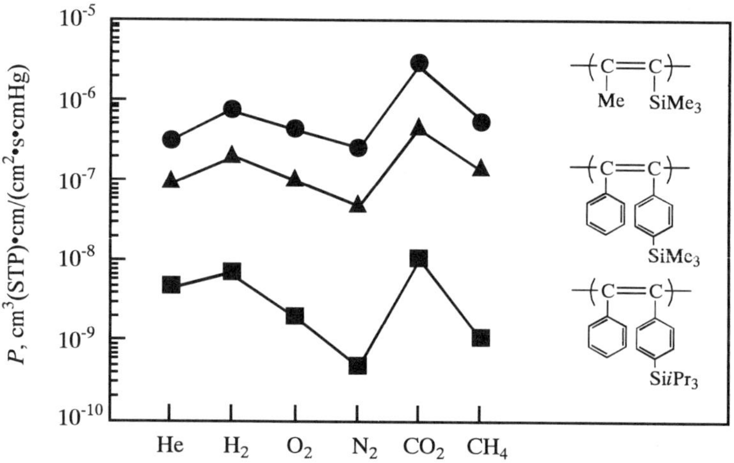


Figure 4. The permeability coefficients ($P$) of poly(**2a**), poly(**2c**), and poly(TMSP) for various gases (25 °C).

16. Shirakawa, H.; Masuda, T.; Takeda, K. *Synthesis and Properties of Acetylenic Polymers*; Supplement C2; Patai, S. Ed., J. Wiley: Chichester, 1994, Chapter 17, pp945–1016.
17. Niki, A.; Masuda, T.; Higashimura, T. *J. Polym. Sci., Part A: Polym. Chem.* **1987**, *25*, 1553.
18. Masuda, T.; Tachimori, H. *J. Macromol. Sci., Pure Appl. Chem.* **1994**, *A31*, 1675.
19. Masuda, T.; Misumi, Y. *Design and Synthesis of New Substituted Polyacetylenes*; Kamachi, M., Nakamura, A. Eds., Springer: Heidelberg, 1996; pp. 59–68.
20. Masuda, T. *The Polymeric Materials Encyclopedia*; Salamone, J. C. Ed., CRC Press: Florida, 1996; Vol. 1, pp 32–39.
21. Kouzai, H.; Masuda, T.; Higashimura, T. *Polymer* **1994**, *35*, 4920.
22. Kouzai, H.; Masuda, T.; Higashimura, T. *J. Polym. Sci., Part A: Polym. Chem.* **1994**, *32*, 2523.
23. Teraguchi, M.; Masuda, T. unpublished results.
24. Tsuchihara, K.; Masuda, T.; Higashimura, T. *J. Polym. Sci., Part A: Polym. Chem.* **1993**, *31*, 547.
25. Tsuchihara, K.; Masuda, T.; Higashimura, T. *Macromolecules* **1992**, *25*, 5816.
26. Ito, H.; Masuda, T.; Higashimura, T. *J. Polym. Sci., Part A: Polym. Chem.* **1996**, *34*, 2925.
27. Teraguchi, M; Masuda, T. *Polym. Prepr. Jpn*. **1997**, *46*, 1291.
28. Isobe, E.; Masuda, T.; Higashimura, T. *J. Polym. Sci., Part A; Polym. Chem.* **1986**, *24*, 1839.
29. Masuda, T.; Isobe, E.; Higashimura, T. *J. Polym. Sci., Part A; Polym. Chem.* **1987**, *25*, 1353.
30. Aoki, T.; Nakahara, H.; Hayakawa, Y.; Kokai, M.; Oikawa, E. *J. Polym. Sci., Part A: Polym. Chem.* **1994**, *32*, 849.
31. Tachimori, H.; Masuda, T.; Kouzai, H.; Higashimura, T. *Polym. Bull.* **1994**, *32*, 133.
32. Tachimori, H.; Masuda, T. *J. Polym. Sci., Part A: Polym. Chem.* **1995**, *33*, 2079.
33. Morisato, A.; Pinnau, I. *J. Membr. Sci*. **1996**, *121*, 243.

# Chapter 3

# Sorption and Diffusion of *n*-Alkanes and Alcohols in Poly(1-trimethylsilyl-1-propyne)

**F. Doghieri, M. Giacinti, L. Nicolini, and G. C. Sarti**

**Dipartimento di Ingegneria Chimica, Mineraria e delle Tecnologie Ambientali, Università di Bologna, Viale Risorgimento 2, 40136 Bologna, Italy**

The sorption of several penetrants in PTMSP has been studied as a function of temperature and pressure. For both solubility and diffusivity isotherms, the experimental results show significant differences between n-alkanes and alcohols. A discussion of the experimental data is presented, considering the glassy matrix as a homogenous phase, and using thermodynamic arguments commonly applied to standard mixtures. It is thus possible to offer a unique description of the thermodynamic properties of the various mixtures examined, in spite of their rather different behaviour. Simple isotherms for the mobility coefficient are also calculated for all penetrants as a function of composition. Remarkably, they show very similar trends for both n-alkanes and alcohols.

During the last decades interest in membrane separation processes and in solutions for packaging problems has given a substantial boost to research on sorption and mass transport in polymeric materials. Several interpretative models for the solubility and diffusivity in rubbery polymers are now available which, at least in principle, allow for the prediction of the permeability of low molecular weight species in polymeric films above their glass transition temperature.

On the other hand, estimation of solubility and diffusivity data in glassy polymers requires direct experimental measurements. The results of these measurements can be correlated through the use of empirical models. The solubility of low molecular weight species in glassy polymers is usually described in terms of the dual mode model (*1*,*2*). The dual mode model is a powerful tool for the representation of most gas solubility isotherms. Its utility has been demonstrated in many studies (*3*-*5*). However, the parameters entering the model are endowed with a physical meaning not always consistent with the experimental observations (*6*). Thus, despite some interesting attempts to interpret the dependence of these parameters on polymer properties, the dual mode model for the solubility in glassy polymers cannot be used in a predictive way.

The proper representation of the thermodynamic properties of out-of-equilibrium systems such as glassy polymers is still an open question. Reliable correlations and predictive expressions for the mass transport properties in polymeric glasses as a function of temperature and concentration are also lacking.

This work offers a contribution to the understanding of some fundamental aspects of sorption and diffusion in glassy polymers. The research focuses on an extensive experimental study of sorption and mass transport in a specific polymeric matrix. A high free volume polymer, (poly 1-trimethylsilyl-1-propyne) [PTMSP], has been used here in order to emphasise aspects of sorption and transport which are peculiar to polymer/penetrant mixtures below the glass transition temperature. The discussion of the experimental data presented in this work permits a clarification of concepts which are of general validity for the interpretation of thermodynamic and mass transport properties in glassy systems.

## Background and Basic Approach

PTMSP is a glassy polymer first synthesised by Masuda et al.(*7*). It gained a reputation as one of the most interesting materials for the production of gas separation or pervaporation membranes. Indeed, the solubility and permeability of organic vapours and gases in PTMSP are orders of magnitude larger than the corresponding properties in conventional glassy polymers and their values are even comparable to or higher than those in many rubbery materials (*8,9*). The extremely high free volume of PTMSP, on the order of 30%, is responsible for the extraordinary properties of the material and makes it a limiting case among glassy polymers (*8,10*). The density and, in turn, the free volume of PTMSP samples, however, may have rather different values according to sample preparation procedures and pretreatments (*11,12*). In a previous study, the effect of various pretreatment protocols on the solubility and diffusivity of ethanol in PTMSP was considered. In particular, it was determined that keeping the polymer sample under vacuum at a relatively low temperature for a long time, results in a decrease in both solubility and mobility of ethanol. On the other hand, only minor changes in solubility and diffusivity were obtained through preswelling treatment protocols (*6*).

In this study an extensive set of experiments was conducted to determine solubility and diffusivity of several vapours in PTMSP. Five different penetrant species were considered, three alkanes of different molecular weight and two polar components. The effect of temperature on solubility and diffusivity was also considered by performing several sorption experiments in the range between 285 and 330 K.

The experimental results are briefly discussed in terms of thermodynamic and mass transport properties in the glassy polymer mixture. The aim of the discussion is to highlight peculiarities of solubility and diffusivity in polymeric systems below the glass transition temperature and to consider possible interpretations. The focus is on the effect of swelling on the thermodynamic and transport properties in glasses. Indeed, it is well known that, contrary to the case of rubbery systems, the solute partial specific volume in dilute glassy mixtures is much lower than the specific volume of the pure solute liquid phase (*13*). From this point of view, PTMSP represents a limiting case among glassy polymers, since practically negligible values of the partial specific volume of both $CO_2$ and $CH_4$ in PTMSP have been measured at solute concentrations up to and higher than 20% by weight (*14*). Therefore, at least for the sorption of limited amounts of penetrant mass, the polymer matrix of PTMSP behaves as a rigid structure with a constant volume. In the sorption experiments performed in this work, only mixtures with relatively low penetrant content are explored, which allows us to neglect in all cases the dilation of the polymer sample under sorption conditions.

## Experimental

In order to measure the solubility and diffusivity of low molecular weight vapours in PTMSP, sequences of differential sorption experiments were performed. The mass uptake as a function of time for each sorption step was measured and recorded using a Cahn microbalance. Details of the experimental equipment are discussed in a previous publication (*15*). Prof. G.Costa of the CNR-IMAG kindly provided the polymer, which was synthesised using $TaCl_4$ as catalyst, and had an intrinsic viscosity of 4.5 *dl/g* in toluene solution at 30°C.

The samples used in this work were cut from the same polymeric membrane, which was 40 μm thick. The membrane was obtained through solvent evaporation from a toluene solution, and kept at temperature around -20°C after its preparation, in order to minimise ageing effect on the polymeric material. The density of the membrane (evaluated from its mass to volume ratio) was slightly higher than 0.80 kg/l. In all cases sorption experiments were performed by contacting the polymer with pure penetrant vapour. This protocol permitted an accurate estimate of the fugacity of the penetrant in the penetrant-PTMSP mixture. In these experiments, the penetrant concentration range was from 0 to approximately 12% by weight. Both sorption and desorption experiments were performed for each penetrant and temperature and no significant difference in the two solubility isotherms was obtained in any case. Indeed, as demonstrated in previous studies (*6,15*), when the analysis is confined to this concentration range, significant dilation of the sample during the sorption experiment is avoided. Therefore, all of polymer/penetrant mixture have, to a very good approximation, the same partial polymer density and this greatly simplifies the data comparison as well as the discussion.

Five different penetrants were considered; n-pentane (n-$C_5$), n-hexane (n-$C_6$), n-heptane (n-$C_7$), ethanol (EtOH) and methanol (MeOH). For the cases of n-pentane, ethanol and methanol, solubility and diffusivity isotherms at three different temperatures 285, 300 and 330 K were examined. Sorption experiments for n-hexane in PTMSP were performed both at 300 and 330 K, while only the highest temperature was used for n-heptane.

The vapour phase pressure was controlled to a fixed value for each sorption step and measured with an accuracy of 0.1 mbar for n-pentane, ethanol and methanol. The use of a more precise manometer was needed for n-hexane and n-heptane sorption in order to measure the vapour phase pressure with an accuracy of 0.01 mbar.

The temperature of the system was controlled to a constant value during the sorption experiments and the thermostat system kept the temperature constant within an accuracy of ±0.1 K. Two different sequences of sorption experiments, with two different samples, were conducted for each penetrant at each temperature. Excellent reproducibility was obtained in all cases. The difference in solubility values obtained from repeated runs was negligible for most of the experimental conditions examined and never exceeded 5%.

## Results and Discussion

**Solubility Isotherms.** In Figure 1, the solubility of n-pentane in PTMSP is reported for three different temperatures. The zero pressure solubility coefficient $S_0$, defined as the limit of the ratio between the penetrant weight fraction $\omega_1$ and the corresponding activity $a_1$, $S_0 = \lim_{a_1 \to 0} \frac{\omega_1}{a_1}$, is significantly high at all temperatures: $S_0 \approx 20$, 15 and 10 at 285, 300 and 330 K, respectively.

As also observed by Morisato et al. (*10*) and by Doghieri and Sarti (*15*), the solubility isotherms of n-pentane in PTMSP have the usual downward curvature, which is typical of the sorption of most gases and vapours in glassy polymers.

From the data in Figure 1, a significant temperature effect on the solubility of n-pentane in PTMSP is also evident at all penetrant activities. As is typical for the sorption in glassy polymers, the solubility of the penetrant increases as the temperature decreases at each activity value. This effect is obviously associated with negative values of the mixing enthalpy, and it appears to be significantly pronounced in the case of n-pentane.

In Figure 2 the solubility isotherms of n-hexane in PTMSP are reported at temperatures of 300 K and 330 K. The behaviour is qualitatively similar to that observed for n-pentane sorption. Both isotherms exhibit a downward curvature and the solubility coefficient in the low pressure limit, $S_0$, is high. For n-hexane, $S_0 \approx 25$ and 16 at 300 and 330 K, respectively. Also for n-hexane sorption, the effect of temperature on solubility indicates a large negative enthalpy of mixing since the solubility decreases substantially with increasing temperature.

Remarkably, upon comparing the solubility of n-pentane and n-hexane in PTMSP, it is observed that, at the same temperature and penetrant activity, the solute mass ratio dissolved in PTMSP is higher for the higher molecular weight component. The same trend is confirmed when the analysis is extended to the case of n-heptane, and it is clearly apparent from Figure 3, where the solubility isotherms for the three alkanes at 330 K are simultaneously reported. From the data shown in Figure 3, the solubility coefficient in the low pressure limit for n-heptane in PTMSP may be evaluated as $S_0 \approx 27$.

In Figure 4 the solubility isotherms at three different temperatures are reported for ethanol in PTMSP. Unlike the case of n-alkanes, the solubility coefficient of ethanol in PTMSP is quite low in the range of low pressures; it increases with penetrant content in the low concentration region and it decreases at higher penetrant fugacity. The

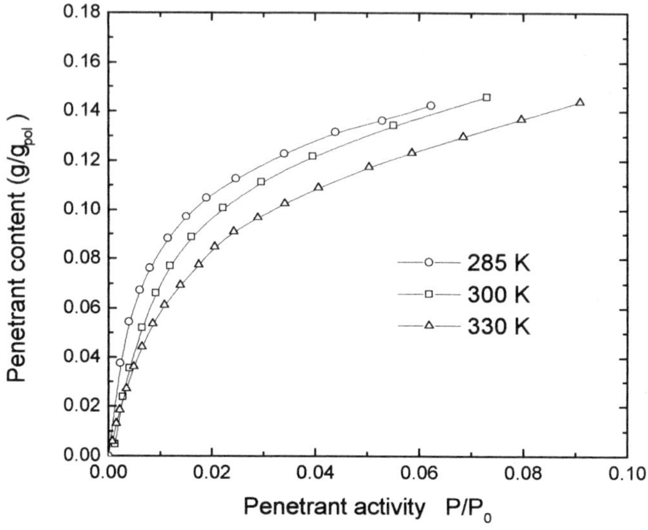


Fig.1 Sorption isotherms for n-pentane in PTMSP.

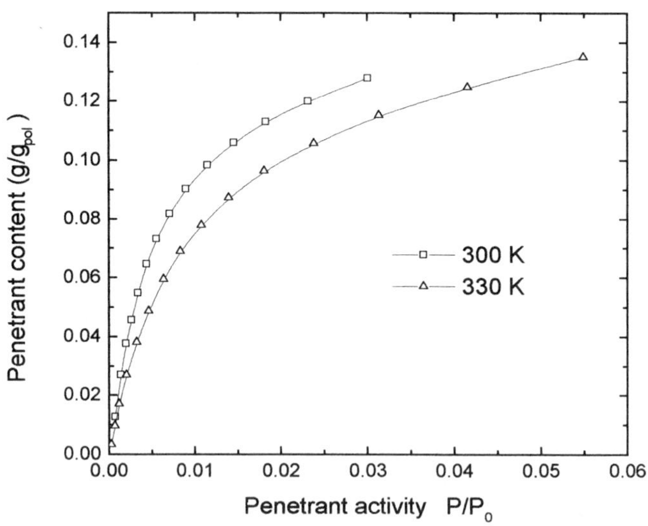


Fig.2 Sorption isotherms for n-hexane in PTMSP.

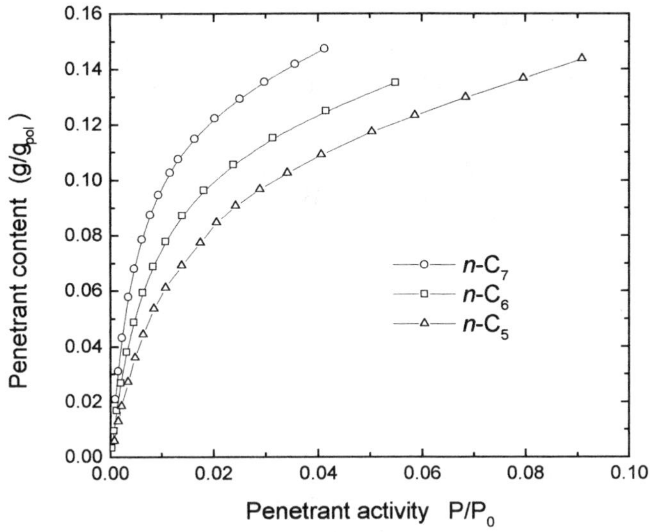


Fig.3 Sorption isotherms for n-alkanes in PTMSP at 330 K.

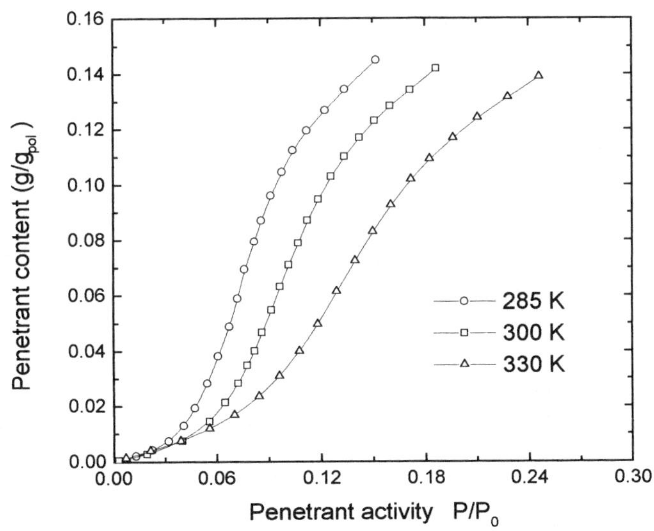


Fig.4 Sorption isotherms for ethanol in PTMSP.

resulting sorption isotherms exhibit an unusual S-shape confirming the behaviour already observed by Nakanishi et al. (*16*) and by Doghieri et al. (*6*).

The peculiar features shown by the sorption isotherms of ethanol in PTMSP are followed also by the solubility curves for methanol in PTMSP, which are presented in Figure 5. As in the case of ethanol, the methanol solubility coefficient in the low pressure limit, $S_0$, is orders of magnitude lower than for alkanes, and the sorption isotherms are characterised by the previously mentioned S shape.

The sorption isotherms for ethanol and methanol reported in Figure 4 and 5 cannot be interpreted on the basis of the well known dual mode model (*1*,*2*). This model assumes that the penetrant content in the glassy polymer matrix may be expressed as function of pressure through the sum of two contributions: the first refers to the penetrant molecules which are considered to be adsorbed onto the surface of microvoids in the interior of the solid polymer, and the second represents the contribution due to penetrant molecules which are strictly dissolved into the solid phase. In the original formulation of the dual mode model the first contribution is expressed as function of pressure in terms of the Langmuir equation and the second through Henry's law.

It must be pointed out that for all the n-alkane sorption isotherms presented in Figures 1 to 3, one can find a temperature dependent Langmuir affinity constant, Langmuir capacity and Henry's law constant which accurately describe the sorption isotherms. On the other hand, as was also observed by Nakanishi et al.(*16*), the S shaped isotherms typical of the alcohols in PTMSP cannot be simply described through the dual mode model. On the contrary, it was already pointed out (*6*) that the Henry's law line drawn through the origin intersects the solubility isotherms measured for alcohols. This behaviour is intrinsically inconsistent with the dual mode basic ideas, since it would require a negative contribution to the sorption isotherm due to adsorption in the low pressure range.

The discussion and interpretation of the difference between n-alkanes and alcohols isotherms will be presented by considering the contributions to the excess partial Gibbs free energy of the penetrants in the polymer phase. In this analysis, rather standard arguments from classical theories of mixture thermodynamics are used.

**Excess Gibbs Free Energy**. The chemical potential of the penetrant in the polymer mixture at T=330 K is reported, in dimensionless terms, in Figure 6 as function of penetrant weight fraction for all the vapours considered in this work. The dimensionless chemical potential $\mu/RT$ is the logarithm of the activity and has been calculated as $\mu/RT = \ln(P/P_0)$ where $P$ is the pressure in the vapour phase at equilibrium and $P_0$ is the penetrant vapour pressure at the same temperature.

When the solubility isotherms are reported in the usual terms of penetrant mass ratio versus activity, qualitative differences are apparent between n-alkanes and alcohols, as already discussed above, which might suggest important differences in the physical mechanism involved in penetrant sorption. However, when the same solubility isotherms are presented as penetrant chemical potential versus penetrant mass fraction for the n-alkanes and alcohols in this study, the isotherms are in all cases qualitatively very similar to one another. When the data are presented in this fashion,

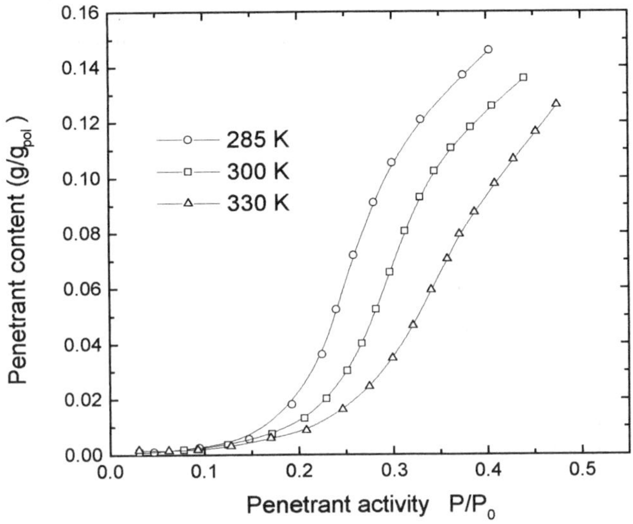


Fig.5 Sorption isotherms for methanol in PTMSP.

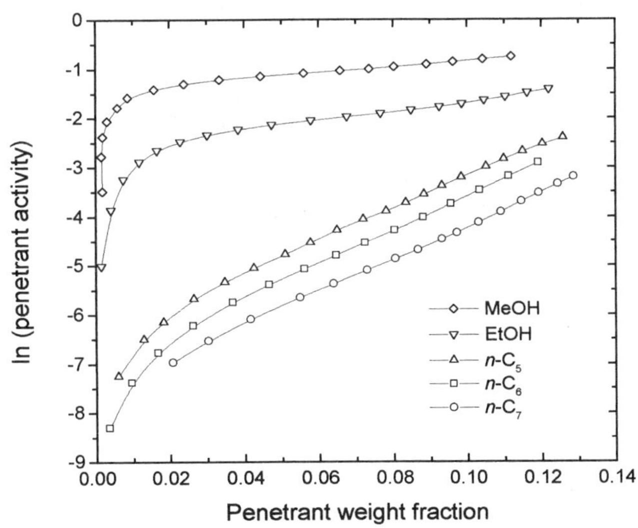


Fig.6 Chemical potential isotherms in PTMSP at 330 K.

only differences in the amount of sorption are evident between n-alkanes and alcohols. That is apparent from Figure 6 where data at 330 K are reported for all the penetrants considered in this study. We simply observe that the excess partial Gibbs free energy for alcohols in PTMSP mixtures is much higher than that for n-alkanes at all penetrant concentration considered. For higher concentrations, the logarithm of activity increases more significantly with penetrant weight fraction for n-alkanes than for alcohols.

The excess chemical potential obviously results from the sum of partial excess enthalpy and entropy as expressed, in dimensionless terms, by the following well known relationship:

$$\frac{\bar{G}_1^{EX}}{RT} = \ln(a_1) = \frac{\bar{H}_1^{EX}}{RT} - \frac{\bar{S}_1^{EX}}{R} \tag{1}$$

where $\bar{H}_1^{EX}$, $\bar{S}_1^{EX}$ and $\bar{G}_1^{EX}$ are the partial excess enthalpy, entropy and free energy of penetrant in the mixture.

The quantitative differences observed between the two families of penetrants are due both to enthalpy and to entropy contributions. Based on a qualitative consideration of the interaction energies between the polymer matrix and the aliphatic or the alcoholic groups of the different penetrants, the enthalpic term in Equation 1 is mostly responsible for the quantitative differences observed in the isotherms. This was also the conclusion drawn in a previous study (*15*) where the solubility of ethanol and n-pentane in PTMSP were qualitatively compared with each other. Based on the solubility isotherms at different temperatures for n-alkanes and alcohols, we can now directly evaluate the excess enthalpy of mixing in Eq.(1). In a previous discussion of the comparison of solubility data for ethanol and n-pentane in PTMSP (*6*), it was assumed that the difference in chemical potential for the two penetrants was essentially due to the enthalpic term. This hypothesis is reasonable since the interaction energy of penetrant molecules with PTMSP segments should be significantly different, relative to interaction energies of pure penetrant molecules, between polar and non polar penetrants.

From the experimental data presented herein, we are now able to estimate the mixing enthalpy for n-alkanes and alcohols in PTMSP and, thus, to evaluate the excess enthalpy term in Eq.(1). The excess enthalpy $\bar{H}_1^{EX}$ for the penetrant in the mixture may be calculated from the solubility data at different temperatures using the well known relation:

$$\bar{H}_1^{EX} = -\left(\frac{\partial \ln a_1}{\partial\, 1/T}\right)_{p,\omega_1} \tag{2}$$

Indeed, the use of Eq. (2) for a nonequilibrium glassy phase deserves further discussion which cannot be presented here and will be reported in a future publication.

**Excess Enthalpy**. From Eq.(2), an estimation of the penetrant excess enthalpy in the concentration range from 3 to 14% by weight was obtained for n-$C_5$, n-$C_6$, ethanol and methanol. In the case of n-pentane, the partial enthalpy of mixing increases with increasing penetrant content. In the concentration interval up to 10% by weight, the n-pentane excess enthalpy ranges from -16 to -12 kJ/mol. The $\bar{H}_1^{EX}$ values obtained are thus approximately half the enthalpy of vaporisation $\Delta \tilde{H}_{vap}^0$ of n-pentane at room temperature ($\Delta \tilde{H}_{vap}^0 \approx 26\ kJ/mol$ as indicated, e.g., in ref. *17*). Lower absolute values of the excess enthalpy are estimated for n-$C_5$ at concentrations higher than 10% by weight, probably due to a slight swelling of a polymeric matrix which is more likely to occur at higher concentrations.

Similarly, the excess enthalpy of n-hexane in PTMSP is evaluated from the data reported in Figure 2. By considering concentrations below 11% by weight, $\bar{H}_1^{EX}$ is estimated to range from -17 to -14 *kJ/mol*. Thus, a relatively narrow range of excess enthalpy is obtained for the case of *n*-hexane, and the absolute value of the excess enthalpy is half the vaporisation enthalpy of the penetrant at room temperature ($\Delta \tilde{H}_{vap}^0 \approx 30\ kJ/mol$ as indicated, e.g., in ref. *17*).

For ethanol, the excess enthalpy $\bar{H}_1^{EX}$ calculated from the temperature dependence of the solubility isotherms, is negligible at very low penetrant concentrations, and it ranges from -7 to -10 *kJ/mol* for the higher concentrations inspected.

For methanol sorption, the excess enthalpy $\bar{H}_1^{EX}$ can be evaluated from the data in Figure 5. Its values range from -3 to -5 *kJ/mol* over most of the concentration interval inspected. It is negligible for very low methanol concentrations, in close analogy with the ethanol case.

For the sake of comparison with the case of *n*-alkanes sorption, for which $\bar{H}_1^{EX}$ is on the order of 1/2 $\Delta \tilde{H}_1^{vap}$, it is worthwhile to notice that the ratio between the absolute value of excess enthalpy and the vaporisation enthalpy at room temperature $\left| \bar{H}_1^{EX} \middle/ \Delta \tilde{H}_1^{vap} \right|$ is approximately equal to 1/4 and 1/8 for ethanol and methanol, respectively.

In summary, the enthalpy of mixing is always negative and thus represents an important contributions in favour of the solubility of penetrants in PTMSP. For alcohols in PTMSP, the excess enthalpy is significantly smaller than for the n-alkanes and even shows negligible values in the low concentration range where the solubility coefficient of alcohols is very small. We can conclude that a relevant contribution to the difference in partial free energy between alcohols and n-alkanes in PTMSP comes from different energetic effects of mixing. Moreover, the very different excess enthalpy values are responsible for the very different solubility isotherms of alcohols and n-alkanes in PTMSP.

**Diffusivity and Mobility Isotherms.** The temperature and penetrant concentration dependence of the diffusion coefficient, $D$, has been measured for binary mixtures of PTMSP with n-pentane, n-hexane, ethanol and methanol. Indeed, for each sorption step in the sequences of isothermal sorption experiments, the kinetics of mass uptake have been analysed and estimates of the corresponding diffusivity have been obtained from both small and long time sorption data, as indicated in a previous study (*6*). The relative difference in the values of the diffusion coefficients obtained through the use of long time and short time sorption data is less than 5% in the great majority of cases and is less than 10% in the worst cases. Indeed, when one compares the diffusion coefficient results obtained from two different experiments performed under the same conditions, the diffusivity values are only determined to within approximately ± 10%.

In the following discussion, the diffusion coefficient data have been measured during the same experiments considered above for the solubility data. The reported diffusivities are obtained as the arithmetic mean between the values calculated from short time and long time data and are plotted as function of the average penetrant concentration in each differential sorption step considered.

In Figure 7, diffusivity isotherms at 300 K are presented for n-pentane, n-hexane, ethanol and methanol. The measured values of $D$ are orders of magnitude higher than the typical values of diffusion coefficients in glassy polymers. The diffusivity of the two alkanes in PTMSP show similar trends as a function of the penetrant content. For both n-pentane and n-hexane, the diffusion coefficient exhibits a maximum variation of about 40% and a rather flat maximum value at intermediate concentrations.

The diffusivity isotherms obtained for ethanol and methanol are similar to each other and quite different from those of the alkanes. Very large diffusion coefficients have been measured in the low penetrant concentration range. Then, the diffusion coefficients decrease to a minimum value at approximately 7% penetrant weight fraction. Afterwards, the diffusivity increases again and shows a local maximum value at $\omega \approx 11\%$. The latter trend is also observed for diffusivity coefficients measured at temperatures of 285 and 330 K. However, for the sake of conciseness, these data are not presented explicitly. The present data also confirm the diffusivity behaviour of ethanol in PTMSP reported previously (*6*).

We observe in this case that, contrary to what could be expected based simply on molecular weight considerations, the values of methanol diffusivity in PTMSP are the lowest among all of the penetrants considered.

Remarkably, the temperature dependence of the alcohol diffusivity in PTMSP does not show a unique trend in the entire concentration range: an increase in temperature results in an increase or a decrease in the diffusivity according to the value of the composition of the polymer/penetrant mixture.

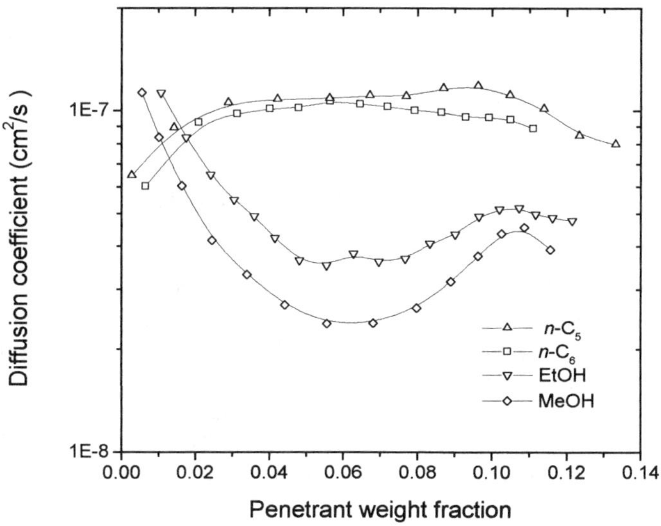


Fig.7 Diffusivity isotherms in PTMSP at 300 K.

A clear interpretation of the concentration and temperature dependence of the mass transport properties of both alkanes and alcohols in PTMSP can be obtained when it is recognised that the diffusion coefficient, $D$, results from the product of the mobility, $L$, and a pure thermodynamic factor, $\alpha$, defined as

$$\alpha = \omega_1 \frac{\partial \ln(a_1)}{\partial \omega_1} \tag{3}$$

where $a_1$ and $\omega_1$ are the penetrant activity and weight fraction, respectively.

Indeed, by considering that the gradient of the chemical potential $\mu$ is the true driving force for mass transport, then the corresponding kinetic conductance is defined in terms of the mobility, $L$, through the following relation:

$$\boldsymbol{J}_1 = -\omega_1 \rho L \nabla(\mu_1/RT) \tag{4}$$

where $\rho$ is the density of the mixture, and $\boldsymbol{J}_1$ is the diffusive mass flux density of the penetrant.

When Eq.(4) is compared with the well known Fick's law, in the form

$$\boldsymbol{J}_1 = -\rho D \nabla \omega_1 \tag{5}$$

it is easily recognised that the $D$, $L$, and $\alpha$ are related as follows:

$$D = \alpha\ L \tag{6}$$

This equation clearly states that the diffusion coefficient results from the product of a purely thermodynamic factor, $\alpha$, and a purely kinetic factor, the mobility, $L$, as long as the chemical potential is the true driving force for the mass transport process.

As the thermodynamic factor $\alpha$ can be determined from the solubility isotherms described in the previous section, we are able to estimate the mobility for all of the penetrant-PTMSP mixtures considered in this study. The thermodynamic factor $\alpha$ at the mean value of the concentration for a given sorption step can be estimated from the following relation

$$\alpha = \frac{\ln(a_{1fin}) - \ln(a_{1in})}{\ln(\omega_{1fin}) - \ln(\omega_{1in})} \tag{7}$$

where the subscripts *in* and *fin* refer to initial and final values for the sorption step, respectively.

The values of mobility are obtained as the ratio between the diffusivity and the thermodynamic factor, estimated using Eq.(7). Mobility values are reported as a function of the average penetrant weight fraction in each sorption step.

In Figure 8, the mobility isotherms at 300 K are reported for binary mixtures of PTMSP with n-pentane, n-hexane, ethanol and methanol. Despite some scatter in the data due to the uncertainty in the values of the diffusion coefficient and to some error propagation due to the use of Eq.(7), it is quite evident that all the isotherms show the same shape. In all cases the mobility coefficient decreases exponentially with increasing penetrant weight fraction. These results confirm the data obtained for

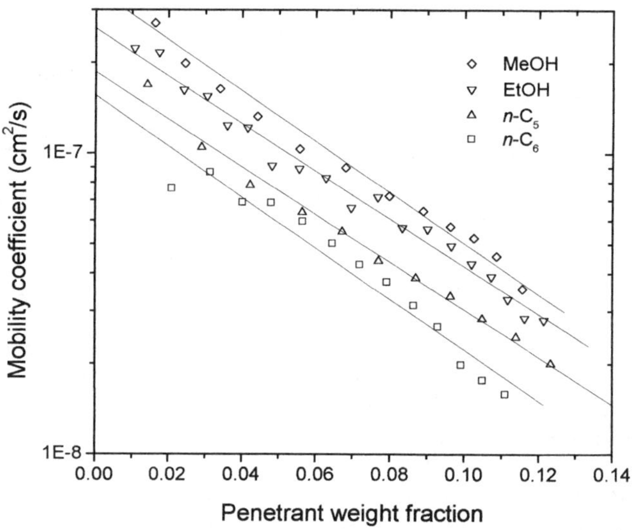


Fig.8 Mobility isotherms in PTMSP at 300 K.

ethanol and n-pentane in a previous study (*15*). Therefore, the difference in the diffusion isotherms between alcohols and n-alkanes is only due to differences in the thermodynamic factor of the diffusion coefficient. The mobility coefficients, which have a pure kinetic meaning, follow very similar trends for these very different families of penetrants.

From the point of view of transport kinetics, it is reasonable to expect that the difference among the penetrants is essentially due to the difference in molecular weight. This point of view is consistent with the data in Figure 8. The mobility coefficient calculated from the data measured in this work is a monotonically decreasing function of the penetrant molecular weight at any fixed value of the penetrant weight fraction. Furthermore, the ratio between the mobility of two given penetrants at fixed temperature and weight fraction is substantially independent of concentration. These observations are confirmed by a comparison of the mobility data at 330 K for the alkanes, which is shown in Figure 9. For the results presented in Figure 9, the mobility is also a decreasing function of penetrant molecular weight, and the ratio of the mobility of one penetrant to another is independent of concentration.

For all the cases considered here, in the concentration range studied, the mobility coefficient almost exponentially decreases with the penetrant weight fraction. The slope, $\partial \ln(L)/\partial \omega$, is between approximately -20 and -16.

The measured concentration dependence of the mobility coefficient for PTMSP is opposite to that commonly observed for rubbery polymers. The reason for this difference is due to the decrease in the specific free volume of PTMSP mixtures as the penetrant concentration increases, at least as long as the swelling of the polymer matrix is negligible (*6*).

Similar isotherms for the mobility coefficient are observed at all temperatures and, moreover, the ratio between the mobility coefficients at different temperatures for the same component are essentially independent of concentration. This observation permits the evaluation of a unique activation energy, $E_a$, for the mobility coefficient of a given penetrant in PTMSP. The values of the activation energy of the mobility coefficient in PTMSP for n-alkanes and alcohols are reported in Table I.

It is worthwhile to notice that the value of the activation energy measured for the penetrants in PTMSP is significantly lower than that in the other glassy polymers used as gas separation membranes. Indeed, this result is consistent with previous results for the activation energy of diffusion coefficients of gases in PTMSP (*18*). The previously reported activation energies of diffusion were evauated based on gas permeability and sorption data.

Tab.I Activation energy for the mobility coefficient

| species | | *n*-$C_5$ | *n*-$C_6$ | EtOH | MeOH |
|---|---|---|---|---|---|
| $E_a$ (kJ/mol) | | 6.0 | ≈6.5 | 14 | 32 |

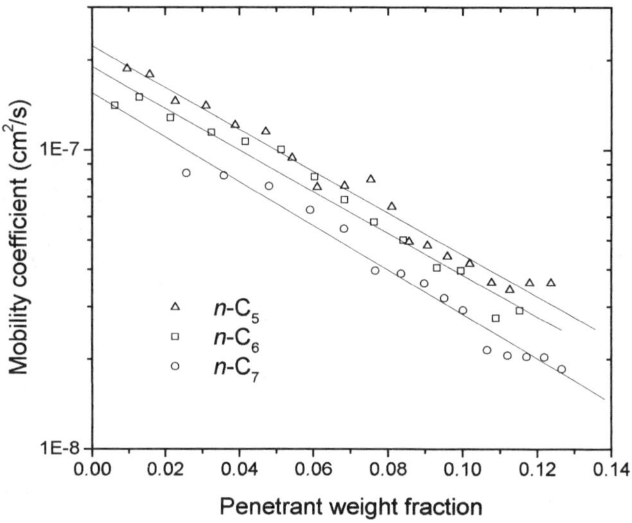


Fig.9 Mobility isotherms in PTMSP of n-alkanes in PTMSP at 330 K.

## Conclusions

The solubility of several penetrants in PTMSP indicate relevant differences between polar and non-polar components. In particular, for the alcohols, unusual S shaped solubility isotherms are observed which cannot be explained by the dual mode model. Variations of solubility with temperature were also considered in this work, and the mixing enthalpy for n-pentane, n-hexane, ethanol and methanol in PTMSP was calculated. Significant negative mixing enthalpies were measured in all cases, and absolute values of the excess enthalpy of the sorption is much higher for the alkanes than for the alcohols.

By considering the solid glassy mixture as a homogeneous phase, the apparent qualitative differences between the solubility of alkanes and alcohols in PTMSP may be easily interpreted as simple quantitative differences in energetic and entropic contributions to the free energy of mixing. Indeed, the chemical potentials of the penetrant species in the mixture show similar isotherms for every case considered. The difference in the value of the chemical potential for alkanes and alcohols in PTMSP is mainly due to differences in energetics of the sorption process, and not to entropic effects.

The diffusivity data also show interesting differences between n-alkanes and alcohols. The diffusivity of alcohols, in particular, give unexpected non monotonic isotherms as a function of penetrant content. This effect is essentially due to the thermodynamic factor of the diffusion coefficient. When the mobility coefficient is calculated, simple monotonic isotherms are obtained in all cases, with similar dependences of penetrant mobility on penetrant content. The mobility of alcohols and alkanes exponentially decrease with increasing penetrant weight fraction. The slope of the corresponding isotherms is similar in all cases.

## Acknowledgements

We wish to thank gratefully Prof. G. Costa of the CNR of Genoa for kindly providing the PTMSP used in this work. This work has been partially supported by the Italian Ministry of University and of Scientific and Technological Research, 40% and 60%.

## Literature Cited

1. Barrer, R.M.; Barrie, J.A.; Slater, J. *J. Polym. Sci.*, **1958**, *27*, 177.
2. Michaels, A.S.; Vieth, W.R.; Barrie, J.A. *J. Appl. Phys.*, **1963**, *34*, 1.
5. Doghieri, F.; Biavati, D.; Sarti, G.C. *I.E.&C. Res.* **1996**, *35*, 2420.
4. Stannett, V.T.; Koros, W.J.; Paul, D.R.; Lonsdale, H.K.; Baker, R.W. *Adv.Polym.Sci.* **1979**, *32*, 69.
5. Koros, W.J.; Paul, D.R.; Rocha, A.A. *J.Polym.Sci., Polym. Phys. Ed.* **1976**, *14*, 687.
6. Koros,W.J.; Paul, D.R. *J.Polym.Sci., Polym. Phys. Ed.* **1978**, *16*, 1947.
7. Masuda, T.; Isobe, E.; Higashimura, T. *Macromolecules*, **1985**, *18*, 841.

8. Platè, N.A.; Bokarev, A.K.; Koliuzhnyi, N.E.; Litvinova, E.G.; Khotimskii, V.S.; Volkov, V.V.; Yampol'skii, Y.P., *J. Membr. Sci.*, **1991**, *60*, 13.
9. Camera-Roda, G.; Bottino, A.; Capannelli, G.; Costa, G.; Sarti, G.C. *Proc. 5th Int. Conf. Pervaporation Proc. in the Chem. Ind.*, Bakish Materials Corporation, Heidelberg (Germany), March 1991, p.495.
10. Morisato, A.; Shen, H.C.; Sankar, S.S.; Freeman, B.D.; Pinnau, I.; Casillas *J.Polym.Sci.,Part B:Polym.Phys.*, **1996**, *34*, 2209.
11. Fusaoka, Y.; Imazu, E.; Fujii, Y. *Proc. of the Vth world Filtration Congress*, Nice (France), June 1990, p.408.
12. Nogase, Y.; Takamura, Y.; Matsui, K. *J. Appl. Polym. Sci.*, **1991**, *42*, 185.
13. Fleming, G.K.; Koros, W.J.; *Macromolecules*, **1986**, *19*, 2285.
14. Pope, D.S.; Koros, W.J.; Hopfenberg, H.B. *Macromolecules*, **1994**, *27*, 5839.
15. Doghieri, F.; Sarti, G.C. *J.Polym.Sci.:Phys.*, **1997**, *35*, 2245.
16. Nakanishi, K.; Odani, H.; Kurata, M.; Masuda, T.; Higashimura, T. *Polymer J.*, **1987**, *19*, 293.
17. Perry's Chemical Engineers' Handbook, $6^{th}$ ed., Mc-Graw Hill, New York (1984).
18. Srinivasan, R.; Auvil, S.R.; Burban, P.M. *J. Membr. Sci.*, **1994**, *86*, 67.

## Chapter 4

# Mixed-Gas Permeation Properties and Physical Aging of Poly(4-methyl-2-pentyne)

**A. Morisato, Z. He, and I. Pinnau**

**Membrane Technology and Research, Inc., 1360 Willow Road, Menlo Park, CA 94025**

Poly(4-methyl-2-pentyne) [PMP] is a glassy, disubstituted, purely hydrocarbon-based polyacetylene. PMP has a density of only 0.78 g/cm$^3$ and a high fractional free volume of 0.28. The polymer has very high hydrocarbon permeabilities; for example, the *n*-butane permeability of PMP in a mixture of 2 mol% *n*-butane in methane is 7,500 x $10^{-10}$ cm$^3$(STP)·cm/cm$^2$·s·cmHg at 25°C. In contrast to conventional, low-free-volume glassy polymer membranes, PMP is significantly more permeable to *n*-butane than to methane in gas mixtures. In this paper, we present the gas permeation properties of PMP in mixtures of *n*-butane with methane. The mixed-gas permeation and physical aging properties of PMP are compared to those of poly(1-trimethylsilyl-1-propyne), the most permeable polymer known.

The recovery of organic vapors from waste gas streams using polymeric membranes is a well established process (*1*). Typically, composite membranes are used for this process. These membranes consist of a thin, selective rubbery layer coated onto a microporous support material. The selectivities of these membranes for organic vapors over nitrogen are typically about 10-100. Currently, commercial vapor separation membrane applications include small systems (10-100 scfm) to recover fluorinated hydrocarbons (Freons) and other high-value solvent vapors from process vent streams to large systems (100-1,000 scfm) for recovery of hydrocarbon vapors in the petrochemical industry (*1*).

The most promising new membrane-based vapor separation applications include: (i) separation of higher hydrocarbons ($C_{2+}$) from methane for dewpoint and heating value control of natural gas, (ii) recovery of hydrogen and/or hydrocarbons from petroleum refinery waste gases, and (iii) separation of olefins (ethylene or propylene) from polyolefin polymerization purge gas streams. In all of these applications, the higher hydrocarbons are the minor components in the feed gas. To minimize membrane area requirements, it is highly desirable to use a selective layer material that is more permeable to the larger, condensable hydrocarbons than to the smaller supercritical gas mixture components, such as methane, hydrogen or nitrogen.

## Background

Gas and vapor transport through polymeric membrane is typically described by the solution-diffusion model. Gas molecules at the high-concentration (high-pressure) side of the membrane dissolve in the membrane material and diffuse through the membrane along a concentration gradient to the low-concentration (low-pressure) side of the membrane. It is assumed that the gas phases on either side of the membrane are in thermodynamic equilibrium with their respective membrane interfaces, and that the interfacial sorption and desorption processes are rapid compared to the rate of diffusion through the membrane. For simple gases, Fick's law leads to the equation:

$$J = \frac{D \cdot S \cdot \Delta p}{\ell} \tag{1}$$

where J is the volumetric gas flux ($cm^3$(STP)/$cm^2$•s), D is the diffusion coefficient of the gas in the membrane ($cm^2$/s), $\ell$ is the membrane thickness (cm), S is the solubility coefficient ($cm^3$(STP)/$cm^3$•cmHg), and $\Delta p$ is the pressure difference across the membrane (cmHg).

The gas permeability, P ($cm^3$(STP)•cm/$cm^2$•s•cmHg) is commonly expressed as:

$$P = D \times S \tag{2}$$

The selectivity $\alpha_{A/B}$ is defined as the ratio of the permeabilities $P_A$ and $P_B$:

$$\alpha_{A/B} = \frac{P_A}{P_B} \tag{3}$$

Equation (3) can also be written as:

$$\alpha_{A/B} = \frac{P_A}{P_B} = \frac{D_A}{D_B} \times \frac{S_A}{S_B} \tag{4}$$

The ratio ($D_A/D_B$) is the ratio of the diffusion coefficients of the two gases and is referred to as the diffusivity selectivity, reflecting the different molecular sizes of the gases. The ratio ($S_A/S_B$) is the ratio of the solubility coefficients of the gases and can be viewed as the solubility selectivity, reflecting the relative condensabilities of the gases. The diffusion coefficient decreases with increasing molecular size, because large molecules interact with more segments of the polymer than small molecules. On the other hand, the solubility coefficient typically increases with molecular size, because large molecules are normally more condensable than smaller ones. For the membranes used for the separation of organic vapors from permanent gases, the balance between the diffusivity selectivity and solubility selectivity determines whether a membrane material is permanent-gas-selective or organic-vapor-selective.

In conventional, low-free-volume, glassy polymers, such as polysulfone, the permeability generally decreases with increasing permeant size. Glassy polymer membranes are used for the separation of permanent gas mixtures ($O_2/N_2$, $H_2/CH_4$, $CO_2/CH_4$). On the other hand, in rubbery polymers, the permeability typically increases with increasing penetrant size and condensability. Rubbery membranes are, therefore, more permeable to large organic vapors than to small permanent gases. The most commonly used rubbery membrane material for commercial membrane-based vapor separation systems is crosslinked polydimethylsiloxane [PDMS], the most permeable rubber known. PDMS has high organic-vapor/permanent-gas selectivity and can easily be formed into thin-film composite membranes.

**Polyacetylenes for Gas and Vapor Separation Applications.** Polyacetylene-based polymers have been studied extensively for gas separation applications (*2-4*). These new amorphous, glassy polymers are characterized by very high glass transition temperatures, typically >200°C, low densities, and high gas permeabilities. In particular, poly(1-trimethylsilyl-1-propyne) [PTMSP], has the highest gas permeabilities of all known polymers (*2-4*). In contrast to conventional glassy polymers, such as polysulfone, PTMSP is significantly more permeable to large organic vapors than to small permanent gases (*5*).

A similar highly permeable, disubstituted polyacetylene, poly(4-methyl-2-pentyne) [PMP], was synthesized more than a decade ago by Masuda *et al.* (6). The chemical structure of PMP is shown in Figure 1. The pure-gas permeation properties of poly(4-methyl-2-pentyne) are similar to those of PTMSP; the gas permeabilities increase as the size, or condensability, of the penetrants increase (*7*). The physical properties, nitrogen permeabilities, and oxygen/nitrogen selectivities of PMP and PTMSP are shown in Table I.

Table I. Physical properties of high-free-volume, glassy poly(4-methyl-2-pentyne) and poly(1-trimethylsilyl-1-propyne).

| Physical Property | PTMSP | PMP |
|---|---|---|
| $T_g$ | > 250 °C | > 250 °C |
| Density | 0.75 g/cm$^3$ | 0.78 g/cm$^3$ |
| Fractional free volume | 0.29 cm$^3$/cm$^3$ | 0.28 cm$^3$/cm$^3$ |
| Nitrogen permeability | 7,000 Barrers* | 1,300 Barrers* |
| Oxygen/nitrogen selectivity | 1.4 | 2.2 |

* 1 Barrer = $1 \times 10^{-10}$ cm$^3$(STP)•cm/cm$^2$•s•cmHg

In this study, we investigated the gas permeation properties of glassy PMP membranes in mixtures of *n*-butane/methane. The influence of feed gas composition and feed temperature on permeability and selectivity is discussed. Furthermore, the

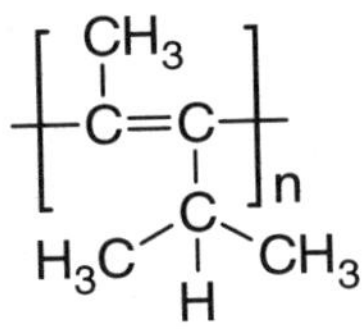

Figure 1. Chemical structure of poly(4-methyl-2-pentyne).

effect of the downstream pressure conditions (atmospheric or vacuum) on the physical aging of PMP membranes was studied using nitrogen as the probe penetrant.

## Experimental

**Polymer Film Preparation.** Isotropic, dense films of PMP were prepared by a ring-casting method from cyclohexane solution (1.5 wt%) onto a glass plate. The films were dried at ambient conditions for 48 h and then under vacuum at 80°C for 3 days to completely remove the solvent. Film samples with thicknesses of 20-40 μm (± 0.5 μm) were used for the permeation measurements.

**Characterization of Mixed-Gas Permeation Properties.** The mixed-gas permeation properties of isotropic PMP films were determined for several *n*-butane/methane mixtures. The permeation experiments were performed using the constant pressure/variable volume method (*8*) at a feed pressure of 150 psig and atmospheric permeate pressure (0 psig). The feed temperature was varied between 0°C and 50°C. The following *n*-butane/methane feed mixtures were used: 1, 2, 4, 6, and 8 mol% *n*-$C_4H_{10}$ in $CH_4$, respectively. The stage-cut, that is, the permeate flow rate to feed flow rate was less than 1%. Thus, the residue composition was essentially equal to the feed composition. The feed, residue, and permeate compositions were determined with a gas chromatograph equipped with a thermal conductivity detector.

The permeate flux, J, measured by the constant pressure/variable volume method is given by (*8*):

$$J = \frac{273 \cdot p_a}{76 \cdot T \cdot A}\left(\frac{dV}{dt}\right) \tag{5}$$

where A is the membrane area ($cm^2$), $p_a$ is the atmospheric pressure (cmHg), T is the gas temperature (K), and dV/dt is the volumetric displacement rate of the soap film in the glass capillary ($cm^3$/s). The mixed-gas permeability coefficient of each gas component was calculated from the relationship:

$$P = \frac{x_2 \cdot J \cdot \ell}{x_1 \cdot p_1 - x_2 \cdot p_2} \tag{6}$$

where $x_1$ and $x_2$ are the mol fractions of the gas components in the feed and permeate streams, respectively, and $p_1$ and $p_2$ are the pressures (cmHg absolute) on the feed and permeate side of the membrane, respectively. The mixed-gas selectivity was calculated with equation (3).

The effects of physical aging on the pure-gas permeation properties of PMP were determined with nitrogen by both the constant pressure/variable volume method (15 psia permeate pressure) and the constant volume/variable pressure method (0 psia permeate pressure). The experiments were carried out at 25°C; the feed pressure was 50 psig.

## Results and Discussion

**Mixed-Gas Permeation Properties of PMP for *n*-Butane/Methane Mixtures.** Recently, we reported the mixed-gas permeation properties of PTMSP, a high-free-volume, glassy polymer (*5,9,10*). Contrary to the permeation behavior of conventional, low-free-volume glassy polymers, such as polysulfone, PTMSP is significantly more permeable to large, condensable hydrocarbons than to small permanent gases, such as methane, nitrogen, or hydrogen. In fact, PTMSP exhibits the highest mixed-gas $C_{2+}$ hydrocarbon/methane and $C_{2+}$/hydrogen selectivity combined with the highest hydrocarbon permeability of any known polymer. One of the most interesting features of mixed-gas transport in PTMSP is blocking of the least condensable gas components in mixtures by co-permeation of condensable gases. For example, in a mixture containing only 2 mol% *n*-butane in methane, the methane permeability in PTMSP is reduced to only 10% of its pure-gas permeability (*5,9*).

The mixed-gas transport behavior of PMP is qualitatively similar to that of PTMSP. The data in Table II show that PMP is significantly more permeable to *n*-butane than to methane. For a feed gas mixture of 2 mol% *n*-butane in methane at a feed pressure of 150 psig and atmospheric permeate pressure at 25°C, the mixed-gas *n*-butane/methane selectivity of PMP is 14 and the *n*-butane permeability is 7,500 x $10^{-10}$ $cm^3(STP)\cdot cm/cm^2\cdot s\cdot cmHg$. This result indicates that the *n*-butane/methane selectivity in PMP is dominated by a high solubility selectivity, similar to the behavior of high-free-volume, glassy PTMSP (*5,10*). The methane permeability of PMP in the mixture was reduced 5-fold by co-permeation of *n*-butane.

Table II. Mixed-gas permeation properties of high-free-volume, glassy PTMSP and PMP. Feed: 2 mol% *n*-butane in methane; feed pressure: 150 psig; permeate pressure: atmospheric (0 psig); temperature: 25°C.

| Polymer | Polymer type | Permeability (Barrers)* $n$-$C_4H_{10}$ | Permeability (Barrers)* $CH_4$ | Mixed-gas selectivity n-$C_4H_{10}$/$CH_4$ | Mixed-gas/ pure-gas $CH_4$ permeability ratio |
|---|---|---|---|---|---|
| PTMSP | glassy | 53,500 | 1,800 | 30 | 0.1 |
| PMP | glassy | 7,500 | 530 | 14 | 0.2 |

* 1 Barrer = $10^{-10}$ $cm^3(STP)\cdot cm/cm^2\cdot s\cdot cmHg$.

The unusual mixed-gas permeation properties of PTMSP are due to its extraordinarily high excess free volume. Srinivasan *et al.* (*11*) and Pinnau *et al.* (*9*) suggested that the free-volume elements in PTMSP may be connected, leading to a structure with interconnected channels throughout the polymer. Such a structure could lead to significant transport of large, condensable gases by sorption and surface diffusion along the interior surface of such channels. Both *n*-butane permeability and *n*-butane/methane selectivity of PMP are lower than those of PTMSP. The lower permeability and selectivity of PMP may be the result of a lower free volume and less connectivity of free volume elements in the polymer matrix.

The permeation properties of PMP were also determined as a function of feed gas composition using mixtures of 1 to 8 mol% *n*-butane in methane. The mixed-gas permeation conditions were the same as those described above. The *n*-butane and methane permeabilities of PMP as a function of the relative *n*-butane pressure are shown in Figure 2. The relative *n*-butane pressure, $p/p_{sat}$, is the partial *n*-butane pressure in the mixture to the *n*-butane saturation pressure at 25°C (35.2 psia). As the relative *n*-butane pressure in the feed gas increased from 0 to about 0.1, the permeability of methane decreased about 5-fold, whereas the *n*-butane permeability was essentially constant. As a result, the *n*-butane/methane selectivity of PMP increased from 11 at a relative *n*-butane pressure of 0.05 to 16 at a relative *n*-butane pressure of 0.38, as shown in Figure 3.

**Influence of Temperature on Mixed-Gas Permeation Properties of PMP.** The permeability coefficients for a mixture of 2 mol% *n*-butane in methane were determined at temperatures ranging from 0 to 50 °C. The feed pressure was 150 psig; the permeate pressure was atmospheric (0 psig). The temperature dependence of gas permeation through polymers is given by the Arrhenius relationship:

$$P = P_0 \exp\left(-\frac{E_p}{RT}\right) \quad (7)$$

where $P_0$ is a pre-exponential factor, $E_p$ is the activation energy of permeation (J/mol), $R$ is the gas constant (8.314 J/mol•K) and T is the gas temperature (K). The activation energy of permeation is the sum of the activation energy of diffusion, $E_d$, and the heat of sorption, $H_s$:

$$E_p = E_d + H_s \quad (8)$$

The activation energy of diffusion depends on the polymer chain mobility and the size of the penetrant relative to the intra- and inter-molecular polymer chain gaps. Diffusion coefficients in polymers always increase with increasing temperature, that is, $E_d$ is always positive. In most polymers, gas solubilities decrease with increasing temperature; therefore, $H_s$ is generally negative.

In pure-gas measurements, PMP showed negative activation energies of permeation to all gases tested (including supercritical gases such as nitrogen or methane), that is, permeabilities increased with decreasing temperature (*7*). On the other hand, in mixed-gas permeation measurements with the binary mixture of 2 mol% *n*-butane in methane, PMP exhibited positive activation energy of permeation for methane (+9.6 KJ/mol) and negative activation energy of permeation for *n*-butane (-7.9 KJ/mol). The Arrhenius plots of permeability for *n*-butane and methane as a function of reciprocal temperature are shown in Figure 4.

The *n*-butane/methane selectivity increased monotonically from 8 to 24 as the feed temperature was decreased from 50 to 2°C, as shown in Figure 5. It is important to note that both *n*-butane permeability and *n*-butane/methane selectivity increased in the PMP membrane as the feed temperature decreased. This temperature/permeability behavior is completely different from that in low-free-volume glassy polymers used

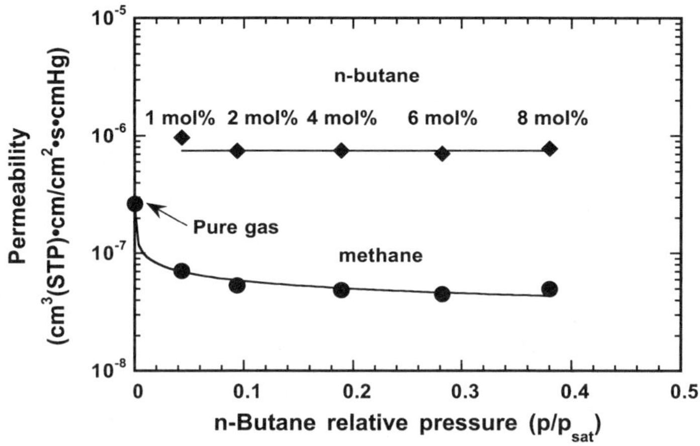


Figure 2. Mixed-gas methane and *n*-butane permeability of poly(4-methyl-2-pentyne) as a function of *n*-butane relative pressure. Feed: 2 mol% n-butane in methane; feed pressure: 150 psig; permeate pressure: atmospheric (0 psig); T=25°C.

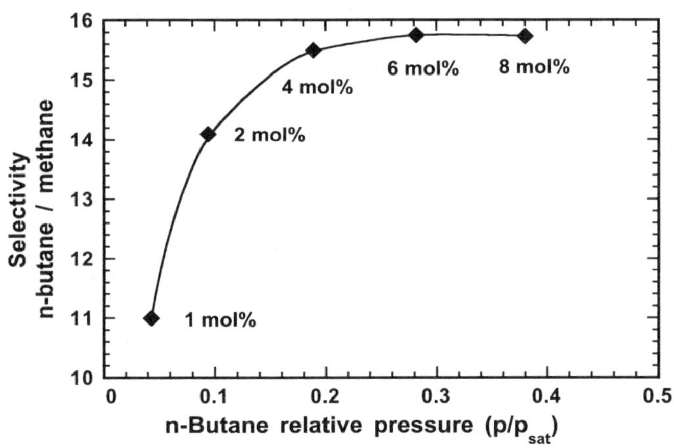


Figure 3. Mixed-gas *n*-butane/methane selectivity of poly(4-methyl-2-pentyne) as a function of *n*-butane relative pressure. Feed: 2 mol% n-butane in methane; feed pressure: 150 psig; permeate pressure: atmospheric (0 psig); T=25°C.

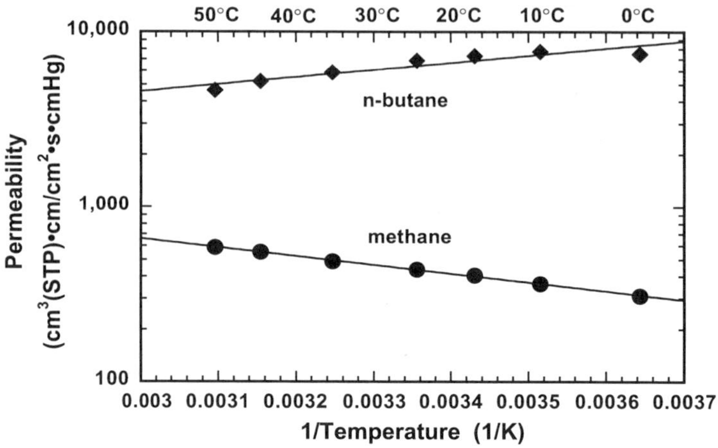


Figure 4. Mixed gas *n*-butane and methane permeability of poly(4-methyl-2-pentyne) as a function of temperature. Feed gas: 2 mol% *n*-butane in methane; feed pressure: 150 psig; permeate pressure: atmospheric (0 psig).

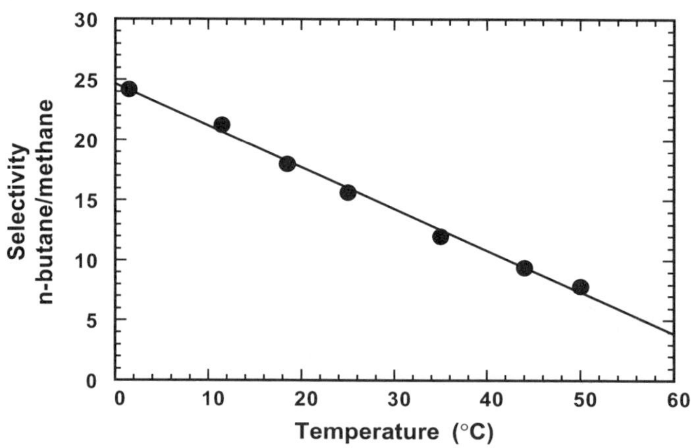


Figure 5. Mixed gas *n*-butane/methane selectivity of poly(4-methyl-2-pentyne) as a function of temperature. Feed gas: 2 mol% *n*-butane in methane; feed pressure: 150 psig; permeate pressure: atmospheric (0 psig).

for separation of permanent gases. In low-free-volume glassy polymers, the permeability decreases for all permanent gas components, whereas the selectivity increases as the temperature decreases.

**Physical Aging of Poly(4-methyl-2-pentyne).** Nagai *et al.* (*12,13*) and Shimomura *et al.* (*14*) reported that the gas permeability of PTMSP decreases rather dramatically with time. This phenomenon has been ascribed to three separate mechanisms: (1) physical aging leading to relaxation of the enormous levels of excess free volume in the polymer, (2) contamination of the sample by sorption of pump oil vapors during permeation experiments, and (3) chemical aging due to oxidation of the double bonds in the polymer chain backbone.

Figure 6 shows the pure-gas nitrogen permeability of PMP as a function of time. The nitrogen permeability of PMP decreased to about 75% of its initial value during 29 days using the constant pressure/variable volume method (15 psia permeate pressure). This result indicates that the high gas permeability of PMP is relatively stable under these test conditions. On the other hand, the gas permeability to nitrogen decreased to only 10% of the initial value under vacuum conditions which utilized a rotary vacuum oil pump to reduce the pressure at the permeate side of the film sample. The higher decrease in nitrogen permeability for the sample tested using the constant volume/variable pressure method is probably due to sorption of pump oil vapor into the excess free volume of the polymer. This hypothesis is supported by a recent study of *Nagai et al.* who showed that the nitrogen permeability in PTMSP also decreased to about 10% of its initial value using a rotary oil vacuum pump, whereas the permeability decreased to only 70% when a sample was tested using an oil-free vacuum pump (*13*).

## Conclusions

Poly(4-methyl-2-pentyne) [PMP] is a high-free-volume, glassy, disubstituted polyacetylene. The mixed-gas permeation properties of PMP are qualitatively similar to those of high-free-volume, glassy PTMSP, the most permeable polymer known. In contrast to conventional low-free-volume glassy polymers, PMP is more permeable to large, condensable penetrants, such as *n*-butane, than to small, permanent gases, such as methane. In gas mixtures, the permeability of methane is reduced to about 20% of its pure-gas value by co-permeation of a highly sorbing hydrocarbon, such as *n*-butane. This reduction in methane permeability is the result of partial blocking of the excess free volume in the polymer by preferential sorption of *n*-butane. The *n*-butane/methane selectivity of PMP increased from 11 to 16 by increasing the relative *n*-butane pressure from 0.05 to 0.38. As the feed temperature decreased, both the mixed-gas *n*-butane/methane selectivity and *n*-butane permeability of PMP increased.

As expected for a high-free-volume, glassy polymer, PMP exhibits significant physical aging. The nitrogen permeability of a polymer sample decreased to about 70% of its initial value after 29 days using the constant pressure/variable volume method. On the other hand, the nitrogen permeability decreased to only 10% of its initial value using the constant volume/variable pressure method, indicating that the

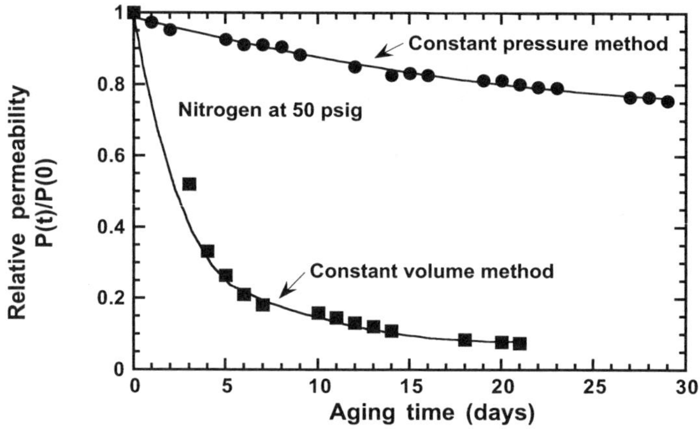


Figure 6. Nitrogen permeability of poly(4-methyl-2-pentyne) as a function of aging time at 25°C.

higher decrease in permeability was caused by sorption of pump oil vapor in the free volume of the polymer. This result is consistent with previous studies on PTMSP.

## Acknowledgments

The authors gratefully acknowledge support of the work by the Department of Energy (SBIR grant number DE-FG03-94ER81811) and the National Science Foundation (SBIR contract number DMI-9661545).

## Literature Cited

1. Baker, R.W.; Wijmans, J.G. *In Polymeric Gas Separation Membranes*; Paul, D.R.; Yampol'skii, Y.; Eds. CRC Press, Boca Raton, FL, 1994, Chapter 8, pp. 354-397.
2. Masuda, T.; Iguchi, Y.; Tang, B.-Z.; Higashimura, T. *Polymer* **1988**, *29*, 2041.
3. Nakagawa, T.; Saito, T.; Asakawa, S.; Saito, Y. *Gas Separation and Purification* **1988**, *2*, 3.
4. Masuda, T.; Isobe, E.; Higashimura, T.; Takada, K. *J. Am. Chem. Soc.* **1983**, 7473.
5. Toy, L. G.; Pinnau, I.; Wijmans, J. G., AIChE Meeting, Miami Beach, FL, 882 (1992).
6. Masuda, T.; Kawasaki, M.; Okano, Y.; Higashimura, T. *Polym. J. (Tokyo)* **1982**, *14*, 371.
7. Morisato, A.; Pinnau, I. *J. Memb. Sci.* **1996**, *121*, 243.
8. Stern, S.; Shah, V.; Hardy, B. *J. Polym. Sci.: Polym. Phys. Ed.* **1987**, *25*, 1263.
9. Pinnau, I.; Toy, L. G. *J. Memb. Sci.* **1996**, *116*, 199.
10. Pinnau, I.; Casillas, C. G.; Morisato, A.; Freeman, B. D. *J. Polym. Sci.: Polym. Phys. Ed.* **1996**, *34*, 2613.
11. Srinivasan, R.; Auvil, S. R.; Burban, P. M. *J. Memb. Sci.* **1994**, *86*, 67.
12. Nagai, K.; Higuchi, A.; Nakagawa, T. *J. Appl. Polym. Sci.* **1994**, *54*, 1651.
13. Nagai, K.; Nakagawa, T. *J. Memb. Sci.* **1995**, *105*, 261.
14. Shimomura, H.; Nakanishi, K.; Odani, H.; Kurata, M.; Masuda, T.; Higashimura, T. *Kobunshi Ronbunshu* **1986**, *43*, 747.

Chapter 5

# Aging of Gas Permeability in Poly[1-(trimethylsilyl)-1-propyne] (PTMSP) and PTMSP–Poly(*tert*-butylacetylene) Blends

**T. Nakagawa[1], T. Watanabe[1], M. Mori[1], and K. Nagai[2]**

[1]**Department of Industrial Chemistry, Meiji University, Higashi-mita, Tama-ku, Kawasaki 214–0033, Japan**
[2]**Department of Chemical Engineering, North Carolina State University, Raleigh, NC 27695**

The effect of aging on gas permeability in glassy poly[1-(trimethylsilyl)-1-propyne] (PTMSP) was investigated in terms of the dual mode sorption and transport theory. The decrease in the gas permeability coefficient depended not only on the decrease in the hole saturation constant of Langmuir adsorption ($C'_H$), which is related to the volume of the microvoids, but also on the decrease in the gas permeability coefficient in the Henry's law mode ($k_D D_D$). The decrease in $C'_H$ originated from the relaxation of nonequilibrium excess volume in the Langmuir mode, which is called physical aging. To stabilize the permeability of PTMSP, poly(*tert*-butyl acetylene) (PTBA), whose solubility parameter resembles that of PTMSP, was synthesized and PTMSP/PTBA blend membranes were prepared. The mixing of each polymer was excellent, and no microphase separation was observed for a blend containing 20 vol % PTBA (PTMSP/PTBA 80/20). The effect of PTBA content on $C'_H$, gas permeability, and also on aging of gas permeability was investigated. Although $C'_H$ values and gas permeability decreased with an increase in PTBA content, the stability of gas permeability is maximized in the PTMSP/PTBA 80/20 blend. The spin-lattice relaxation time ($T_1$) of carbon atoms was also a maximum for the PTMSP/PTBA 80/20 blend, which means that molecular motion is slowest for this composition. The $T_1$ value of the carbons in the methyl groups of the trimethylsilyl group was very low compared with $T_1$ values of carbons in the backbone. These $T_1$ values did not change during aging for the PTMSP/PTBA 80/20 blend. The oxygen permeability of this 80/20 blend membrane is 6,140 Barrers at 30°C.

Poly[1-(trimethysilyl)-1-propyne] (PTMSP) is a glassy polymer which was first synthesized by Masuda and Higashimura's group *(1,2)*. PTMSP has the highest gas permeability, diffusivity, and solubility of all nonporous polymeric membranes, though the selectivity is very low. Both the high diffusivity and solubility coefficients of gases come from the large excess free volume of PTMSP. The biggest problem, however, is a decrease in its gas permeability with time (*i.e.* physical aging) and upon absorption of low volatility organic vapors (*i.e.*, pump oil vapor). The absorption of low volatility organic vapors can be prevented by careful experimental protocol. However, the effect of physical aging on gas permeability must be minimized by modification of PTMSP.

In this paper, physical aging of gas permeability properties of PTMSP is described using the dual mode sorption and transport model. This model is also used to rationalize the increase in stability of gas permeation properties as a result of blending PTMSP with poly(*tert*-butyl acetylene) (PTBA).

## Experimental

**Polymer Synthesis.** PTMSP was synthesized according to Masuda's method(*1,2*). The polymerization was performed in toluene under a dry nitrogen atmosphere for 24 h at 80°C. $TaCl_5$ was used as a catalyst.

The polymerization of *tert*-butyl acetylene was performed in toluene under a dry nitrogen atmosphere at 30°C for 24 h. The catalyst was $MoCl_5$. The chemical structure of the monomers, 1-trimethylsilyl-1-propyne and t-butylacetylene, and the corresponding polymers, PTMSP and PTBA, are shown in Figure 1.

**Membrane Preparation.** All PTMSP, PTBA and PTMSP/PTBA blend polymer membranes were cast on a horizontal glass plate from polymer solutions in toluene. They were immersed in methanol until just before beginning experimental measurements to prevent hysteresis of the membranes.

**Gas Sorption and Permeability.** Gas sorption isotherms at 35°C and elevated pressure were determined by a gravimetric method using a CAHN 2000 electronic micro-balance. Gas permeabilities were measured using an adaptation of the constant volume, variable pressure technique, which employed an MKS Baratron model 310HBS-100SP pressure transducer for upstream pressures below 1 atm and at temperatures in the 30-90°C range. An MKS Baratron model 370HA pressure transducer was used for permeability measurements at 35°C and upstream pressures as high as 40 atm.

**NMR Measurements.** High resolution, solid-state NMR measurements were performed using a JEOL JNM400 spectrometer at 30°C. The spin-lattice relaxation time ($T_1$) of $^{13}C$ and $^{29}Si$ were obtained using the CP $T_1$ pulse sequence at a spinning speed of 6 kHz (*3*).

## Theoretical Background

A quantitative description of penetrant solution and diffusion in microheterogeneous media has evolved over the past forty years and has become known as the dual mode sorption theory (*4*). Originally, this theory postulated that two concurrent modes of sorption are operative in a microheterogeneous medium. Nonlinear sorption isotherms can be decomposed into a linear part that accounts for normal solution (Henry's law-type domain) and a nonlinear Langmuir-type domain that accounts for immobilization of the penetrant molecules at fixed sites in the medium.

Meares concluded that the glassy state contains a distribution of microvoids frozen into the structure as the polymer is cooled through its glass transition temperature (*5-7*). In the glassy state, segmental motion of the polymer chains is restricted, resulting in fixed microvoids or "holes" at the molecular level as shown in Figure 2. These microvoids in the glassy state immobilize a portion of the penetrant molecules by entrapment or adsorption.

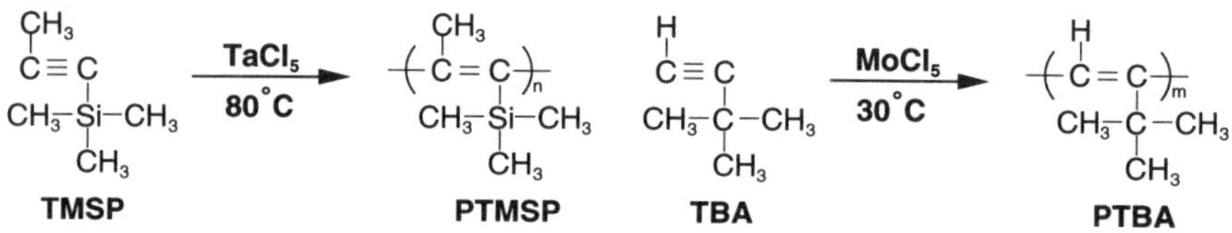


Figure 1. Chemical structure of the monomers, 1-trimethylsilyl-1-propyne and t-butyl acetylene, and the corresponding polymers, PTMSP and PTBA.

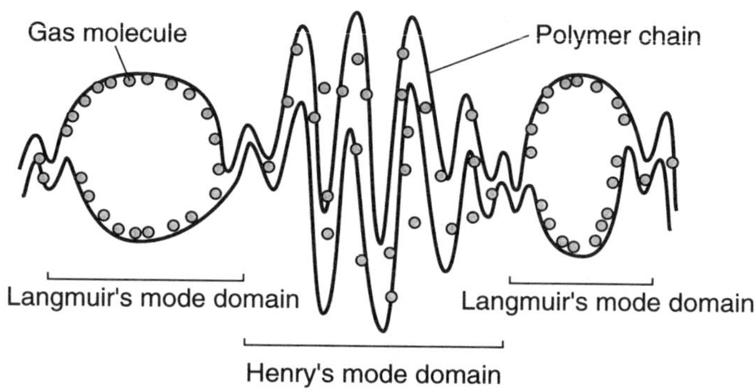


Figure 2. Schematic pictorial representation of dual-mode sorption.

Based on this model, sorption isotherms are described using the following equation (*4*):

$$C = C_D + C_H = k_D p + \frac{C'_H bp}{1+bp} \quad (1)$$

where $C$ is the total concentration of gas dissolved in the polymer [$cm^3$(STP)/$cm^3$(polymer)], $C_D$ is the gas concentration in the Henry's law mode, and $C_H$ is the gas concentration in the Langmuir mode. The parameters of the model are $k_D$, which is the Henry's law solubility constant, $b$, which is the hole affinity constant, and $C'_H$, which is the hole saturation constant. The gas pressure in contact with the polymer sample is $p$.

The three parameters of the dual mode sorption theory, $k_D$, $C'_H$, and $b$, are determined by analyzing an isothermal plot of $C$ versus $p$. For gas sorption at low pressure, where $bp \ll 1$, the sorption isotherm in Equation 1 reduces to the following linear expression,

$$C = (k_D + C'_H b)p \quad (2)$$

Generally, solubility coefficients of the same gas in different amorphous polymers do not vary widely. Therefore, $C'_H$ is roughly proportional to the amount of microvoids in the polymers. At sufficiently high pressure, the microvoids become saturated and will no longer sorb additional penetrant molecules. When $bp \gg 1$, sorption in the microvoids reaches the saturation limit, $C'_H$, and Equation 1 again reduces to a linear form,

$$C = k_D p + C'_H \quad (3)$$

As shown by Equation 3, $k_D$ can be obtained from the slope of the sorption isotherm at high pressure, and $C'_H$ can be determined from the $y$ intercept of the high pressure data. $b$ can then be determined from the slope of the low pressure data as indicated in Equation 2.

If the penetrants sorbed in both the Henry's law and Langmuir modes diffuse independently through both domains, the permeability coefficient of the glassy membrane is expressed as follows (*8-10*):

$$P = k_D D_D + \frac{C'_H b D_H}{1+bp} \quad (4)$$

where $D_D$ is the diffusion coefficient in the Henry's law mode, and $D_H$ is the diffusion coefficient in the Langmuir mode.

Plots of permeability coefficients versus $(1 + bp)^{-1}$ should be linear. The intercept and slope of the line correspond to $k_D D_D$ and $C'_H b D_H$, respectively. Once the equilibrium parameters from the dual mode sorption theory are known, the two diffusion coefficients, $D_D$ and $D_H$, can be determined.

## Results and Discussion

**Effect of Physical Aging on Gas Permeability.** Literature values of the oxygen permeability coefficient of PTMSP are strikingly variable. At room temperature, the minimum and maximum values are 3,000 and 12,000 Barrers (*11,12*). The unusually wide variation in $O_2$ permeability makes PTMSP a remarkable material. In this regard, the effect of storage time in vacuum at 30°C on gas permeability is shown in Figure 3. The permeability coefficients of each gas decreased by 50-60 % over the course of 2 weeks. This phenomenon is typical for aging of PTMSP membranes whose thicknesses are 100-200 μm. There are two reasons for the decrease in gas permeability coefficients with time: (1) absorption of low volatility pump oil vapor in the PTMSP membrane and (2) physical aging. Even if a liquid nitrogen trap is used between the permeation cell and an oil rotary vacuum pump or oil diffusion pump in the gas permeation apparatus, PTMSP still absorbs oil vapor and permeability coefficients decrease. A plasticizer such as dioctylphthalate, which is contained in the rubber gasket of the permeation cell, also decreases the gas permeability (*13*). The effect of catalysts (*14*), absorbed oil vapor (*12,15,16*), temperature (*17*) and storage in air or under vacuum (*12,17,18*), on the decrease in PTMSP permeability has also been reported.

**Characterization of Aged PTMSP membranes.** Physical aging is a very special phenomenon in glassy polymers. It is well known that annealing below the glass transition temperature is effective for reducing the nonequiliblium excess volume of glassy polymers. Annealing causes a decrease in the nonequiliblium excess volume, and this decrease then causes a reduction in diffusivity and solubility (*19-22*). The nonequiliblium excess volume is the volume of the Langmuir mode, which can be evaluated using $C'_H$ values. The $C'_H$ value of PTMSP calculated from the sorption isotherm of $CO_2$ at 35°C, which was first reported by Ichiraku *et al.* (*24*), was 112.9 $cm^3$(STP)/$cm^3$(polymer). This value is fifteen times higher than that of glassy polystyrene (*23,24*) and is the highest $CO_2$ $C'_H$ value of any known polymer.

One of the authors reported decreases in *d*-spacing (*13*) and $C'_H$ (*12*) of PTMSP membranes upon aging. Upon aging, based on density measurements, wide-angle X-ray diffraction, and positron annihilation lifetime spectroscopy studies, Yampol'skii and co-workers also inferred a decrease in the free volume of PTMSP(*18*).

The $CO_2$ permeability coefficient measured at 35°C for a PTMSP membrane annealed at 75°C for varying lengths of time is shown in Figure 4. The measurement was performed by increasing the upstream pressure of $CO_2$ from 1 to 40 atm, and then decreasing it from 40 to 1 atm. The permeability coefficient was lower at higher pressures, which is consistent with the dual sorption and partial immobilization transport theory expressed by Equation 4. Figure 4 shows no plasticization of PTMSP by $CO_2$ even at the highest pressure (40 atm), but a small hysteresis effect was observed. The relationship between $CO_2$ permeability coefficient and $(1 + bp)^{-1}$ is shown in Figure 5. As already mentioned, physical aging derives from relaxation of the nonequiliblium excess volume. This causes a decrease in $C'_H$ and $D_H$. It is quite clear, however, that the decrease in the *y*-intercept in Figure 5 signifies a decrease in $k_D D_D$ in the Henry's mode. Therefore, the relaxation of the nonequiliblium excess free volume in glassy PTMSP not only affects the Langmuir mode but also the segmental motion in the Henry's law mode.

We previously characterized the molecular motion of carbon atoms in the main chain and side chain of PTMSP using $^{13}C$ spin-lattice relaxation time (*3*). During aging, changes in $T_1$ values of the side chain carbons were smaller than those of the backbone chain carbons. This result suggests that molecular motions, such as twisting, of the backbone carbons were most strongly influenced by aging.

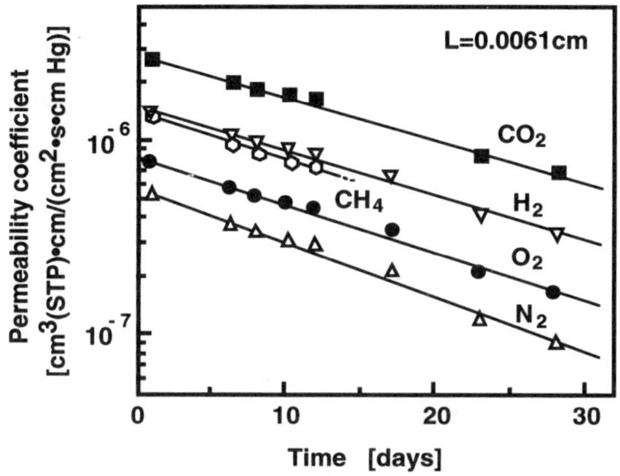


Figure 3. Effect of aging time at 30°C on the gas permeability coefficient of PTMSP. Upstream pressure for each of these measurements is 20-30 cmHg. Aging protocol: stored under vacuum at 30°C.

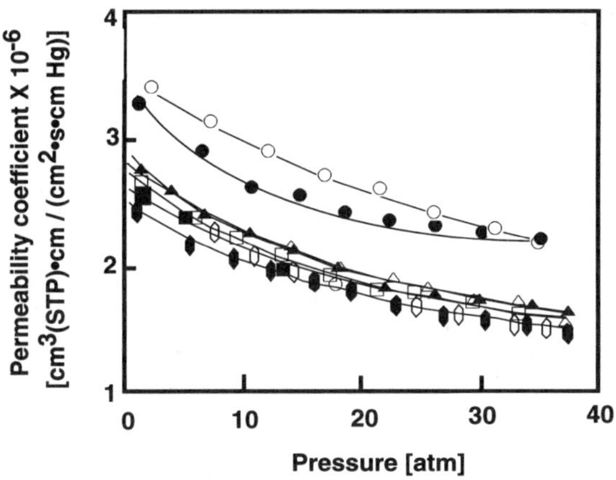


Figure 4. Effect of thermal treatment time at 75°C in nitrogen on a weekly basis for one month on the $CO_2$ permeation isotherm of PTMSP at 35°C.

○● : 1 week △▲: 2 weeks □■ : 3 weeks

◇◆ : 4 weeks.

The open and closed symbols represent data taken during increasing and decreasing pressure, respectively.

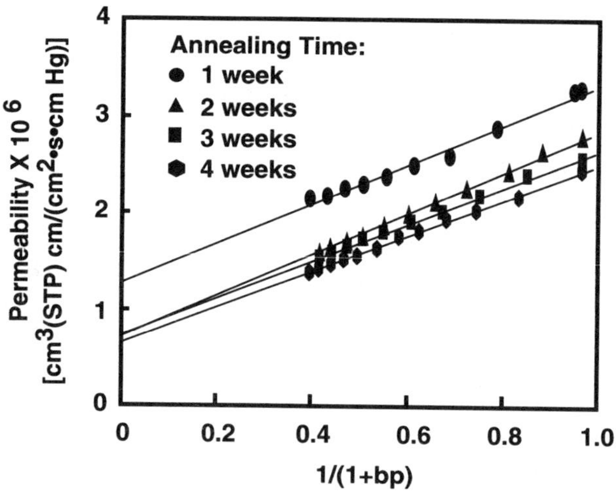


Figure 5. $CO_2$ permeability coefficient versus $(1+bp)^{-1}$. These data are replotted from the closed symbols in Figure 4.

**Gas Permeability and Aging of Blend Membranes of PTMSP and PTBA.** Although the gas permeability of PTMSP is the highest of all polymeric materials, its decrease over time is remarkable. Therefore, stabilization of gas permeability in PTMSP is required. Many papers have reported modifications of PTMSP to control physical aging. Such methods include filling PTMSP with poly(dimethylsiloxane) oil (*25*), halogenation (*26-28*), copolymerization with phenylpropyne (*29*), blending with poly(phenylpropyne) (*30*), and laminating with poly(dimethylsiloxane) film (*16*). In this paper, the gas permeability and molecular motions of blend membranes composed of PTMSP and PTBA are reported.

**Characterization of PTMSP/PTBA Blend Membranes.** Figure 6 shows wide-angle X-ray diffraction patterns of the blend membranes. The content of PTBA (in volume %) was gradually increased from 0 % (PTMSP) to 100 % (PTBA). The PTBA membrane exhibits a peak at 17.8 degrees 2θ in blend membranes containing 20 % or more PTBA. PTBA also exhibits a peak at 9.8 degrees. The 9.8 and 17.8 degree peaks suggest the average intersegmental and intrasegmental distances, respectively. PTBA is an amorphous, glassy polymer, like PTMSP. PTMSP exhibits an amorphous halo peak centered at 11.2 degrees 2θ.

Figure 7 shows the effect of PTBA content on $C'_H$ (obtained from $CO_2$ sorption) and oxygen permeability coefficient at 30°C. As expected, both $C'_H$ and $O_2$ permeability decrease monotonically upon increasing the concentration of PTBA since PTBA has a lower fractional free volume than PTMSP.

**Gas Permeability and Oxygen/Nitrogen Selectivity.** Figure 8 shows the effect of PTBA content on the permeability coefficient of oxygen and the ideal separation factor for oxygen and nitrogen at 30°C. The PTBA oxygen permeability coefficient is 160 Barrers, which is about 25% of the value for poly(dimethylsiloxane), the most permeable rubbery polymer. It is very helpful to understand the effect of the difference in side chain bulkiness on sorption and transport properties. In this regard, the trimethylsilyl group and methyl group in PTMSP enlarge the intersegmental distance and, therefore, free volume relative to that in PTBA. From Figure 7, the decrease in the permeability coefficient of oxygen is almost linear with the decrease in $C'_H$. The compatibility of PTMSP with PTBA is evaluated by the relationship between the logarithm of gas permeability of the blends and the vol % of PTBA (*31-33*). To calculate the volume fraction of PTBA in the blend requires the density of each component. In this study, the density of PTMSP and PTBA were determined to be 0.74 and 0.83 g/cm$^3$, respectively. Typical representative equations for the permeability of homogeneous and microphase separated blends are expressed by Equations 5 and 6, respectively:

$$\ln P = \varphi_1 \ln P_1 + \varphi_2 \ln P_2 \quad (5)$$

$$P = P_1 \frac{P_2 + 2P_1 + 2\phi_2(P_1 - P_2)}{P_2 + 2P_1 + \phi_2(P_1 - P_2)} \quad (6)$$

where $P_1$ is the permeability coefficient of component 1, $P_2$ is that of component 2, and $\varphi_1$ and $\varphi_2$ are the volume fractions of components 1 and 2, respectively. In Equation 6, components 1 and 2 refer to the continuous and discontinuous phases, respectively. The straight dashed line in Figure 8 is the theoretical line corresponding to Equation 5 and the solid curve corresponds to Equation 6, when PTMSP and

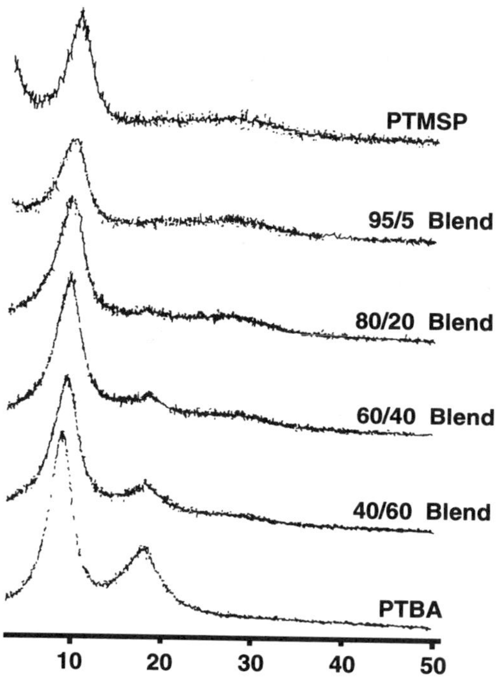


Figure 6. Wide-angle X-ray diffraction patterns of PTMSP/PTBA blend membranes. Composition is given in volume percent.

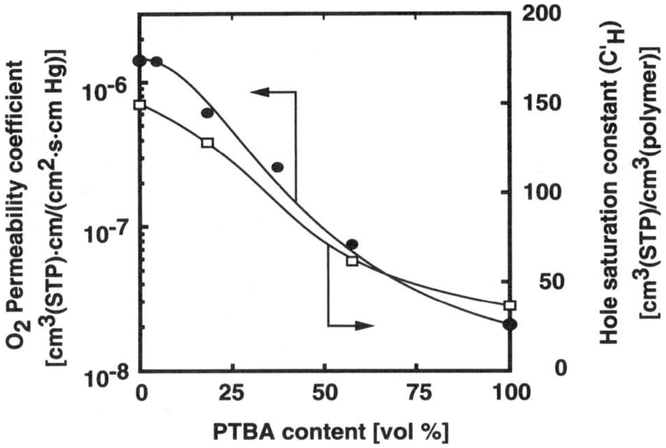


Figure 7. Effect of PTBA content on the $O_2$ permeability coefficient at 30°C and the hole saturation constant ($C'_H$) of $CO_2$ in the blend membranes at 35°C. Upstream pressure for $O_2$ permeation experiments is 20-30 cm Hg. $C'_H$ is calculated from the $CO_2$ isotherms up to 30 atm.

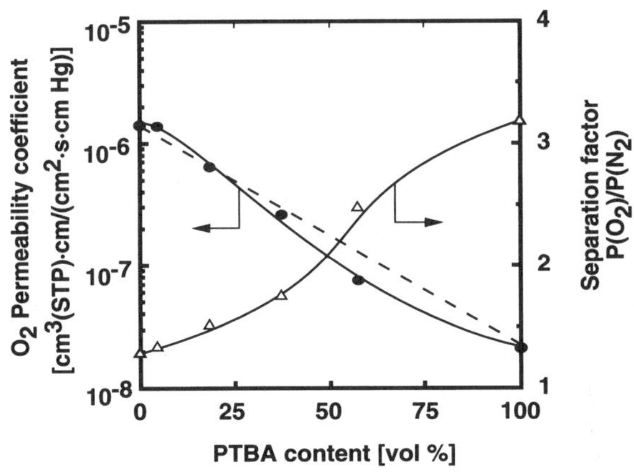


Figure 8. Effect of PTBA content on the $O_2$ permeability coefficient and the $O_2/N_2$ ideal separation factor at 30°C. Upstream pressure for permeation experiments is 20-30 cm Hg.

PTBA are the continuous phase and discontinuous phase, respectively. The solubility parameters of PTMSP and PTBA are 7.74 and 7.45 $(cal/cm^3)^{1/2}$, respectively (*34*), and the degree of polymerization of PTBA is known to be slightly less than that of PTMSP. Thus the formation of a phase-separated blend composed of PTBA domains in a PTMSP matrix was not expected. When the blend ratio of PTMSP/PTBA is 80/20, gas permeability follows the straight line, which is consistent with a homogeneous (*i.e.*, non-phase separated) blend. Figure 9 shows the time dependence of the permeability coefficients of $N_2$, $O_2$, and $CO_2$ of the PTMSP/PTBA 80/20 blend membrane, which was kept in an evacuated vessel at 30°C between permeability measurements. The decrease in permeability coefficients with time has been reduced relative to that of pure PTMSP. The influence of aging time on oxygen permeability coefficients of the blend membranes is shown in Figure 10 as a function of the initial permeability of the blend. The 80/20 PTMSP/PTBA membrane is the best based on the balance of the permeability coefficient and its stability.

$CO_2$ sorption isotherms for the blend membranes at 35°C are shown in Figure 11, and the dual-mode sorption parameters are summarized in Table I. $C'_H$ decreases with an increase in PTBA content. The $C'_H$ value does not directly correlate with the stability of permeability, because the $C'_H$ value is a measure of the total volume of the nonequiliblium excess volume (*i.e.*, the microvoids), and it is the size distribution of the microvoids which is important for stabilization. The bigger microvoids easily decay with aging. The initial $C'_H$ value of 128 $cm^3(STP)CO_2/cm^3$polymer for the 80/20 blend membrane was not changed after aging this blend for 2 weeks.

Table I Dual-mode sorption parameters for $CO_2$ in PTMSP, PTBA, and PTMSP/PTBA blend membranes

| Membrane | $k_D$ [1)] | $C'_H$ [2)] | b [3)] |
|---|---|---|---|
| PTMSP | 0.937 | 149 | 0.0347 |
| 80/20 Blend | 0.673 | 128 | 0.0440 |
| 40/60 Blend | 0.738 | 61.6 | 0.0728 |
| PTBA | 0.714 | 36.4 | 0.0924 |

1) $cm^3(STP)/(cm^3(polymer)\cdot atm)$
2) $cm^3(STP)/\ cm^3(polymer)$
3) $atm^{-1}$

**Solid-state NMR of PTMSP/PTBA Blend Membranes.** The CP/MAS $^{13}C$ NMR spectrum of PTBA is shown in Figure 12. The peak assignments are also shown in this figure. The two peaks at 60-90 ppm and 180-210 ppm are spinning side bands (SSB). In the liquid state, the NMR signal of the methyl groups ($\alpha$) in the t-butyl moiety is split into two parts due to the difference in the configuration of the polymer chains. However, a single peak is observed in the solid state spectrum presented in Figure 12. In Figure 13, the CP/MAS $^{13}C$ NMR spectra of several PTMSP/PTBA blends are shown. Each chemical shift coincides with the CP/MAS spectra of the PTMSP and PTBA membranes. No changes in chemical shift values were observed with variations in PTBA content. The intensity of the peak in the

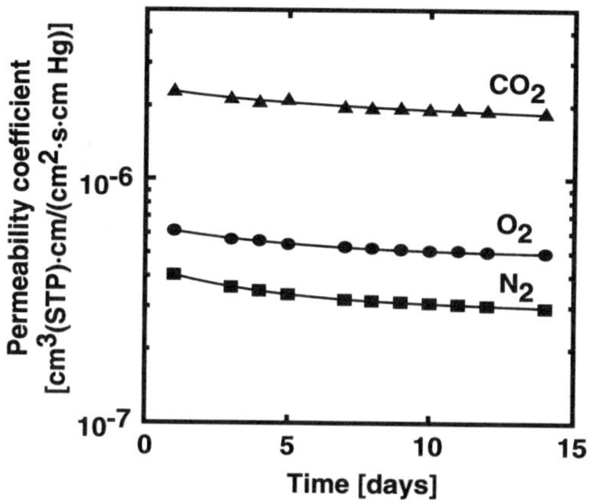


Figure 9. Effect of time under vacuum on the gas permeability coefficients at 30°C of PTMSP/PTBA 80/20 membrane. Upstream pressure is 20-30 cm Hg.

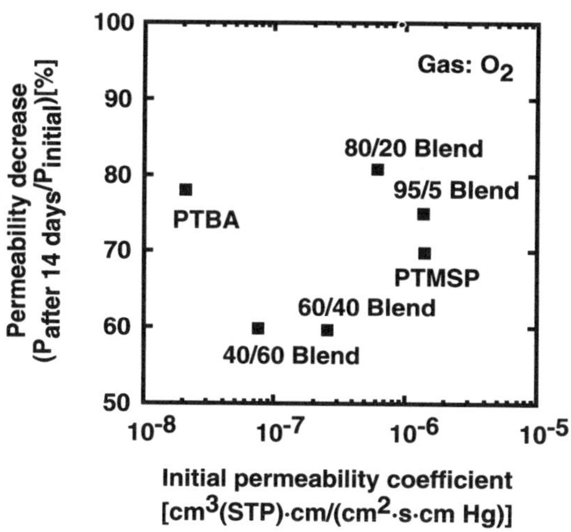


Figure 10. Effect of time under vacuum on the decrease of oxygen permeability of blend PTMSP/PTBA membranes at 30°C. Upstream pressure is 20-30 cm Hg.

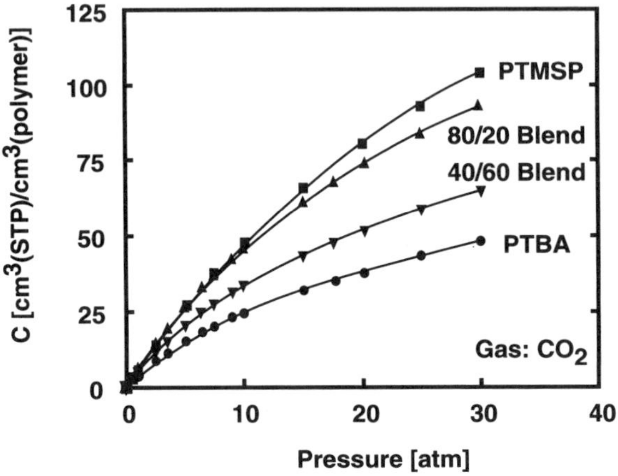


Figure 11. Sorption isotherms for $CO_2$ in PTMSP, PTBA, and PTMSP/PTBA blend membranes at 35°C.

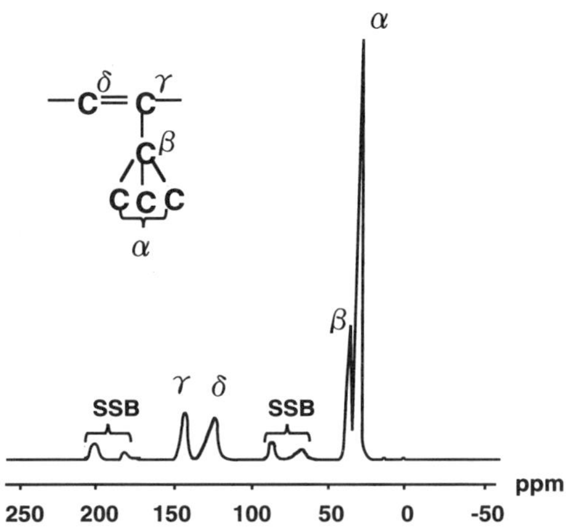


Figure 12. Solid-state CP/MAS $^{13}C$ NMR spectrum of PTBA membrane. SSB=Spinning side band.

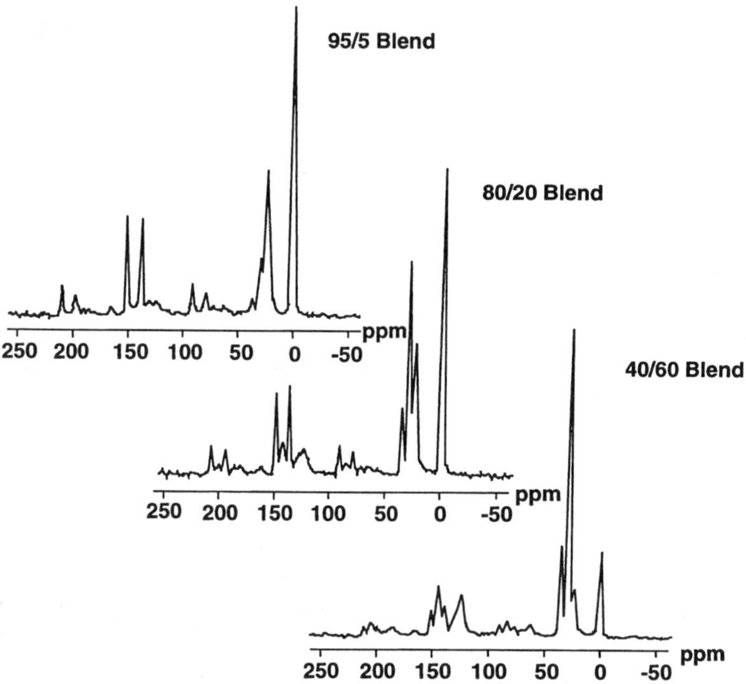


Figure 13. Solid-state CP/MAS $^{13}C$-NMR spectra of PTMSP/PTBA blend membranes.

backbone chain is very weak for the 40/60 blend membrane because of the higher PTBA content.

**Molecular Motion of PTMSP/PTBA Blends Before and After Aging.** The authors have already reported the influence of aging on the molecular motion of PTMSP, copolymer membranes and blend membranes composed of 1-trimethylsilyl-1-propyne and 1-phenylpropyne (*35*). The motions of the carbons in the main chains changed upon aging, but those of the side chain groups did not change.

Figures 14 and 15 show the relationship between PTBA content and carbon spin-lattice relaxation time ($T_1$) of the PTMSP side-chain and backbone carbon atoms, respectively. The effect of PTBA content on $C'_H$ is also shown in these figures. $T_1$ of the methyl group carbon atoms (a) in the side-chain of PTMSP became longer with increases in PTBA content. This result indicates that the molecular motion of the trimethylsilyl groups became slower. On the other hand, $T_1$ values for the carbon atoms of the other methyl group (b) and the backbone chain show a maximum at 20 vol % PTBA, which suggests perfect mixing of PTMSP and PTBA. These results coincide with the maximum stability of the gas permeability coefficient (*cf.* Figure 10).

## Conclusions

PTMSP exhibits significant decreases in gas permeability with age. During aging, the nonequilibrium excess volume of glassy PTMSP relaxes. The reasons for the decrease in gas permeability were not only relaxation of the nonequilibrium excess volume in the Langmuir mode, but also reduction of the permeability coefficient in the Henry's law mode. This latter effect appears to be correlated with relaxation of the nonequilibrium excess volume.

Blending PTMSP with PTBA is a novel strategy to improve the stability of the gas permeability of PTMSP membranes by controlling the excess free volume, which is characterized by $C'_H$. The compatibility of PTMSP and PTBA is excellent. The 80/20 blend exhibited perfect mixing based on the relationship between gas permeability and volume content of PTBA and based on the molecular motions of the blend as characterized by $^{13}C$ $T_1$. The stability of gas permeability was also best for the 80/20 blend, and the oxygen permeability coefficient of this blend is 6,140 Barrers at 30°C. The nonequilibrium excess volume hardly decreased with age. The molecular motions of the trimethylsilyl group were not influenced by aging. However, motions of the carbon atoms in the backbone were reduced after aging. The spacing between the polymer segments in PTMSP/PTBA blend membranes is wide enough to permit motions of the side-chain after aging.

## Literature Cited


1. Masuda, T.; Isobe, E.; Higashimura, T.; Takada, K. *J. Amer. Chem. Soc.*, **1983**, *105*, 7473.
2. Masuda, T.; Isobe, E.; Higashimura, T. *Macromol.* **1985**, 18, 841.
3. Nagai, K.; Watanabe, T.; Nakagawa, T. *Polymer J.*, **1996**, 28, 933.
4. Vieth, W. R.; Howell, J. M.; Hsieth, J. H. *J. Memb. Sci.*, **1976**, 1, 177.
5. Meares, P. *J. Amer. Chem. Soc.*, **1954**, 76, 3415.
6. Meares, P. *Trans. Farad. Soc.,* **1957**, 53, 101.
7. Meares, P. *Trans. Farad. Soc.*, **1958,** 54, 40.

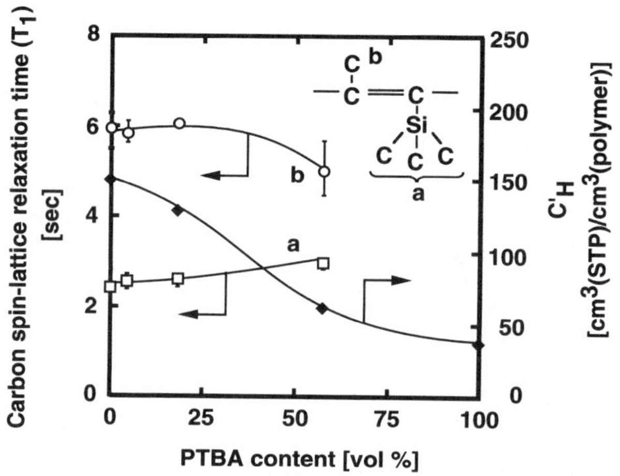


Figure 14. Relationship between PTBA content and side chain $^{13}C$ $T_1$ and $C'_H$ in PTMSP/PTBA blend membranes.

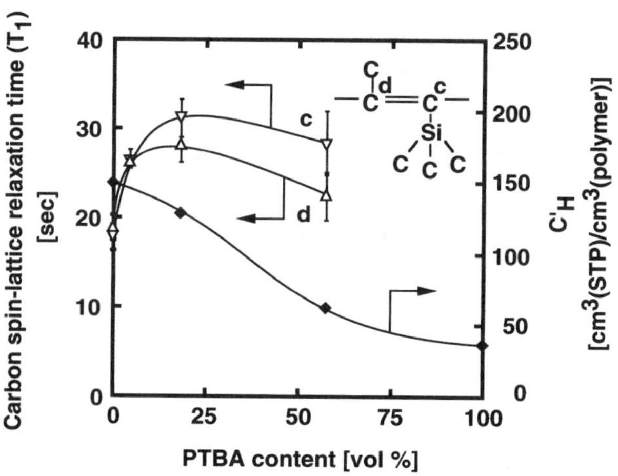


Figure 15. Relationship between PTBA content and backbone $^{13}C$ $T_1$ and $C'_H$ in PTMSP/PTBA blend membranes.

8. Paul, D. R.; Koros, W. *J. Polym. Sci.: Part B: Polym. Phys.*, **1976**, 14, 675.
9. Koros, W. J.; Paul, D. R.; Rocha, A. A. *J. Polym. Sci.: Part B: Polym. Phys.*, **1976**, 14, 687.
10. Koros, W. J.; Patton, C. J.; Felder, R. M. ; Finder, S. J. *J. Polym. Sci.: Part B: Polym. Phys.*, **1980**, 18, 1485.
11. Odani, H.; Masuda, T. In *Polymers for Gas Separation*; Toshima, N. Ed., VCH Publishers, Inc.: New York, **1992**; Chap. 4, pp107-144.
12. Nagai, K.; Nakagawa, T. J. *Memb. Sci.*, **1995**, 105, 261.
13. Nakagawa, T.; Saito, T.; Asakawa, S.; Saito, Y. *Gas Sep. Purif.*, **1988**, 2, 3.
14. Nagai, K.; Nakagawa, T. *J. Memb. Sci.*, **1995**, 105, 261.
15. Witchey-Lakshmanan, L. C.; Hopfenberg, H. B.; Chern, R. T. *J. Memb. Sci.*, **1990**, 48, 321.
16. Asakawa, S.; Saito, Y.; Waragi, K.; Nakagawa, T. *Gas Sep. Purif.*, **1989**, 3, 117.
17. Shimomura, H.; Nakanishi, K.; Odani, H.; Kurata, M.; Masuda, T.; Higashimura, T. *Kobunshi Ronbunshu,* **1986**, 43, 747.
18. Ysampol'skii, Y. P.; Shantorovich, V. P.; Chernyakovskii, F. P.; Kornilov, A. I.; Plate, N. A. *J. Appl. Polym. Sci.*, **1993**, 47, 85.
19. Chen, A. H.; Paul, D. R. *J. Appl. Polym. Sci.*, **1979**, 24, 1539.
20. Wonders, A. G.; Paul, D. R. *J. Memb. Sci.*, **1979**, 5, 63.
21. Hachisuka, H.; Kito, H.; Tsujita, Y.; Takizawa, A.; Kinoshita, T. *Polym. J.*, 1**988**, 35, 1333.
22. Moe, M. B.; Koros, W. J.; Paul, D. R. *J. Polym. Sci.: Part B: Polym. Phys.*, **1988,** 26, 1931.
23. Hachisuka, H.; Tsujita, Y.; Takizawa, A.; Kinoshita, T. *Polym. J.*, **1989**, 21, 681
24. Ichiraku, Y.; Stern, S. A.; Nakagawa, T. *J. Memb. Sci.*, **1987**, 34, 5.
25. Nakagawa, T.; Fujisaki, S.; Nakano, H.; Higuchi, A. *J. Memb. Sci.*, **1994**, 94, 183.
26. Nagai, K.; Higuchi, A.; Nakagawa, T. *J. Appl. Polym. Sci.*, **1994**, 54, 1207.
27. Nagai, K.; Mori, M.; Watanabe, T.; Nakagawa, T. *J. Polym. Sci.: Part B: Polym. Phys.*, **1997**, 35, 119.
28. Langsam, M.; Anand, M.; Karwacki, E. J. *Gas Sep. Purif.*, **1988**, 2, 162.
29. Nagai, K.; Higuchi, A.; Nakagawa, T. *J. Appl. Polym. Sci.*, **1994**, 54, 1353.
30. Nakagawa, T.; Watanabe, T.; Nagai, K. *ACS Symposium Series*, **1998**, in press.
31. Paul, D. R. *J. Memb. Sci.,* **1966**, 22, 360.
32. Barrie, J. A.; Ismail, J. B. *J. Memb. Sci.*, **1983**, 13, 197.
33. Barnabeo, A. E.; Creasy, W. S.; Robeson, L. M. *J. Polym. Sci.: Part A: Polym. Chem.,* **1975**, 13, 1979.
34. Fedors, R. F. *Polym. Eng. Sci.*, **1974**, 14, 143.
35. Nagai, K.; Mori, M.; Watanabe, T.; Nakagawa, T. *J. Polym. Sci.: Part B: Polym. Phys.*, **1997**, 35, 119.

## Chapter 6

# Physical Aging Phenomena in Poly(*tert*-butylacetylene) Solvent Cast Films: Influence of Operating Conditions

**G. Costa[1], C. Iurilli[2], B. Cavazza[1], and A. Turturro[2]**

[1]Istituto di Studi Chimico-Fisici di Macromolecole Sintetiche e Naturali, IMAG—CNR, via De Marini 6–16149 Genova, Italy
[2]Dipartimento di Chimica e Chimica Industriale, University of Genova, via Dodecaneso 31–16146 Genova, Italy

The influence of the experimental conditions used to prepare films of cis/trans poly(*tert*-butylacetylene) [PTBA] on density and physical aging phenomena have been investigated. Films with different free volumes were obtained by casting from solvents with different thermodynamic characteristics. In order to choose the most appropriate solvents, the solubility parameter of PTBA was determined both through a calculation based on group contribution methods and by viscosity measurements in different solvents. Films from solvents with different solubility parameters have been prepared under controlled conditions. On increasing evaporation time and polymer-solvent affinity, higher density values and, consequently, higher chain packing and lower free volume are achieved. Aging phenomena, studied by dilatometric and calorimetric techniques, occur whenever the films undergo thermal treatment, indicating that densification takes place as a result of enhanced chain mobility at the elevated temperatures of the thermal treatment protocol.

Most properties of the substituted polyacetylenes (PA), such as permeability to penetrants, solubility, and thermal stability are strongly related to the number and/or steric hindrance of the substituents (*1-4*). However, these properties cannot be related in a straightforward fashion to the kind of substituents or to the mono- or double-substitution along the main chain. In fact, the configuration of the repeat units and, consequently, the chain conformation and packing have to be taken into account and might be responsible for the properties of different polymers. For instance, it has been shown that structural differences originating from different cis/trans tactic ratio in poly(1-trimethylsilyl-1-propyne) [PTMSP] and poly(*tert*-butylacetylene)[PTBA] films, lead to large changes in solubility (*5,6*), density (*7*), oxygen-permeability and oxygen/nitrogen permselectivity (*1*), as well as in chain packing and sorption properties (*7,8*).

One of the major problems which limits the practical application of PTMSP in the field of polymer-based membranes, both for gas-separation and pervaporation processes, is the physical aging which occurs even at room temperature despite its high glass transition temperature (*3,6,9-14*). This has also been observed in PTBA (7). Indeed, recent investigations in our laboratory on solvent cast films of PTBA samples with different tacticity have shown that chain packing and physical aging phenomena induced by thermal treatments are strongly related to the geometric structure of the polymer (7). It can be reasonably foreseen that films with different properties (due to the modified organization of the polymer chains) can also be obtained by varying the experimental conditions used to prepare them. For example, by casting films from various solvents, the arrangement of the chain segments during the film formation (solvent evaporation) can lead to different chain packing and, consequently, to a more or less permeable membrane.

In the present work the specific volume of cast films, obtained by evaporation under controlled conditions from solutions of PTBA in solvents with different thermodynamic characteristics, is reported. In order to make a careful choice of the systems, the solubility parameter of PTBA has been determined both through a calculation based on group contribution methods and by viscosity measurements. A correlation between thermodynamic characteristics of the solvents, free volume of the resulting membranes and extent of chain packing which occurs when the membranes undergo thermal treatments, is also given.

## Experimental

**Materials.** The monomer, *tert*-butylacetylene, was distilled twice over $CaH_2$ at atmospheric pressure and under nitrogen atmosphere. Analytical grade solvents were refluxed over a proper drying agent and distilled in an inert atmosphere. $MoCl_5$ (purity >99.5%) and $(C_6H_5)_4Sn$ were obtained from Aldrich, and used as received.

**Polymerizations.** Polymers were prepared as described in the literature (*15*). Typically, polymerization reactions to prepare PTBA with a mixture of cis and trans isomers (cis/trans PTBA) were carried out in toluene at 30°C for 24 h in the presence of $MoCl_5$ as a catalyst. 100% cis PTBA was prepared in benzonitrile, using the same reaction time and temperature mentioned previously. For 100% cis PTBA, $MoCl_5/(C_6H_5)_4Sn$ was used as the catalytic system. Monomer conversion was determined by gas chromatography using chlorobenzene as an internal standard.

**Polymer Characterization.** Viscosity measurements were carried out at 30°C using an automatic Schott-Gerate viscometer. Calorimetric analysis was performed under nitrogen in a Perkin-Elmer DSC 7; the experiments were done in the temperature range from 25 to 200°C. This range was chosen on the basis of the thermogravimetric analysis. Heating and cooling rates were always 10 K/min. The microstructure of the polymers was determined by solution $^{13}C$ NMR spectroscopy, in $CDCl_3$ (with tetramethylsilane as an internal standard) at room temperature on a Bruker AM-270 operating at 67.93 MHz for carbon.

**Film Preparation.** Solvent cast films ~200μm thick were prepared on a teflon plate from solutions (concentration 2 g/dl) in different solvents. Solvent evaporation was controlled by using an appropriate plate cover, according to the evaporation time selected. The solvent evaporation rate was followed by determining the weight of the polymer solution as a function of time. The evaporation times, $\tau_{ev}$, were 6, 70, and 145 h, and they refer to the time required to evaporate 90% of the solvent.

**Dilatometric Analysis.** The density of the films at room temperature and as a function of temperature was determined using a capillary dilatometer with mercury as the filling liquid (*16*). Dilatometric runs were performed in the temperature range from 20 to 90°C (heating rate 0.25 K/min) to avoid degradation of the polymer. Degradation has been reported in samples kept at 100°C for a long time (*1*). More details related to these experiments can be found in our previous paper (*7*).

## Results and Discussion

**Calculation of the Solubility Parameter.** The solubility parameter of PTBA, $\delta_{PTBA}$, has been calculated on the basis of group contribution methods, described by Hoy (*17*), Van Krevelen (*18*) and Small (*19*). These methods provide an estimate of the solubility parameter based on the primary chemical structure of the repeat unit which, for PTBA, is given in Figure 1.

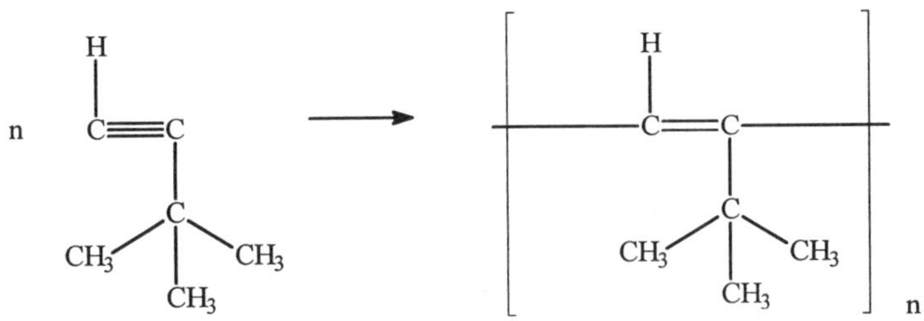

Figure 1 Chemical structure of poly(*tert*- butylacetylene)

Taking into account the molar volume group contributions for glassy polymers (*18*) and the group molar attraction constants at 25°C, the calculated values of $\delta_{PTBA}$ are: 16.4, 15.7 and 15.6 $MPa^{1/2}$ according to Van Krevelen, Hoy and Small, respectively.

**Solubility Parameter from Viscosity Measurements.** Based on the calculated values of $\delta_{PTBA}$, several solvents (with solubility parameters ranging from 14.5 to 19 $MPa^{1/2}$) were selected in order to perform viscosity measurements. These experiments were performed on both 100% cis PTBA samples and PTBA with cis/trans microstructure (55% cis). The results are summarized in Table I. Some significant differences have been noticed in the intrinsic viscosity behavior of the two samples in different solvents. For instance, $CHCl_3$ / toluene and n-heptane/ n-hexane present opposite solvent characteristics for 100% cis PTBA and for PTBA with lower cis content. The viscosity is higher in heptane than in hexane for 100% cis PTBA and lower in heptane than in hexane for cis/trans PTBA. The same trend is observed for toluene and chloroform.

Table I. Relative viscosity of PTBA samples in solvents with different solubility parameters, δ

| Solvents | $\delta^{\circ}$ $MPa^{1/2}$ | $\ln\eta_{rel}^{*}$ 100% cis PTBA | cis/trans PTBA |
|---|---|---|---|
| n-hexane | 14.9 | 0.23 | 0.11 |
| n-heptane | 15.3 | 0.26 | 0.10 |
| methylcyclohexane | 15.9 | 0.35 | 0.15 |
| cyclohexane | 16.8 | 0.38 | 0.14 |
| carbontetrachloride | 17.6 | 0.35 | 0.13 |
| toluene | 18.2 | 0.21 | 0.11 |
| chloroform | 18.7 | 0.26 | 0.10 |

° Hoy (*17*)
* determined at 30°C; concentration: 0.25 g/dl

Similar behavior has been described in the literature for other polyacetylenes. Indeed, different solubility properties, originating from differences in the polymer microstructure, have been reported for PTMSP (*5,6,20,21*) and poly(trimethylsilylacetylene) (*5*).

A plot of ln $\eta_{rel}$ at c=0.25 g/dl against the solubility parameter of the solvents is given in Figures 2 and 3 for solutions of 100% cis and cis/trans PTBA samples, respectively.

The best fit of the experimental data gives two very similar curves; the maximum in these curves occurs with the best solvent, which is cyclohexane for 100% cis PTBA and a solvent with δ between cyclohexane and methylcyclohexane for cis/trans PTBA. Generally, it is assumed that the polymer has the same solubility parameter as that of the best solvent (*19,22*). Therefore, $\delta_{PTBA}$ values deduced from our experiments are almost the same for the two samples: 16.8 $MPa^{1/2}$ for the 100% cis sample and 16.3 $MPa^{1/2}$ for the cis/trans PTBA sample. These data are in very good

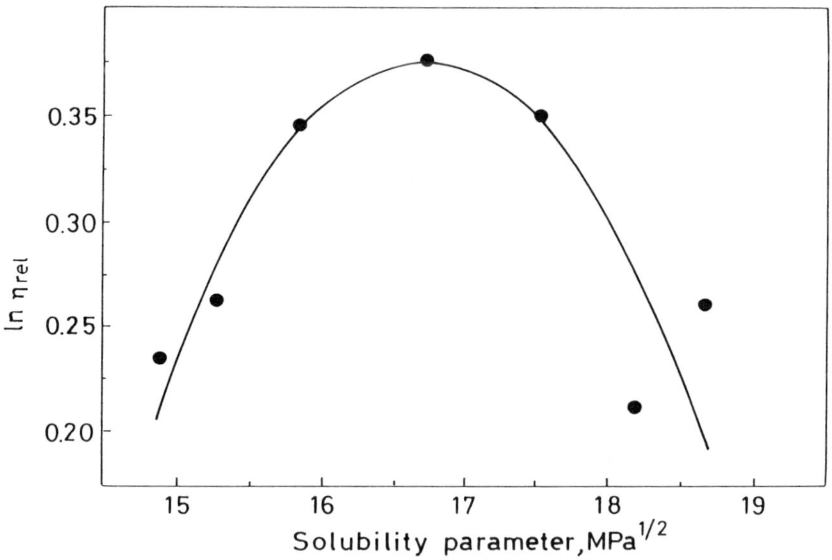


Figure 2. ln$\eta_{rel}$ at c=0.25g/dl as a function of the solubility parameters of different solvents; cis/trans PTBA

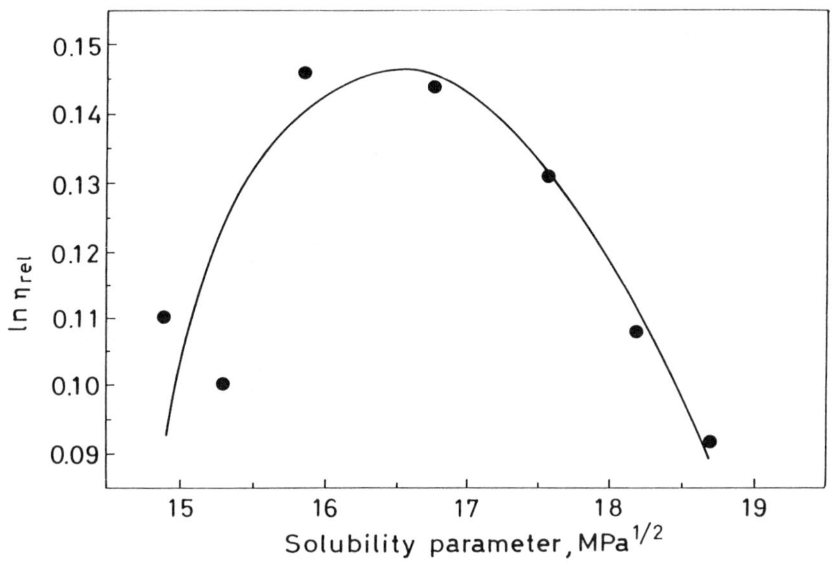


Figure 3. ln$\eta_{rel}$ at c=0.25g/dl as a function of the solubility parameters of different solvents; 100% cis PTBA

agreement with the calculated PTBA solubility parameters. Deviations between measured and calculated values are only about 5%.

**Density.** The density at room temperature of cis/trans PTBA films cast from toluene and cyclohexane and prepared at different evaporation times, $\tau_{ev}$, were obtained from dilatometric measurements. The specific volume, $v_{sp}$, of the polymer as a function of temperature was calculated using the following relationships (*16*):

$$v_{sp}(T) = [\, v_d(T)-(m_{Hg}-\Delta m_{Hg})\, v_{sp,Hg}(T)]/m_p \quad (1)$$

$$\Delta m_{Hg} = (\Delta h \cdot \sigma)/v^0_{sp,Hg} \quad (2)$$

where $m_p$ and $m_{Hg}$ are the mass of polymer and mercury in the dilatometer, respectively; $v_d$ and $v_{sp,Hg}$ are the volume of the dilatometer and the mercury, respectively, at the measurement temperature; $\Delta h$ is the change of the mercury height in the dilatometer capillary with temperature; $\sigma$ is the inner cross-section of the capillary; and $v^0_{sp,Hg}$ is the specific volume of mercury at 20°C.

A typical dilatometric curve is shown in Figure 4 for a film obtained from toluene solution. Run I and II refer to two successive experiments performed on the same sample: after run I was carried out on as-cast films up to 80°C, the dilatometer was cooled (cooling rate 0.5 K/min) and kept overnight at room temperature. Run II was performed with the same heating rate as for run I. As already found for the PTBA/$CHCl_3$ system (*7*), a temperature, $T_P = 51$°C, at which the packing phenomenon begins is present only during the first heating run, both for PTBA/toluene and PTBA/cyclohexane systems. A linear increase of specific volume with temperature is observed from room temperature up to $T_p$. Then, a deviation occurs and the volume increases at a lower rate. This is due to the normal dilation of the polymer with temperature and to a simultaneous densification process induced at high temperature by slow molecular motions which lead to a better packing of the chains. This phenomenon is discussed in our previous work (*7*). Density data given in Table II refer to films as-cast and after the thermal treatment in the dilatometer from room temperature up to 80°C (run I).

Table II. Density values for cis/trans PTBA cast from toluene and cyclohexane before and after heat treatment in dilatometer

| Solvent | $\tau_{ev}$, h | $\rho^{25°C}$, g/cm$^3$ As-cast | $\rho^{25°C}$, g/cm$^3$ Heat-treated | $\Delta\rho$ |
|---|---|---|---|---|
| Toluene | 6 | 0.9030 | 0.9145 | 0.0115 |
| | 145 | 0.9309 | - | - |
| Cyclohexane | 6 | 0.9189 | 0.9253 | 0.0064 |
| | 70 | 0.9298 | 0.9357 | 0.0059 |

The results show that by increasing the evaporation time better chain packing is achieved independent of the solvent and in agreement with our previous findings (*7*). Moreover, when the best solvent, cyclohexane, is used, a much lower evaporation time (70 h) is needed to reach the same density as that of a film prepared from toluene solution (145 h). A comparison of these data with our previous results on chloroform cast films (*7*) clearly shows that, under the same preparation conditions, density increases as the affinity between the polymer and the solvent increases. That is, going from chloroform to toluene to cyclohexane as the solvent, density of the as-cast films increases.

These results are interpreted as follows: the polymer concentration at which film formation begins, during the solvent evaporation, increases on increasing the affinity between polymer and solvent. Thus, when a good solvent is used, the polymer is homogeneously dissolved in solution up to very high concentrations and the chains have enough time to slowly organize to a more compact structure than that which is obtained when polymer precipitation occurs from a more dilute solution. As a result, the sample will be characterized by a low specific volume, i.e. a low free volume content, when the polymer is cast from thermodynamically better solvents.

Different density values have also been reported for PTMSP films prepared from toluene and n-hexane (*6*). We believe that, also in this case, the higher density values measured for films cast from n-hexane solution can be ascribed to the better thermodynamic characteristics of this solvent. Indeed, based on results from our laboratory the solubility parameter of n-hexane, as compared to $\delta_{toluene}$, is nearer to that of PTMSP.

The density of heat-treated PTBA films is always higher than that of as-cast films, which strengthens the notion that physical aging always occurs when specimens undergo thermal treatment.The amount of density increase which takes place as a result of heat treatment decreases when the initial density is higher, that is, when films have been obtained from solutions of the polymer in good solvents. This trend is similar to the one observed for $CHCl_3$ cast PTBA films prepared with different evaporation times (*7*). However, the amount of increase in density upon heat treatment, $\Delta\rho$, in that system does not correlate with $\Delta\rho$ values reported in the present work. This finding cannot be easily explained at present.

**Thermal Behavior**. Calorimetric measurements confirmed that a packing phenomenon takes place during the first heating scan. DSC profiles exhibit a very broad exothermic peak, which starts at about 100°C (Figure 5), which is much higher than the temperature $T_p$ revealed by dilatometry (Figure 4). This discrepancy originates from the different scanning rates, 10 and 0.25 K/min, used for the calorimetric and dilatometric experiments, respectively. The associated packing enthalpy, ranging from 11 to 5 J/g, can be related to the solvent evaporation time used to prepare the films, 6 to 145 h. In other words, a higher exothermic effect is observed in films with lower density. Second DSC scans did not exhibit any exothermic phenomenon.

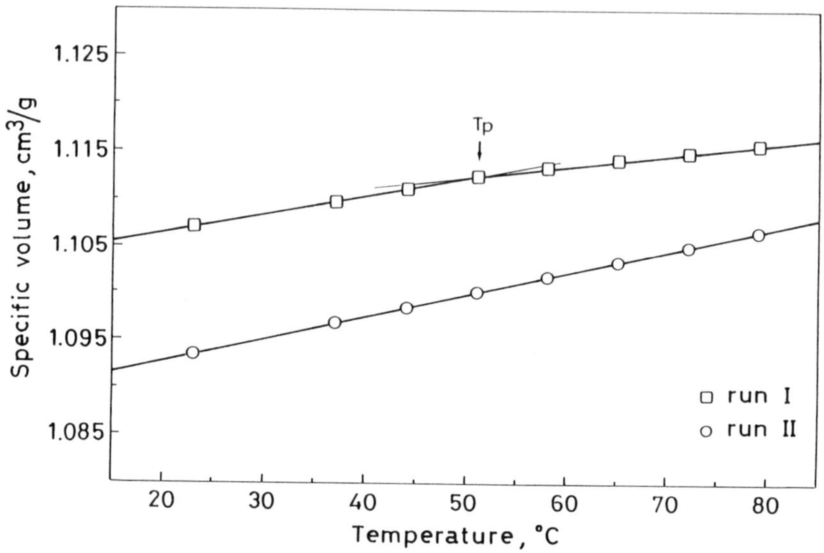


Figure 4. Specific volume of cis/trans PTBA film cast from toluene as a function of temperature ($\tau_{ev} = 6$ h)

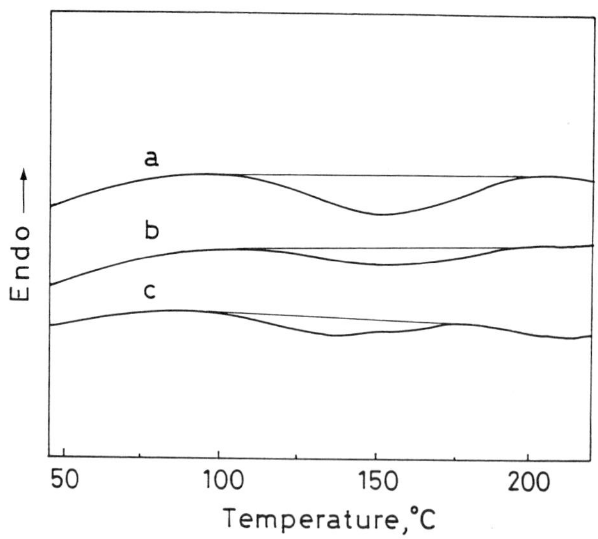


Figure 5. DSC thermograms of cis/trans PTBA films cast from toluene; $\tau_{ev}$ : a) 6 h; b) 70 h; c) 145 h

## Concluding Remarks.

Films from cis/trans PTBA prepared with lower solvent evaporation rates are always characterized by higher density values, independent of the thermodynamic characteristics of the solvent.

When the specimens undergo thermal treatment during either dilatometric analysis or calorimetric measurements, a packing process takes place as a result of enhanced chain mobility. This process occurs at relatively low temperature not too far above room temperature.

The cubic thermal expansion coefficient, determined from the specific volume vs. temperature plot, both before and after the densification process, (run II), is $\alpha_p = 2.0 \cdot 10^{-4}\ ^\circ C^{-1}$ (*7*), typical of that for glassy polymers.

The density increase upon ordering is larger for initially less dense samples, but the final state is not identical and retains memory of the initial packing. The thermal treatment does not overcome the effect of the solvent evaporation rate on chain packing.

The most interesting result from the present study is that it is possible to prepare membranes with different levels of free volume and, consequently, with different physical aging stability, by choosing the most suitable operating conditions (solvent, evaporation time, etc.).

**Acknowledgments.** The authors wish to thank Mr. G. Zannoni, Istituto di Chimica delle Macromolecole-CNR, Milano, for $^{13}C$ NMR experiments and Mr. V. Trefiletti for thermal characterization. Financial support *from Progetto Strategico CNR and MURST 60%* is gratefully acknowledged.

## Literature Cited

1. Takada, K.; Matsuya, H.; Matsuda, T.; Higashimura, T. *J. Appl. Polym. Sci.* **1985**, *30*, 1605.
2. Masuda, T.; Tang, B.-Z; Higashimura, T *Macromolecules* **1985**, *18*, 2369.
3. Masuda, T.; Iguchi, Y.; Tang, B.-Z.; Higashimura, T. *Polymer* **1988**, *29*, 2041.
4. Masuda, T; Higashimura, T. *J. M. S.-Pure Appl. Chem.* **1994**, *A31*, 1675.
5. Costa. G.; Grosso, A.; Sacchi, M.C.; Stein, P.C.; Zetta, L. *Macromolecules* **1991**, *24*, 2858.
6. Tanaka, A.; Nitta, K.; Mackawa, R.; Masuda, T.; Higashimura, T. *Polym.J.* **1992**, *24*, 1173.
7. Costa, G.; Chikhaoui, S.; Turturro, A; Carpaneto, L. *Macromol. Chem. Phys.* **1997**, *198*, 39.
8. Morisato, A.; Miranda, N.R.; Freeman, B.D.; Hopfenberg, H.B.; Costa, G.; Grosso, A.; Russo, S. *J. Appl. Polym. Sci.* **1993**, *49*, 2065.
9. Masuda, T.; Higashimura, T. *Adv. Polym. Sci.* **1987**, *81*, 121.
10. Langsam, M.; Robeson, L.M. *Polym. Eng. Sci.* **1989**, *29*, 44.
11. Witchey-Lakshmanan, L.C.; Hopfenberg, H.B.; Chern, R.T. *J. Membr. Sci.* **1990**, *48*, 321.

12. Tasaka, S.; Inagaki, N.; Igawa, M. *J. Polym. Sci. Part B: Polym. Phys.* **1991**, *29*, 691.
13. Yampol'skii, Yu.P.; Shishatskii, S.M.; Shantorovich, V.P.; Antipov, E.M.; Kuzmin, N.N.; Rykov, S.V.; Khodjaeva, V.L.; Platè, N.A. *J. Appl. Polym. Sci.* **1993**, *48*, 1935.
14. Nagai, K.; Watanabe, T.; Nakagawa, T. *Polym. J.* **1996**, *28*, 933.
15. Okano, Y.; Masuda, T.; Higashimura, T. *Polym.J.(Tokyo)* **1982**, *14*, 477
16. Turturro, A.; Bianchi, U. *Chim.Ind.(Milan)* **1967**, *49*, 362.
17. Hoy, K.L. *J.Paint Technol,* **1965**, *42*, 236.
18. Van Krevelen D.W.; Hoftyzer, P.Y. *Properties of Polymers: their Estimation and Correlation with Chemical Structure, 2nd Ed.*, Elsevier, Amsterdam, **1976.**
19. Barton, A.M.F. *Handbook of Solubility Parameters and other Cohesion Parameters*, CRC Press, Boca Raton, **1983**.
20. Masuda, T.; Isobe, E.; Higashimura, T. *Macromolecules* **1985**, *18*, 841.
21. Izumikawa, H.; Masuda, T.; Higashimura, T. *Polym.Bull.* **1991**, *27*, 193.
22. Bristow, G.M.; Watson, W.F. *Trans.Faraday Soc.* **1958,** *54*, 1742.

## Chapter 7

# Effects of Physical Aging on Gas Permeability and Molecular Motion in Poly(1-trimethylsilyl-1-propyne)

**Kazukiyo Nagai[1], B. D. Freeman[1], Tetsuya Watanabe[2], and Tsutomu Nakagawa[2]**

**[1]Department of Chemical Engineering, North Carolina State University, Box 7905, Raleigh, NC 27695–7905**
**[2]Department of Industrial Chemistry, Meiji University, Higashi-mita, Tama-ku, Kawasaki 214, Japan**

The effect of physical aging on gas permeability and molecular motion of poly(1-trimethylsilyl-1-propyne) (PTMSP) membranes synthesized using various catalysts was studied. During aging, the gas permeability of PTMSP membranes synthesized using $TaCl_5$ and $TaCl_5$-$Ph_3Bi$ was dramatically reduced, but the gas permeability of PTMSP synthesized using $NbCl_5$ was stable. The decrease in gas permeability of PTMSP membranes synthesized using various catalysts was correlated with changes in the non-equilibrium excess free volume in the polymer matrix. In the case of PTMSP synthesized using $NbCl_5$, no change in $^{13}C$ $T_1$ and $^{29}Si$ $T_1$ (NMR spin-lattice relaxation times) was observed during aging. However, $^{13}C$ $T_1$ of the backbone chain carbons of PTMSP membranes synthesized using $TaCl_5$ and $TaCl_5$-$Ph_3Bi$ increased upon aging. This change probably accompanied the relaxation of non-equilibrium excess free volume of PTMSP membranes synthesized these catalysts.

Polymeric membranes must have good permeabilities, permselectivities, and long-term stability. Moreover, they must be compatible with the process environment in which they will be used. However, polymers with high permeability tend to have low selectivity and vice versa (*1*). Poly(1-trimethylsilyl-1-propyne) (PTMSP) has the highest intrinsic gas permeability of all known synthetic polymers. It was first synthesized by Masuda and Higashimura in the 1980's (*2*). However, PTMSP has the lowest selectivities of all known polymers and exhibits unstable gas permeability (*3*).

PTMSP has a glass transition temperature of more than 250°C (*2*). In all glassy polymers, small-scale polymer segmental motions lead to relaxation of non-equilibrium excess free volume and, as a result, the physical properties of glassy polymers drift over time. Because PTMSP has more free volume than other glassy polymers, a dramatic decline in gas permeability occurs when the non-equilibrium excess volume in PTMSP relaxes (*3,4*). Membrane contamination via absorption of organics (such as pump oil vapor) also decreases gas permeability of PTMSP membranes (*4*). In the absence of such contaminants, the decrease in gas permeability in PTMSP membranes is due to a relaxation of non-equilibrium excess free volume, a process known as physical aging (*4*). As the amount of excess free volume is reduced during physical aging, the constraints on molecular motion increase, and polymer segmental mobility decreases.

The catalyst used for polymerization influences several properties such as the solubility of PTMSP in various solvents (*7*). Even though the cis-trans content of PTMSP has not been quantitatively determined, Costa *et al.* (*5*) and Izumikawa *et al.* (*6*) reported that the polymerization catalyst $NbCl_5$ produced a more regular chain configuration than $TaCl_5$. In addition, PTMSP synthesized using $NbCl_5$ had a cis-rich structure relative to that prepared with $TaCl_5$. In this study, the effects of physical aging on gas permeability and molecular motion of PTMSP membranes synthesized using various catalysts are reported.

## Experimental

**Polymers.** The PTMSP samples were previously synthesized (*4,8*), utilizing Masuda *et al.*'s methods (*2,7*).

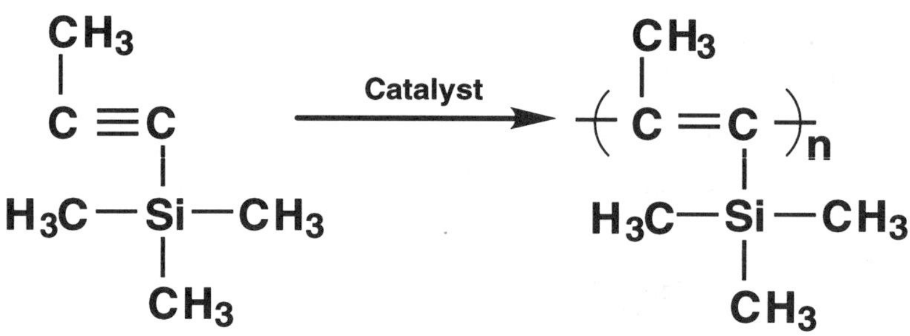


Poly(1-trimethylsilyl-1-propyne) (PTMSP)

Polymerization catalysts included tantalum (V) chloride ($TaCl_5$), niobium (V) chloride ($NbCl_5$), and $TaCl_5$-triphenyl bismuthine ($Ph_3Bi$) as a cocatalyst ($TaCl_5$:$Ph_3Bi$=1:1). The weight-average molecular weights ($M_w$, g/mol) of the PTMSP samples were $86\times10^4$ ($TaCl_5$), $260\times10^4$ ($TaCl_5$-$Ph_3Bi$), and $35\times10^4$ ($NbCl_5$). The intrinsic viscosities ([$\eta$], $dLg^{-1}$) were 6.0 ($TaCl_5$), 9.3 ($TaCl_5$-$Ph_3Bi$), and 0.7 ($NbCl_5$), and were measured in toluene at 30°C. No thermal events were recorded in differential scanning calorimetry scans at temperatures up to 250° for all PTMSP samples. These samples were, therefore, glassy under the conditions used in this study.

**Membranes.** PTMSP membranes with uniform thicknesses of 60 to 200 μm were prepared by casting on a horizontal glass plate from a solution of the polymer dissolved in toluene. To prevent physical aging and contamination, all membranes were immersed in methanol until just before beginning gas permeation, gas sorption, and NMR experiments. The methanol treatment was performed at least 48 hours at room temperature in order to attain the equilibrium sorption of methanol in the PTMSP membranes (*i.e.*, to prepare samples with the same initial condition). PTMSP membranes swell when immersed in methanol. The experiments were started after the macroscopic contraction of the swollen membrane was finished. The detailed procedures before beginning the experiments were reported previously (*8*). Aged membranes were prepared by storing them in a vacuum vessel at 30°C.

**Gas Permeation and Sorption Measurements.** The gas permeability was determined by the vacuum-pressure method using an MKS Baratron pressure transducer. The upstream pressure was up to 1 atmosphere. The gases used were hydrogen ($H_2$), nitrogen ($N_2$), oxygen ($O_2$), carbon dioxide ($CO_2$), and methane ($CH_4$). Sorption isotherms for propane ($C_3H_8$) were determined at 35°C using a gravimetric quartz spring balance operating at pressures up to 1 atmosphere.

**NMR Measurements.** High resolution, solid-state NMR measurements were performed at 30°C using a JEOL JNM 400 spectrometer (*9*). The spin-lattice relaxation times ($T_1$) were obtained by the CP $T_1$ pulse sequence. The same $T_1$ measurements were performed on at least three separate membranes prepared at the same time from the same casting solution.

## Results and Discussion

**Gas Permeability.** Figure 1 presents $N_2$ permeability coefficients for as-cast and aged PTMSP membranes synthesized using various catalysts. The initial value of an as-cast membrane was measured between 24 and 48 hours after the membrane was removed from methanol.

Gas permeabilities of PTMSP membranes synthesized using $TaCl_5$ and $TaCl_5$-$Ph_3Bi$ were dramatically lower than those of PTMSP membranes synthesized using $NbCl_5$. For other gases (*e.g.*, $H_2$, $O_2$, $CO_2$, and $CH_4$), the same trend was observed.

Figure 2 presents the relationship between kinetic diameter (*10*) and aging ratio of PTMSP membranes synthesized using various catalysts. The aging ratio is the permeability coefficient of a membrane stored for 12 days relative to its initial permeability [P(12 days)/P(as-cast)]. The PTMSP synthesized using $NbCl_5$ had stable gas permeability coefficients, but the gas permeability of PTMSP synthesized using $TaCl_5$ and $TaCl_5$-$Ph_3Bi$ was reduced upon aging. With increasing kinetic diameter, the aging ratio slightly decreased.

The ideal separation factors changed with age because the aging ratio of each gas was different. Table I summarizes the ideal separation factors of as-cast and aged PTMSP membranes synthesized using various catalysts. PTMSP synthesized using $TaCl_5$-$Ph_3Bi$, which exhibited the most distinct decline in gas permeability upon aging, showed a clear increase in separation factor after aging. However, in the case of $P(O_2)/P(N_2)$, where there is a very small difference in kinetic diameter, the change in separation factor with aging was small. PTMSP synthesized using $NbCl_5$, which had more stable gas permeability values than PTMSP synthesized using $TaCl_5$ or $TaCl_5$-$Ph_3Bi$, showed the smallest change in separation factor.

Based on these gas permeation results, aging of gas permeability appears to depend on the chain configuration of PTMSP since the polymerization catalyst, which is believed to be important in determining chain configuration, influenced aging behavior. With the aging conditions used, the gas permeability of PTMSP synthesized using $NbCl_5$ was more stable than that of PTMSP membranes synthesized using $TaCl_5$ or $TaCl_5$-$Ph_3Bi$. This result suggests that the non-equilibrium state of PTMSP synthesized using $NbCl_5$ is more stable than that of polymers prepared with $TaCl_5$ or $TaCl_5$-$Ph_3Bi$.

In general, the permeability can be represented as the product of gas solubility and gas diffusivity. In previous studies of light gases (*4,11*), the reduction in gas permeability of PTMSP upon aging depended on gas diffusivity more than gas

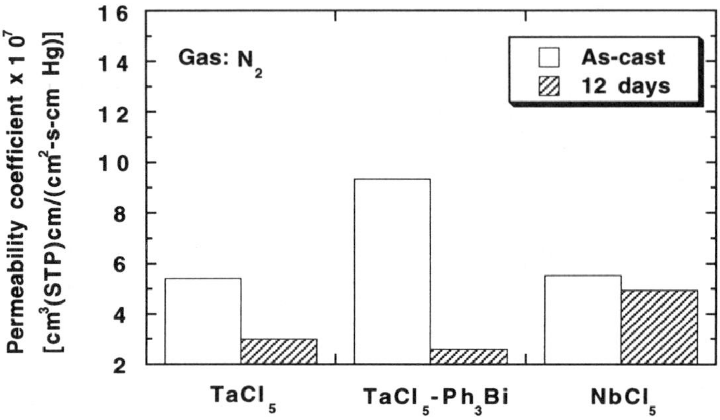


**Figure 1.** Permeability coefficients of nitrogen in as-cast and aged PTMSP membranes at 30°C.

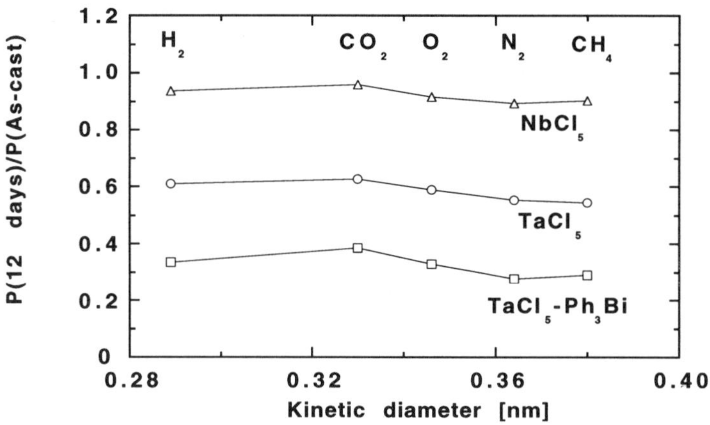


**Figure 2.** Relationship between aging ratio and kinetic diameter of the PTMSP membranes synthesized using various catalysts. Aging ratio: Permeability coefficient (12 days)/ Permeability coefficient (as-cast).

solubility. Gas diffusivity, in turn, depends on free volume and molecular motion of the polymer chains.

**Table I. Ideal Separation Factors of As-cast and Aged[a] PTMSP Membranes Synthesized Using Various Catalysts**

| Catalyst | $H_2/N_2$ | | $O_2/N_2$ | | $CO_2/N_2$ | | $CO_2/CH_4$ | |
|---|---|---|---|---|---|---|---|---|
| | Initial | Aged | Initial | Aged | Initial | Aged | Initial | Aged |
| $TaCl_5$ | 2.7 | 2.9 | 1.5 | 1.6 | 4.9 | 5.6 | 1.9 | 2.2 |
| $TaCl_5$-$Ph_3Bi$ | 2.5 | 3.0 | 1.4 | 1.6 | 4.7 | 6.5 | 1.9 | 2.5 |
| $NbCl_5$ | 2.6 | 2.8 | 1.4 | 1.5 | 4.9 | 5.3 | 1.9 | 2.0 |

[a] Aging protocol: stored in a vacuum vessel for 12 days at 30°C.

**Free Volume.** The Langmuir sorption capacity, $C'_H$, of the dual-mode sorption model (*12,13*) characterizes gas sorption in the non-equilibrium excess free volume of a glassy polymer. The total sorbed gas concentration in a glassy polymer, which is a summation of Henry's law dissolution ($C_D$) and Langmuir type hole filling ($C_H$), is given by:

$$C = C_D + C_H = k_D p + \frac{C'_H bp}{1+bp} \tag{1}$$

where $k_D$ is the Henry's law coefficient, and b is the hole-affinity parameter.

Based on parameters estimated from sorption isotherms for $C_3H_8$ in as-cast and aged PTMSP membranes, only $C'_H$ showed a distinct change as a result of the aging protocol. Figure 3 presents $C'_H$ values for $C_3H_8$ in as-cast and aged PTMSP membranes. The $C'_H$ values for PTMSP synthesized using $TaCl_5$-$Ph_3Bi$ was dramatically reduced upon aging, indicating a decrease in the non-equilibrium excess free volume during the aging process. On the other hand, the $C'_H$ value for PTMSP synthesized using $NbCl_5$ was practically independent of aging, suggesting that the polymerization catalyst strongly influences aging behavior in PTMSP.

When the aging protocol includes thermal annealing at various temperatures, the same trend in $C'_H$ is observed (*4*). That is, both gas permeability and $C'_H$ of PTMSP are reduced by thermal treatment (*4*). Furthermore, aging-induced decreases in gas permeability of PTMSP synthesized using various catalysts are correlated systematically with changes in $C'_H$. Therefore, aging appears to reduce the non-equilibrium excess free volume of PTMSP which, in turn, reduces the Langmuir sorption capacity and the gas permeability.

**Molecular Motion.** Spin-lattice relaxation times, $T_1$, determined using solid-state NMR provide a practical method to characterize molecular motion of polymers in the solid state. In this regard, Figure 4 presents $^{29}Si$ $T_1$ values for as-cast and aged PTMSP membranes. Based on the results in this table, the spin lattice relaxation time is not sensitive to aging for PTMSP synthesized with the three catalysts considered in this study.

Table II summarizes the $^{13}C$ $T_1$ of each distinct carbon in as-cast and aged PTMSP membranes. As previously described (*9*), for PTMSP synthesized using $NbCl_5$, no changes in $T_1$ values were observed as a result of aging. In contrast, the $T_1$ values of the backbone chain carbons ($C_c$ and $C_d$) of PTMSP synthesized using $TaCl_5$

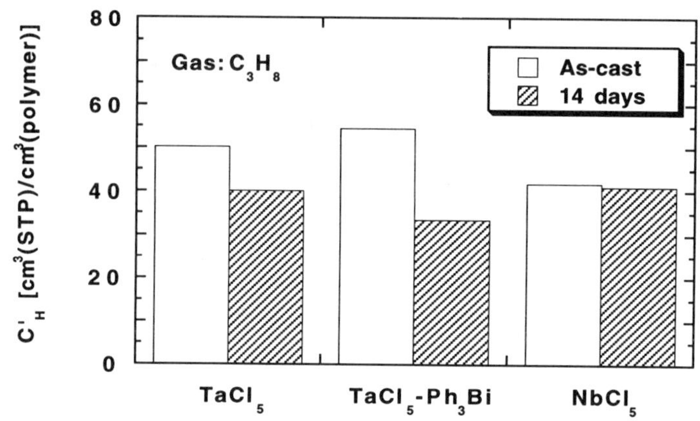


**Figure 3.** $C'_H$ values for propane in as-cast and aged PTMSP membranes.

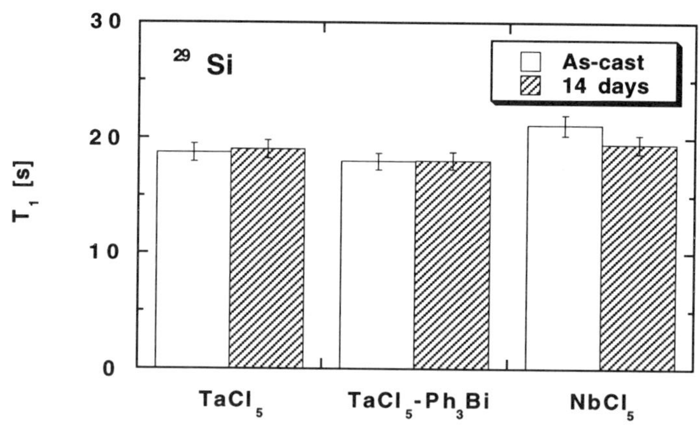


**Figure 4.** $^{29}Si$ $T_1$ of as-cast and aged PTMSP membranes.

and $TaCl_5$-$Ph_3Bi$ decreased upon aging. The $T_1$ values of side chain carbons ($C_a$ and $C_b$) were almost constant. The reduction in backbone chain carbon spin lattice relaxation time is probably closely coupled to the relaxation of non-equilibrium excess free volume of PTMSP synthesized with $TaCl_5$ and $TaCl_5$-$Ph_3Bi$.

**Table II.** $^{13}C$ $T_1$ **of** $C_a$-$C_d$[a] **in the As-cast and Aged**[b] **PTMSP Membranes**

| Catalyst | $C_a$ [s ] | | $C_b$ [s] | | $C_c$ [s] | | $C_d$ [s] | |
|---|---|---|---|---|---|---|---|---|
| | Initial | Aged | Initial | Aged | Initial | Aged | Initial | Aged |
| $TaCl_5$ | 2.4 | 2.4 | 5.9 | 6.4 | 18.9 | 26.1 | 17.8 | 27.9 |
| $TaCl_5$-$Ph_3Bi$ | 2.3 | 2.8 | 5.8 | 5.6 | 22.2 | 27.4 | 21.9 | 25.2 |
| $NbCl_5$ | 2.6 | 2.6 | 5.2 | 5.0 | 25.6 | 25.9 | 27.7 | 26.2 |

[a] Repeat Unit Structure: $\mathbf{C_b}H_3$-$\mathbf{C_d}$=$\mathbf{C_c}$-Si($\mathbf{C_a}H_3)_3$

[b] Aging protocol: stored in a vacuum vessel for 14 days at 30°C.

## Conclusions

The effect of physical aging on gas permeability of PTMSP synthesized using various catalysts was studied. With the aging protocol used in this study, the gas permeability of PTMSP synthesized using $TaCl_5$ and $TaCl_5$-$Ph_3Bi$ was dramatically reduced after aging. However, the gas permeability of PTMSP synthesized using $NbCl_5$ was stable. The $C'_H$ values of PTMSP prepared with different catalysts also exhibited the same behavior. Based on $T_1$ measurements, only backbone chain carbons of PTMSP synthesized using $TaCl_5$ and $TaCl_5$-$Ph_3Bi$ decreased as a result of aging. This decrease in molecular motion is probably closely related to the relaxation of non-equilibrium excess free volume of PTMSP synthesized with $TaCl_5$ and $TaCl_5$-$Ph_3Bi$. Based on these composite data, the non-equilibrium state of PTMSP prepared using $NbCl_5$ appears to be more stable than that prepared using $TaCl_5$ and $TaCl_5$-$Ph_3Bi$.

## Literature Cited


1. Robeson, L. M. *J. Membrane Sci.* **1991**, *62*, 165-185.
2. Masuda, T.; Isobe, E.; Higashimura, T.; Takada, K. *J. Am. Chem. Sci.* **1983**, *105*, 7473-7474.
3. For example, Shimomura, H.; Nakanishi, K.; Odani, H.; Kurata, M.; Masuda, T.; Higashimura, T. *Kobunshi Ronbunshu* **1986**, *43*, 747-753.
4. For example, Nagai, K.; Nakagawa, T, *J. Membrane Sci.* **1995**, *105*, 261-272.
5. Costa, G.; Grosso, A.; Sacchi, M. C.; Stein, P. C., Zetta, L. *Macromolecules* **1991**, *24*, 2858-2861.
6. Izumikawa, H.; Masuda, T.; Higashimura, T. *Polym. Bull.* **1991**, *27*, 193-199.
7. Masuda, T.; Isobe, E.; Higashimura, T. *Macromolecules* **1985**, *18*, 841-845.
8. Nagai, K.; Higuchi, A.; Nakagawa, T. *J. Polym. Sci.: Part B: Polym. Phys.* **1995**, *33*, 289-298.
9. Nagai, K.; Watanabe, T.; Nakagawa, T. *Polym. J.* **1996**, *10*, 933-935.
10. Breck, D. W. Zeolite Molecular Sieves; Wiley; NY, 1974; p. 636.
11. Nakagawa, T.; Saito, T.; Asakawa, S., Saito, Y. *Gas Sep. Purif.* **1988**, *2*, 3-8.
12. Barrer, R. M.; Barrie, J. A.; Slater, J. *J. Polym. Sci.* **1958**, *27*, 177-197.
13. Michaels, A. S.; Vieth. W. R.; Barrie, J. A. *J. Appl. Phys.* **1963**, *34*, 1-13.

Chapter 8

# Free Volume in Polymeric Membrane Materials as Determined by Positron Annihilation Lifetime Spectroscopy

**Yuri P. Yampolskii[1] and Victor P. Shantarovich[2]**

**[1]Institute of Petrochemical Synthesis, 29 Leninsky Pr., 117912 Moscow, Russia**
**[2]Institute of Chemical Physics, 4 Kosygina Str., 117334, Moscow, Russia**

Positron annihilation lifetime spectroscopy (PALS) is an efficient tool for measuring free volume and sizes of free volume elements in polymeric materials. This is particularly important for studies of membrane materials, since free volume determines the permeation rate of small molecules. Free volume was studied by means of PALS in polymers characterized by extremely high permeability: poly(1-trimethylsilyl-1-propyne) and copolymers of 2,2-bistrifluoromethyl-4,5-difluoro-1,3-dioxole and tetrafluoroethylene. The results obtained were compared with those observed for conventional glassy polymers. For the first time, the size distribution of free volume has been determined for these membrane materials.

The free volume (FV) in polymer systems is of great interest because the size and concentration of its elements (holes) affect numerous transport and other physicochemical properties of polymers. Positron annihilation lifetime (PAL) spectroscopy is now one of the most efficient approaches for investigations of FV. The foundations of this method for probing polymers were based in particular on Walker-Brandt-Berko's free volume model (*1*). According to this model, Positronium, Ps, (a bound atomic system, which consists of an electron and the positron) tends to be localized or trapped before its annihilation in FV or, in other words, in areas with reduced electron density. Accordingly, annihilation characteristics (lifetimes and intensities of longer lifetime components of annihilation radiation) provide information regarding the concentration and sizes of FV elements. (*2-5*)

PAL spectra have been extensively used for probing free volume in membrane polymeric materials. Correlations have been reported between free volume parameters and transport properties of polymers (*4,6-9*). It is generally accepted that PAL distributions in most polymers consist of three components. The long-lived component characterized by lifetime $\tau_3$ (ns) and intensity $I_3$ (%) is sensitive to free volume. This work focuses on free volume as measured by the PALS method in polymers. Three of the polymers investigated in this work have chemical structures that lead to unusually loose chain packing: poly(1-trimethylsilyl-1-propyne) (PTMSP) and copolymers of 2,2-bistrifluoromethyl-4,5-difluoro-1,3-dioxole and tetrafluoroethylene known by their trademark names of AF2400 and AF1600. The mole fraction of the dioxole comonomer is approximately 90% and 65% in AF2400 and AF1600, respectively. In these materials a fourth component of the PAL spectrum appears. It is distinguished by a lifetime $\tau_4$, which is much longer than that observed in other polymers.

Until recently, most of the PAL data were analyzed in a finite-term lifetime approach. A computer program PATFIT, which represents annihilation lifetime distribution in a discrete manner, i.e. as a sum of several exponents, is employed for this purpose. As an alternative, Gregory and Jean (*10,11*) proposed using a continuous lifetime analysis. In this approach the Laplace inversion program CONTIN, originally developed by Provencher (*12*) for analysis of fluorescence spectra, is used to obtain a continuous probability density function of annihilation lifetimes from PAL spectra. In this way one can obtain size distributions of FV in polymers.

One of the purposes of this work is to compare the results of application of both programs, PATFIT and CONTIN, for the treatment of positron annihilation characteristics obtained in studies of polymer samples distinguished by unusually long Ps lifetimes or, according to existing notions, substantially larger sizes of FV. Dense films of PTMSP and both dense films and porous membranes prepared from poly(2,6-diphenylphenylene oxide) (PPO) were the objects of the investigation using the CONTIN method. Additionally, some new results are presented which shed light on Ps behavior in polymers and are discussed in terms of the importance of proper methodology for the application of PALS as a way for probing free volume.

## Experimental Section

PTMSP was synthesized by polymerization in the presence of the catalytic system $TaCl_5$/triisobutylaluminum. The polymer had an intrinsic viscosity of 5.7 dl/g and $M_w$=2,000,000 (GPC). The films were cast from solution in toluene and dried in a vacuum oven at 40-50° C for several days until a constant weight was achieved.

Commercial forms of the amorphous glassy perfluorinated polymers AF2400 and AF1600 purchased from DuPont Co in powder form were used as received. Some of the properties of these materials are described in (*13*). The films were cast from 2 wt.% solutions in perfluorotoluene and dried in the same way as the PTMSP films.

Dense films of PPO were cast from solution in chloroform and prepared in the manner just described. Porous PPO membranes obtained by the phase inversion method were provided by Dr.F.P. Cuperus (The Netherlands) and used as-received. Two samples of different porosity were prepared. Those samples were characterized by surface areas of 70 and 200 $m^2$/g.

Isotactic poly(vinyl cyclohexane) (PVCH) films having varying degree of crystallinity were obtained by ion-coordination suspension polymerization in the presence of $TiCl_4$/tri(isobutyl)aluminum as the catalytic system. The product was treated with boiling n-hexane. Thus the isotactic fraction was obtained in the solution while the atactic polymer remained as an insoluble residue. Films of isotactic PVCH were cast from cyclohexane solutions. The polymer had an intrinsic viscosity of 2.2 dl/g in cyclohexane. Crystallinity in the range of 5-35% was obtained by annealing at 60-140°C for different times. Crystalline content was determined by X-ray diffraction.

The PAL measurements were performed using an ORTEC standard "fast-fast" coincidence circuit of a positron lifetime spectrometer with time resolution of 230 psec (full width at half maximum (FWHM) of the prompt coincidence curve). Integral statistics for each PAL spectrum included at least $10^7$ counts. A high purity and defect free single crystal Silicon sample was used as a reference. Its PAL spectrum consisted of a single component with a lifetime of 220 psec. Measurements were performed either at ambient conditions, i.e. in contact with air, or in nitrogen at 1 atm in a special cell.

## Results and Discussion

**High Free Volume Polymers.** It has been shown that PAL distribution of glassy polymers distinguished by extremely high gas permeability and diffusion coefficients cannot be fit by a 3 component model spectrum; on the other hand, a 4 component model provide excellent fits (*13,14*). Table I gives PAL spectra for PTMSP and the two perfluorinated glassy polymers. These materials are endowed with permeability coefficients to light gases on the order of $10^3$-$10^4$ Barrers (*13,15*).

Table I. Positron annihilation lifetime spectra components $\tau_i$, ns and $I_i$, % in high permeability polymers

| Polymer | $\tau_1$ | $I_1$ | $\tau_2$ | $I_2$ | $\tau_3$ | $I_3$ | $\tau_4$ | $I_4$ |
|---|---|---|---|---|---|---|---|---|
| PTMSP* | 0.21 | 41 | 0.58 | 25 | 2.5 | 5.0 | 6.7 | 30 |
| AF2400 | 0.20 | 22 | 0.42 | 62 | 1.7 | 2.4 | 6.0 | 13 |
| AF1600 | 0.18 | 21 | 0.39 | 56 | 0.68 | 8.5 | 4.7 | 15 |
| PS** | 0.23 | 41 | 0.54 | 26 | 2.1 | 33 | – | – |

*PTMSP sample obtained by polymerization in the presence of $TaCl_5$
**PS: atactic polystyrene.

The PAL spectrum of polystyrene, which is considered as a conventional glassy polymer, is also presented in Table I for comparison. It is seen that much longer lifetimes (5-7 ns for $\tau_4$) are characteristic of high permeability materials. Interestingly, such long lifetimes have been observed as well in silicagels and zeolites (*16*) or even in porous UF membranes as is shown below. All of this information is in line with the assumption that high free volume polymers like PTMSP are akin to porous sorbents (probably with open or closed porosity).

An equation was proposed and extensively used for the calculation of equivalent radii $R_i$ of FV elements based on PALS lifetime data (*2,3,17*):

$$\tau_i = 0.5\left[1 - \frac{R_i}{R_0} + \frac{1}{2\pi}\sin\left(\frac{2\pi R_i}{R_0}\right)\right]^{-1} \quad (1)$$

where $\tau_i$ (i=3 or 4) is expressed in nanoseconds, $R_i$ and $R_0$ are in Angstroms, and $R_0=R_i + \Delta R$, which is an empirical parameter. The value of $\Delta R$ is 1.656 Å. An application of this equation shows that the radius of FV elements as sensed by the PAL method in high permeability materials like PTMSP or AF2400 can be as large as about 6-7 Å, whereas in conventional glassy polymers it is equal to 3-4 Å. Below, it is shown that, in fact, these values can be a lower limit of FV element sizes in these high permeability materials.

**Aging of PTMSP.** An unusual feature of PTMSP is that this glassy material is prone to aging behavior at ambient temperatures, i.e. at temperatures over 200 K below its glass transition, resulting in a decrease in gas permeation parameters (*18*). This effect is observed in vacuum or in inert atmosphere. Therefore, it is a consequence of structure or morphological transformations which proceed rapidly in this material. Another type of slower process takes place in the presence of oxygen during long term storage of this polymer. This process involves partial oxidation of the main chain (*19,20*). Table II presents $CO_2$ permeability and solubility coefficients of as-cast PTMSP and the material aged after storage for several years at ambient conditions.

Table II. $CO_2$ transport and sorption parameters, PALS parameters and d-spacing of original and aged PTMSP at 25°C.

| Parameter | Original PTMSP* | Aged PTMSP |
|---|---|---|
| P, Barrer** | 18,000 | 1060 |
| S, $cm^3(STP)/cm^3 \cdot atm$*** | 10 | 3.9 |
| $C'_H$, $cm^3(STP)/cm^3$*** | 139 | 18 |
| $\tau_3$, ns | 2.5 | 1.7 |
| $I_3$ ,% | 5.0 | 7.8 |
| $\tau_4$, ns | 6.7 | 3.8 |
| $I_4$, % | 30 | 15 |
| $R_4$, Å | 7.1 | 5.8 |
| d-spacing, Å (WAXD)# | 10 | 8.9 |

* The polymer was synthesized using $TaCl_5$ catalyst

**1 Barrer = $10^{-10}$ $cm^3(STP) \cdot cm/cm^2 \cdot s \cdot (cm\ Hg)$

*** According to Bondar et al. (21).

# WAXD = wide angle X-ray diffraction

It is seen that both parameters decrease markedly thanks to partial oxidation (appearance of C=O bonds according to (*19,20*)) of the main chains that makes the chains more flexible. Infrared and nuclear magnetic resonance spectroscopy reveal the presence of the C=O bonds. Simultaneously, the PAL spectrum is changed, reflecting a significant decrease in FV element size (from 7.1 to 5.8 Å). Approximately the same change is observed in interchain distance (so called d-spacing) as revealed by WAXD.

**Effect of Oxygen Present During the PAL Measurement.** Another peculiarity of PAL spectra in high surface area materials like silica gels is that an additional mechanism for annihilation of o-Ps appears which involves the reaction with molecular oxygen on pore walls (*22*). Recently it was shown that this is also true for high permeable (high surface area) material such as PTMSP (*23*). If the PAL measurement is performed for this polymer in inert atmosphere instead of air the lifetime parameter $\tau_4$ increases from 6.6 to 13.8 ns. This result means that the real sizes of the free volume elements in PTMSP are even larger than the values given in the previous paragraph.

Table III gives the results of a similar study of the effects of replacement of atmospheric oxygen by pure nitrogen on the PAL parameters of another highly permeable glassy polymer, AF2400. For comparison, another perfluorinated membrane material, Nafion, was also studied. It should be noted that Nafion is distinguished by low gas permeability: the permeability coefficient of oxygen in Nafion is only 0.6 Barrer at ambient conditions (*24*).

Table III. Effect of gas atmosphere during measurement on the observed PALS parameters of AF2400 and Nafion

| Parameter | AF2400 | | Nafion | |
|---|---|---|---|---|
| | Air | Nitrogen | Air | Nitrogen |
| $\tau_1$, ns | 0.196±0.007 | 0.202±0.015 | 0.173±0.032 | 0.166±0.025 |
| $I_1$, % | 22.6±1.8 | 21.9±3.4 | 10.9±4.2 | 13.1±3.3 |
| $\tau_2$, ns | 0.415±0.006 | 0.422±0.010 | 0.389±0.015 | 0.399±0.012 |
| $I_2$, % | 60.6±1.5 | 60.8±3.1 | 78.2±2.4 | 77.4±2.4 |
| $\tau_3$, ns | 1.59±0.12 | 2.12±0.37 | 0.95±0.25 | 1.28±0.49 |
| $I_3$, % | 3.4±0.2 | 2.8±0.2 | 4.3±2.2 | 2.8±0.7 |
| $\tau_4$, ns | 6.02±0.04 | 8.00±0.14 | 3.30±0.07 | 3.29±0.20 |
| $I_4$, % | 13.4±0.13 | 14.5±0.3 | 6.7±0.3 | 6.6±0.8 |

Although the PAL spectrum of Nafion can be described by 4 components, the $\tau_4$ lifetime and corresponding intensity are relatively small. Table III shows that the lifetime $\tau_4$ of AF2400 is increased from 6.0 ns in air to about 8 ns in the inert atmosphere without any change of the corresponding intensity. Some changes also occur in the $\tau_3$ lifetime. On the contrary, oxygen exerts virtually no effect on the PAL spectrum of Nafion, a result that agrees with the concept of a low "surface area" in conventional glassy polymers that do not exhibit exceptionally long o-Ps lifetimes.

**Continuous PAL Free Volume Size Distribution.** The results described above were obtained by a finite-term treatment of the experimental lifetime distribution. However, an application of the CONTIN program allows calculation of the lifetime distribution, which can be converted into the size distribution of FV elements using Equation 1.

Table IV shows a comparison of PAL spectra in PTMSP and a dense PPO film (crystallinity of about 40%) obtained using PATFIT and CONTIN programs. It is seen that PPO has a PAL spectrum typical of most polymeric materials: the intermediate positron lifetime component $\tau_2$ is close to 0.4 ns, whereas the o-Ps component is in the range of 2-3 ns.

Table IV. Comparison of lifetime distributions of positron annihilation in dense films of PTMSP and PPO using PATFIT and CONTIN programs.

| Parameter | PTMSP* | | PPO | |
|---|---|---|---|---|
| | PATFIT | CONTIN | PATFIT | CONTIN |
| $\tau_1$, nsec | 0.136±0.006 | 0.13 | 0.141±0.008 | 0.15 |
| $\tau_2$, nsec | 0.360±0.010 | 0.40 | 0.370±0.008 | 0.35 |
| $\tau_3$, nsec | 1.520±0.012 | 1.75 | 1.780±0.140 | – |
| $\tau_4$, nsec | 5.580±0.020 | 5.63 | 3.060±0.130 | 2.30 |
| $I_1$, % | 23.90±1.40 | 24.5±0.4 | 18.35±1.50 | 14.6±1.2 |
| $I_2$, % | 29.40±1.30 | 26.7±0.7 | 50.46±1.21 | 54.0±2.0 |
| $I_3$, % | 5.58±0.02 | 5.4±0.8 | 15.38±2.65 | – |
| $I_4$, % | 40.50±0.30 | 42.9±3.0 | 15.80±2.97 | 31.0±0.6 |

*Sample obtained by polymerization in the presence of the catalytic system $TaCl_5$/triisobutylaluminum

Although the PATFIT program allows calculation of two longer lifetime components $\tau_3$ and $\tau_4$ in PPO, these lifetimes do not differ markedly and have similar intensities of about 15%. The CONTIN program, which was used in the range of annihilation rates from 0.1 to 15 ns$^{-1}$ (or lifetimes from 0.06 to 10 ns) with 75 points in the solution, describes these two lifetime components as one fairly broad peak with a maximum at 2.3 ns and an intensity of 31%. Alternatively, the PATFIT and CONTIN analyses give similar results for the long lived components of PTMSP ($\tau_3$, $\tau_4$, $I_3$, $I_4$). The PATFIT treatment of annihilation data for PTMSP gave a lifetime $\tau_4$=5.58 ns with a very large intensity of 40.5%. In conclusion, the PATFIT and CONTIN data exhibit very good agreement for PAL spectra of PTMSP, which can be considered as evidence of an appropriate selection of the resolution function.

The lifetime distribution for PTMSP at ambient temperature is shown in Figure 1. Using this distribution, the size distribution of FV was calculated by means of Equation 1. Naturally, the o-Ps size distribution consists of two peaks (Figure 2). The dashed line in Figure 2 shows the calculated dependence of the annihilation rate ($\lambda_i=1/\tau_i$ ns$^{-1}$) versus FV radius used in computing the size distribution by means of Equation 1.

The CONTIN method has been extensively used for determination of the size

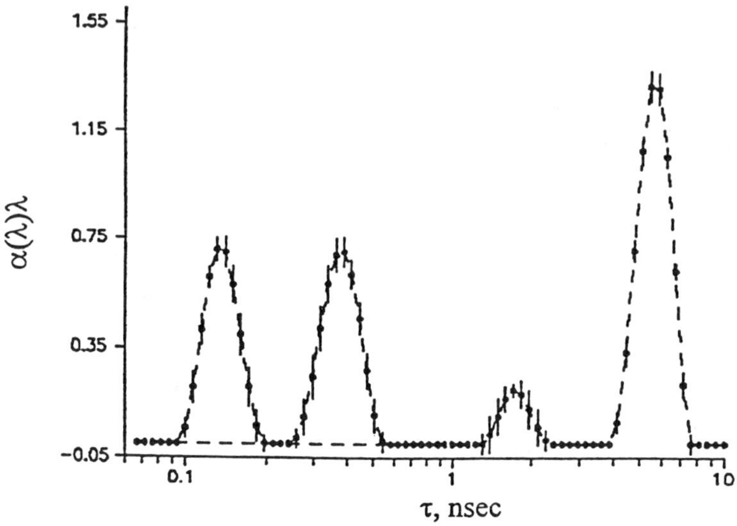


Figure 1. Lifetime distribution of PTMSP obtained using the Laplace inversion program CONTIN. The value α(λ) is the probability density function of the decay with the rate λ.

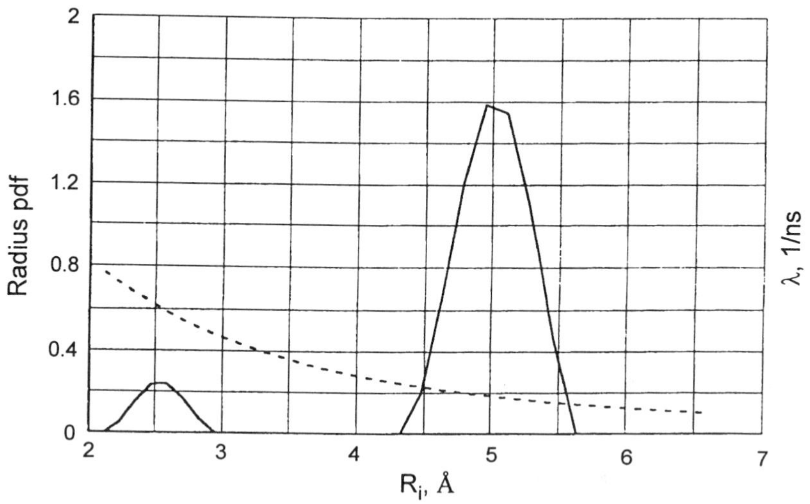


Figure 2. Free-volume hole radius distribution R pdf(R) (relative units) of PTMSP (pdf(R) is probability density function) obtained from the two right-hand peaks in Fig. 1. The dashed line shows the calculated dependence (3) of annihilation rate, (in $ns^{-1}$) versus radius of FV elements used in computing the size distribution.

distribution of FV in different polymers (*9,11,25-27*), and the present work gives an opportunity for a comparison of FV distribution in various materials. Recently, the performance of the CONTIN method and the maximum entropy method (MELT) were compared (*28*). It was shown that both the CONTIN and MELT approaches provide excellent estimates of the intensities and lifetimes of Ps components. Therefore, we can compare the widths of o-Ps peaks in the cases of monomodal FV size distributions (polymers such as polystyrene, polycarbonates, etc. with "normal" free volume) and bimodal size distributions (polymers such as PTMSP with unusually large free volume). For the latter material, the full width at half maximum, $\delta$, of the right-hand peak (Figure 2) corresponding to the fourth component $\tau_4$ is equal to 0.7 Å. The corresponding $\delta$ value for the third component $\tau_3$ is 0.46 Å. The values of $\delta$, which correspond to FV size distributions described by the $\tau_3$ and $\tau_4$ components of various polymers, are given in Table V. It can be concluded that the widths of the peaks of FV size distributions do not show a systematic variation when the average size $R_i$ of FV is increased from 2.5 to about 5 Å.It could be assumed that a transition from the microvoids having the sizes typical for conventional glassy polymers (2-4 Å) to unusually large microcavities charasteristic for PTMSP would be accompanied by corresponding extension of FV size distribution. Table V shows that this is not the case. It will be interesting to study the size distribution of FV in other materials which show extralong lifetime components in their PAL spectra. This work is now in progress.

Table V. The width of peaks $\delta$ in CONTIN size distribution of FV in polymers at ambient temperature.

| Polymer | $R_i$, Å | $\delta$ (Å) | Reference |
|---|---|---|---|
| Polystyrene | 2.88 | 0.77 | 25 |
| Polyepoxide | 2.48 | 0.89 | 11 |
| Polycarbonate | 2.94 | 0.64 | 9 |
| Tetramethyl polycarbonate | 3.20 | 0.61 | 9 |
| Hexafluoro polycarbonate | 3.38 | 0.37 | 9 |
| Hexafluorotetramethyl polycarbonate | 3.69 | 0.68 | 9 |
| PTMSP | 2.5 | 0.46 ($\tau_3$) | This work |
| | 5.0 | 0.70 ($\tau_4$) | |

**A Model of Ps and $e^+$ Behavior in Polymers.** Let us consider now the results of the investigation of porous PPO membranes. Figure 3 gives the probability density function $\lambda\alpha(\lambda)$ of annihilation with rate $\lambda$ for the Ps part of the PAL spectra. These results were obtained by CONTIN analysis of the experimental PALS data for samples with surface areas of 70 and 200 $m^2/g$. Increasing the surface area to 200 $m^2/g$ causes the right-hand peak (Figure 3, curve 2) to shift to longer times $\tau_4$, and its intensity becomes much larger. The long lifetime component of the Ps spectrum of the sample with a surface area of 200 $m^2/g$ and having an intensity (according to PATFIT

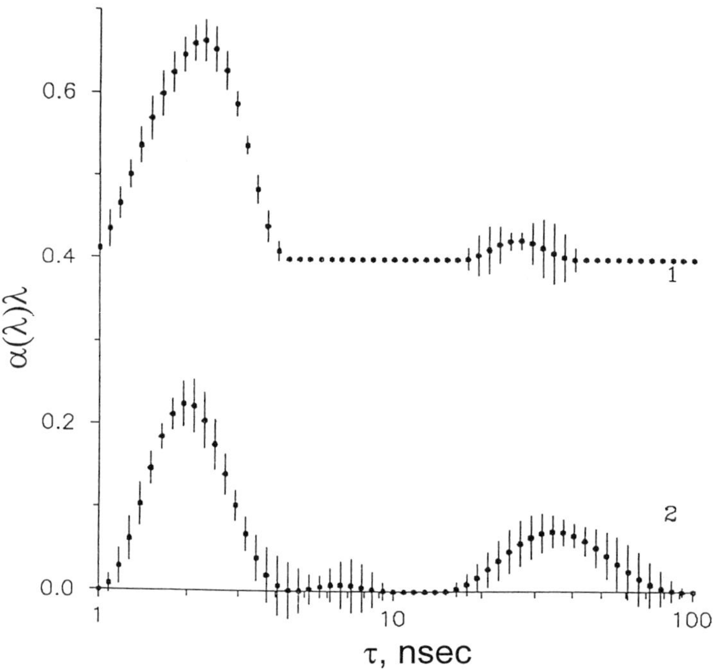


Figure 3. Positronium lifetime distribution of the two porous samples of PPO: 1 surface area is 70 $m^2/g$, 2 surface area is 200 $m^2/g$.

analysis) of 6.5% is characterized by an average lifetime of 35 ns even at atmospheric conditions, i.e. in the presence of oxygen. It should be noted that the application of Equation 1 at such long lifetimes is not accurate (*5*), because neglecting the intrinsic annihilation rate of triplet Ps ($\lambda$=0.007 $nsec^{-1}$) in comparison with Ps annihilation in the FV elements is not fully justified. A correction proposed by us (*5*) leads to somewhat larger estimated pore radii (approximately by a factor of 1.3) relative to the prediction of Equation 1. Therefore we must conclude that the PPO sample with a surface area of 200 $m^2/g$ contains pores with a mean radius of about 14 Å or more.

It would be interesting to find a relation between PAL results and pore volume (and/or pore size distribution and/or surface area) as determined by standard methods like Helium and Mercury porosimetry or the BET method. Indeed, Venkateswaran et al. (*29*) found a positive correlation connecting surface area in porous materials and the intensity of the long-lived Ps component ($I_4$ in the nomenclature of the present work). Interestingly, this correlation does not involve the other parameter of the PAL spectrum, the long-lived lifetime $\tau_4$. The polymeric materials studied in the present work encompass a wide range of surface area. Since the surface area of PTMSP dense films has been reported (*30*) to be as large as 550 $m^2/g$, specific surface area in the materials studied varied by nearly one order of magnitude. Hence, further exploration of this correlation is possible. As seen from the data given in Table VI, an increase in surface area is accompanied by a growth of intensity $I_4$.

Table VI. Effect of the specific surface areas in "porous" samples on the intensities in the PAL spectra $I_4$.

| Material | Surface area, $m^2/g$ | $I_4$,% |
|---|---|---|
| Porous PPO membrane | 70 | 2 |
| Porous PPO membrane | 220 | 7.1 |
| PTMSP dense film | 550 | 40.5 |

However, there is no quantitative agreement with the empirical correlation reported in (*29*). It can be assumed that surface area should correlate with both $I_4$ and $\tau_4$ parameters of PAL spectra.

The study of these samples permits one to estimate as well the diffusion coefficient of Ps in the polymer matrix before its localization in a FV element. This knowledge is important for the methodology of PAL spectroscopy applied to probing free volume structure in polymers. A strong variation of the intensity of the long-lived peak due to variation of the surface area implies that the Ps diffusion length, $l_{Ps}$, is comparable to the average distance between pores in the PPO membrane material. The assumption made above leads (*31*) to the following inequality for $l_{Ps}$: $3\cdot10^{-7}<l_{Ps}<5\cdot10^{-7}$ cm. Bearing in mind that $l_{Ps} = (6D_{Ps}\tau)^{1/2}$, with $\tau$ representing the nonlocalized Ps lifetime, one can obtain the following estimate for the diffusion coefficient of nonlocalized Ps:

$$5\cdot10^{-5}<D_{Ps}<1.5\cdot10^{-4}\ cm^2\cdot sec^{-1}$$

The lower limit is rather close to the value reported in (*29*) for cross-linked epoxy resins ($1.5 \cdot 10^{-5}$ $cm^2 \cdot sec^{-1}$) and somewhat lower than $D_{Ps}$ in the amorphous part of ice (*32*). The $D_{Ps}$ values found by us, Venkataswaran et al. (*29*), and Eldrup et al. (*32*) are significantly larger that the diffusion coefficient of localized Ps ($3 \cdot 10^{-6}$ $cm^2 \cdot sec^{-1}$) which was determined using quenching experiments in different glassy polymers (polycarbonate, polysulfone, and polystyrene) (*33*). These results favor the notion that it is Ps, and not free positrons, which have a much larger diffusion coefficient ($D_+$=0.1 $cm^2 \cdot sec^{-1}$ according to Maurino and Brandt (*34*)) that diffuses before annihilation. An alternative hypothesis, which should be, therefore, discarded, is that free positrons, after diffusion, are localized in a pore where they form Ps. It can also be noted that the accuracy of diffusion coefficient estimations could be increased in studies with oxygen-free samples.

A concept being developed by us is based on an assumption (*35*) that in the PAL method, preferential localization of Ps in disordered areas of a polymer and of free positrons in ordered areas takes place. Semi-crystalline materials with varied crystallinity can be used in order to test this concept. In Table VII, the variation of $I_3$ intensities is shown for poly(vinyl cyclohexane) samples with crystallinity in the range from 5 to 35%.

Table VII. Influence of crystallinity of PVCH on the intensities $I_3$ and $I_3'$.*

| Crystallinity, c % | $I_3$, % | $I_3'$, % |
|---|---|---|
| 5 | 23.0 ± 0.2 | 23.0 |
| 10 | 23.4 ± 0.2 | 20.7 |
| 20 | 20.3 ± 0.3 | 18.4 |
| 35 | 19.6 ± 0.3 | 14.9 |

*The value of $I_3'$ is defined as $I_3' = I_3(5\%)[1 - c]$, where $I_3(5\%)$ is the value of I at the crystallinity c = 5%.

The value $I_3'$ has been calculated on the basis of an additivity rule valid for the model (*35*). In this model, o-Ps formed in the crystalline and amorphous phases annihilate in these phases respectively. Therefore, the observed weaker dependence of $I_3$ on crystallinity favors redistribution of Ps from ordered (crystalline) to disordered areas, where o-Ps can find FV elements in which to localize before annihilation. These results also agree with the estimate of $D_{Ps}$ made above.

## Conclusions and Future Directions

Thus, this work contains the first report of the lifetimes and FV (hole) size distributions in polymers distinguished by an unusually loose structure (high free volume): PTMSP and microporous PPO membranes. It was shown that PATFIT and CONTIN descriptions of the annihilation data are in reasonable agreement: maxima of CONTIN peaks coincide with $\tau_i$ components in the PATFIT finite-term spectra. The o-

Ps peak corresponding to the larger FV elements in PTMSP has a width comparable to those of smaller FV elements in conventional glassy polymers.

A fundamental result of the PALS studies of high free volume polymers such as PTMSP, amorphous glassy teflons AF, as well as microporous PPO membranes is that the size distribution of FV in these materials is bimodal and not monomodal as in conventional glassy and rubbery polymers. It will be useful in the future to compare the results of the application of different programs (CONTIN, MELT) for hole size distribution in samples with well defined microstructure, such as zeolites, in order to check the findings of the PALS method.

Moreover, it will be especially interesting to test these results also by molecular modeling. The creation of well equilibrated, realistic amorphous packing models for PTMSP and the polymers having similar high FV is the major prerequisite to determine if there is, in fact, a bimodal size distribution of FV elements, whether the FV elements form an infinite or, at least, large, in the microscopic scale, cluster, and, therefore, why PTMSP and similar materials are so permeable.

## Acknowledgment

The work was supported by the Russian Foundation of Basic Research (Grant 95-03-08325). The research described in this publication was made possible in part by Award No RC2-347 of the U.S. Civil Research and Development Foundation for the Independent States of the Former Soviet Union (CRDF). The authors are grateful to Dr. V.S. Khotimsky who prepared the PTMSP sample, Dr.V.I. Kleiner for the synthesis of PVCH and Dr. F.P. Cuperus (ATO-Agrotechnology, The Netherlands) for the porous PPO samples. The CONTIN program was kindly provided by Prof. R.B. Gregory.

## Literature Cited

(1) Brandt W.; Berko S.; Walker W.W. *Phys.Rev.* **1960**, *120*, 1289
(2) Tao, S.J. *J.Chem.Phys.* **1972**, *56*, 5499
(3) Eldrup, M.; Lightbody, D.; Sherwood, J.N. *Chem.Phys.* **1981**, *63*, 51
(4) Gol'danskii, A. V.; Onishchuk, V. A.; Shantarovich, V. P.; Volkov, V. V.; Yampolskii, Yu. P. *Khim.Fiz.* **1988**, *7*, 616
(5) Shantarovich, V. P.; Yampolskii, Yu. P.; Kevdina, I. B. *Khim.Vys.Energ.* **1994**, *28*, 53
(6) Okamoto, K.; Tanaka, K.; Katsuba, M.; Kita, H.; Sueoka, O.; Ito, Y. *Polymer J.* **1993**, *25*, 275
(7) Tanaka, K.; Okamoto, K.; Kita, H.; Ito, Y. *Polymer J.* **1993**, *25*, 577
(8) Kobayashi, Y.; Haraya, K.; Hattori, S.; Sasuga, T. *Polymer* **1994**, *35*, 925
(9) Jean, Y. C.; Yuan, J-P.; Liu, J.; Deng, Q.; Yang, H. *J.Polym.Sci.: Part B: Polym.Phys.* **1995**, *33*, 2365
(10) (a) Gregory, R.B.; Yongkang Zhu In: *Positron and Positron Chemistry*; Jean, Y. C., Ed.; World Scientific: Singapore, 1991; p.136. (b) Gregory, R. B. *J.Appl.Phys.* **1991**, *70*, 4665

(11) Deng, Q.; Jean, Y.C. *Macromolecules* **1993**, *26*, 30

(12) Provencher, S.W., *Comput. Phys. Commun.* **1982**, *27*, 229

(13) Alentiev, A. Yu.; Yampolskii, Yu. P.; Shantarovich, V. P.; Nemser, S. M.; Plate, N. A. *J.Membr.Sci.* **1997**, *126*, 123

(14) Yampolskii, Yu. P.; Shantarovich, V. P.; Chernyakovskii, F. P.; Kornilov, A. I.; Plate, N. A. *J.Appl.Polym.Sci.* **1993**, *47*, 85

(15) Plate, N. A.; Yampolskii, Yu. P. In: *Polymeric Gas Separation Membranes*; Paul, D. R.; Yampolskii Yu. P. Eds.; CRC Press, Boca Raton, 1994, p.155-207

(16) Ito, Y.; Takano, T.; Hasegawa, M. *Appl.Phys. A*, **1988**, *45*, 193

(17) Nakanishi, H.; Wang, S.J.; Jean Y.C. In *Positron Annihilation Studies of Fluids;* Sharma, S. C , Ed.; World Sci.: Singapore, 1988, p.292.

(18) Nakagawa, T.; Saito, T.; Asakawa, S.; Saito, Y. *Gas Sep.Purif.* **1988**, *2*, 3

(19) Yampolskii, Yu. P.; Shishatskii, S. M.; Shantarovich, V. P.; Antipov, E. M.; Kuzmin, N. N.; Rykov, S. V.; Khodjaeva, V. L.; Plate, N. A. *J. Appl. Polym. Sci.* **1993**, *48*, 1935

(20) Nagai, K.; Nakagawa, T. *J. Appl. Polym. Sci.* **1994**, *54*, 1651

(21) Bondar, V.; Alentiev, A.; Masuda, T.; Yampolskii, Yu. *Macromol.Chem.Res.* **1997**, *198*, 1701

(22) Goldanskii, V. I.; Mokrushin, A.D.; Tatur, A. O.; P.Shantarovich, V. P. *Appl.Phys.* **1975**, *5*, 379

(23) Consolati, G.; Genco, I.; Pegoraro, M.; Zanderighi, L. *J.Polym.Sci.:Part B: Polym.Phys.* **1996**, *34*, 357

(24) Timashev, S. F.; Vorobiev, A. V.; Kirichenko, V. I.; Popkov, Yu. M.; Volkov, V. I.; Shifrina, R. R.; Lyapunov, A. Yu.; Bondarenko, A. G.; Bobrova, L. P. *J.Membr.Sci.* **1991**, *59*, 117

(25) Liu, J.; Deng, Q.; Jean, Y. C. *Macromolecules* **1993**, *26*, 7149

(26) Dlubek, G.; Clarke, A. P.; Fretwell, H. M.; Dugdale, S. B.; Alam, M. A. *Phys.Stat.Solid A* **1996**, *157*, 351

(27) Gregory, R. B.; Chai, K.-J. *J. de Physique, IV*, **1993**, *3*, 305

(28) Wang, C.-L.; Maurer, F.H.J.; *Macromolecules* **1996**, *29*, 8249

(29) Venkateswaran, K.; Cheng, K. L.; Jean, Y. C. *J. Phys.Chem.* **1984**, *88*, 2465

(30) Nakanishi, K.; Odani, H.; Kurata, M.; Masuda, T.; Higashimura, T. *Polymer J.* **1987**, *19*, 393

(31) Shantarovich, V. P., Novikov, Yu. A.; Azamatova, Z. K.; Yampolskii, Yu. P. *Vysokomol. Soed.* **1998**, . *40*, 983

(32) Eldrup, M.; Vehanen, A.; Schultz, P. J.; Lynn, K. G. *Phys. Rev. Lett.* **1983**, *51*, 2001

(33) Hirata, K.; Kobayashi, Y.; Ujihira, Y.; *J.Chem. Soc. Faraday Trans.* **1996**, *92*, 985

(34) Maurino, M.; Brandt, W. *Bull.Amer.Phys.Soc.* **1979**, *24*, 72

(35) Gol'danskii, A. V., Onishuk, V. A.; Shantarovich, V. P.; Musaelyan, I. N. *Khimiya Vysokikh Energii* **1985**, *19*, 13

## Chapter 9

# Facilitated Transport Membranes: Separation of Ethene from Ethane with Silver Ion-Exchanged Nafion® Hollow Fibers

**Odd Ivar Eriksen[1], Ivar M. Dahl[1], Ingrid B. Vik[1], and Mark L. Posey[2]**

**[1]SINTEF Applied Chemistry, P.O. Box 124, Blindern, 0314 Oslo, Norway**
**[2]Phillips Petroleum Company, Bartlesville, OK 74004**

Nafion® hollow fibers were ion-exchanged in aqueous sodium hydroxide, mounted in a steel permeator and tested with a humidified 1:1 feed mixture of ethene and ethane. Humidified helium was used as a purge gas at atmospheric pressure. The permeability of ethene and ethane was about 24 and 9 Barrers, respectively. The separation factor of about 3 corresponded to the solubility ratio of the feed components in water. The mounted fibers were afterwards ion-exchanged in aqueous 2 M silver nitrate solution, and steady-state fluxes of ethene and ethane were measured at feed pressures of 1-5 atm and with a downstream helium purge at atmospheric pressure. The permeability of ethene through the silver ion-exchanged Nafion® hollow fibers was in the range of 2900-1700 Barrers with calculated facilitation factors in the range of 115-66 at feed pressures of 1-5 atm. From the flux data, the estimated values for the equilibrium constant K´ for co-ordination of ethene to silver ions within the fibers was comparable to values reported in bulk aqueous solutions. The permeate purity was determined as a function of feed pressure and ethene recovery and a long-term separation test was performed to evaluate stability.

In a facilitated transport membrane process, carriers react or coordinate reversibly with a solute which is transported through the membrane. The principles for such a membrane process for separation of ethene from ethane are shown in Figure 1. On the feed side of the membrane, ethane and ethene are dissolved in the membrane surface. Only ethene can form a complex with the silver ions within the membrane. Diffusion of the ethene-silver ion complex across the membrane to the permeate side takes place according to a concentration gradient. On the permeate side the partial pressure of ethene must be low so that a decomplexation reaction occurs and ethene is released from the membrane surface. Ethane can only diffuse through the membrane according to the concentration gradient across the membrane.

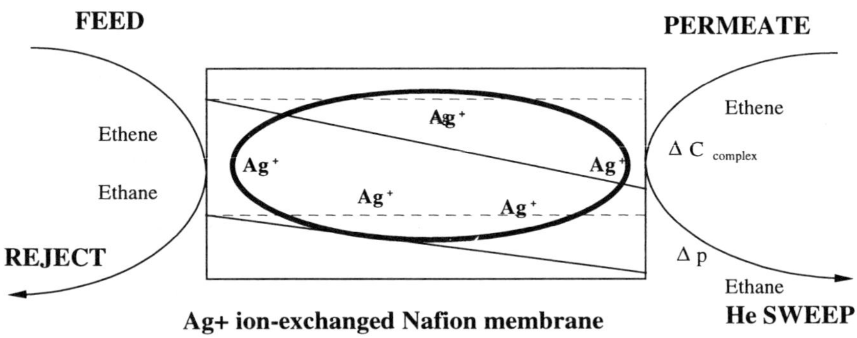


Figure 1. Facilitated transport membrane for separation of ethene from ethane.

A drawback of immobilized liquid membranes (ILM) *(1)* is that the liquid phase used as a solvent for the carrier molecules may be forced out of the pores at high pressure gradients. In particular, this can be a severe problem for thin separating barriers which are necessary for gas separation membranes in order to obtain high fluxes, and thereby an economical process. Another drawback is that the carrier molecules may leak out over time.

Water-swollen, silver ion-exchanged hollow fibers would be an interesting alternative to porous hollow fibers for which aqueous solutions of $Ag^+$ can be immobilized. Such fibers should offer a facilitated transport membrane separation system for gaseous olefins with improved stability and performance compared to immobilized liquid membranes (ILM) for two main reasons:

1) A large portion of the carriers are counter-ions to the negatively charged polymeric network, and carrier loss caused by solvent condensation on the membrane surfaces is diminished.
2) The membrane is a non porous polymer which is swollen by the solvent. If the membrane dries out during operation, it can be swollen again by proper humidification of the feed.

Flat films are the most convenient configuration for laboratory tests of membranes, and in previous work we reported that silver-exchanged Nafion® N-117 flat films gave a very good selectivity to ethene over ethane for a humidified feed at atmospheric pressure *(2),(3)*. Table I shows that the water content in these membranes can be adjusted by a swelling procedure, and that this affects the diffusivity and the equilibrium constant for the ethene-silver ion co-ordination within the polymeric membrane.
In this work silver ion-exchanged hollow fibers were tested instead of flat films. The procedures and techniques presented here are also discussed in a Phillips Petroleum patent *(4)*. We demonstrate a very simple procedure to obtain silver ion-exchanged hollow fiber modules that are very convenient for studying the separation of ethene from ethane at feed pressures up to 5 bars. Some preliminary flux data are presented to show the very high permeability and selectivity to ethene that can be obtained with these fibers. Based on the flux data, the equilibrium constant for the co-ordination of ethene to the silver ions within the Nafion® hollow fibers was estimated to predict the saturation feed pressure for the fibers.

## Experimental

**Preparation of ion-exchanged hollow fibers.** Nafion® hollow fibers (a poly (perfluorosulfonic) ionomer supplied by Perma Pure Products, Inc. under the trademark of Nafion, tubing size (030), having an inner diameter of 0.61 mm and an outer diameter of 0.84 mm), were refluxed 4 hours in aqueous 0.2 M NaOH solution. The $Na^+$-exchanged Nafion® fibers were washed and stored in distilled water. The Nafion® fibers were then mounted and glued in the inlet and outlet gaskets of a stainless steel permeator having a length of 24.5 cm and an inner diameter of 1 cm.

The effective membrane area on the inside of the hollow fibers in the permeator was 25 $cm^2$ for a 7 fiber module and 350 $cm^2$ for a 50 fiber module. The

Table I. Summary of Test Results Obtained for Nafion® Flat Film Membranes.

| Nafion® N-117 Membrane | L µm | $J_1$ ethane | $J_2$ ethene | $P_1$ ethane | $P_2$ ethene | Water content weight % | S.F. $J_2/J_1$ |
|---|---|---|---|---|---|---|---|
| untreated [(1)] | | | | | | | |
| $Na^+$-membrane | 212 | $4.2x10^{-6}$ | $1.1x10^{-6}$ | 2 | 6 | | 3 |
| $Ag^+$-membrane | 212 | $3.9x10^{-7}$ | $1.8x10^{-4}$ | 2 | 1010 | 15 | 460 |
| preswollen [(2)] | | | | | | | |
| $Na^+$-membrane | 318 | $4.3x10^{-6}$ | $1.4x10^{-5}$ | 36 | 120 | | 3 |
| $Ag^+$-membrane | 316 | $2.2x10^{-6}$ | $1.1x10^{-3}$ | 19 | 9740 | 55 | 500 |
| preswollen [(3)] | | | | | | | |
| $Na^+$-membrane | 347 | $4.9x10^{-6}$ | $1.5x10^{-5}$ | 45 | 136 | | 3 |
| $Ag^+$-membrane | 332 | $2.4x10^{-6}$ | $1.3x10^{-3}$ | 21 | 12510 | 67 | 540 |

Fluxes, J, of ethene and ethane are in $cm^3(STP)\cdot cm^{-2}\cdot s^{-1}$ through $Na^+$ - and $Ag^+$ - membranes of Nafion with a 1:1 molar ratio of ethene and ethane as feed gas at atmospheric pressure. The corresponding permeabilities, P, are given in Barrers ($10^{-10}$ $cm^3(STP)\cdot cm\cdot cm^{-2}\cdot s^{-1}\cdot cm\ Hg^{-1}$) The separation factor (S.F.) is given as the flux ratio.

(1) no heat pretreatment.

(2) preswollen in glycerine by heating to 225 °C.

(3) heated dry in oven to 340 °C and quenched, preswollen in glycerine by heating to 225 °C. All the $Ag^+$-membranes were immersed in aqueous 2 M $AgBF_4$ solution 4 days before testing.



shell compartment of the permeator, which was equipped with an inlet and outlet for the purge gas, was filled with distilled water during the mounting procedure. After the fibers had been mounted, the water in the shell compartment was removed, and the $Na^+$-exchanged Nafion® fibers were tested with the humidified feed gas. After measuring the steady-state fluxes of the feed gases through the sodium ion-exchanged hollow fibers, the shell compartment was filled with an aqueous 2 M $AgNO_3$ solution for 3 days. After the aqueous $AgNO_3$ solution had been removed from the shell compartment, the silver ion-exchanged Nafion® fibers were tested with the humidified feed gas. All tests were performed at room temperature for total feed pressures between 1 and 5 atm and with helium as a purge gas at 1 atm.

**Characterization of the fibers.** The $Na^+$-fibers were 100% ion exchanged and had a water content of 16 weight %. The $Ag^+$ exchanged fibers had an Ag content corresponding to 80% ion exchange and a water content of 14 weight %.

**Analysis of water content.** Dry sodium or silver ion-exchanged fibers were cut into samples of different lengths and weighed before immersing in distilled water for 3 days. Excess water inside the fibers was removed by blowing briefly through the samples with air. Excess water on the outside was removed with an absorbing paper. The samples were then immediately placed in closed vessels and weighed.

**Analysis of sodium and silver content.** The Nafion®-$Na^+$ fiber samples were digested in a nitric acid/hydrogen peroxide solution at high temperature and pressure, and the content of sodium or silver was determined by means of Inductively Coupled Plasma Atomic Emission Spectrometry (ICP). The Nafion®-$Ag^+$ fiber samples were digested with nitric acid and dry-ashed at 600 °C. The residue was dissolved in nitric acid and diluted, and the silver content was determined by ICP analysis.

**Flux measurements.** The test apparatus and the method used for flux calculations were described in a previous paper *(2)*. Figure 2 shows a schematic of the apparatus used for testing the modules. Both the sodium and silver ion-exchanged hollow fibers were tested with custom-mixed 50:50 mole % mixture of ethane and ethene. The feed stream was humidified through a steel cylinder packed with Chromosorb P adsorbent (200 g) impregnated with water (200 ml). The humidified feed flowed through the mounted fibers at a flow rate of about 17-38 $cm^3$(STP)/min. Helium (99.995 %) was humidified by bubbling through water at atmospheric pressure and flowed through the shell compartment as purge gas at a rate of 10 $cm^3$(STP)/min. The flow of feed and purge in the fibers was co-current. The relative humidity of the feed and the purge gases to the fibers was not measured, but due to the steady-state flux conditions, it was assumed that the relative humidity of both the purge and the feed gases was near 100 %. The flow rates along the feed side of the permeator (inside the fibers) were controlled with a mass flow controller on the reject stream from the permeator. During testing, both the permeate and reject streams were analyzed by gas chromatography, and the fluxes of ethane and ethene were calculated from calibration data as previously reported *(2)*. The relative standard deviation in the flux values was below 4 % for at least five measurements at steady-state conditions. Mass balances were performed using the compositions of the purge and reject streams exiting the permeator to calculate the average partial pressures on both sides of the hollow fibers. All gases were supplied by Hydro Gas.

## Results and Discussion

**$Na^+$-exchanged modules.** Nafion® polymers in the $Na^+$-exchanged form have a low intrinsic ethene/ethane separation factor. The permeability through the $Na^+$-exchanged fibers, based on linearized flux data, was about 24 and 9 Barrers for ethene and ethane respectively, using the wall thickness of the dry fibers. The calculated permeabilities were confirmed for two other modules prepared and tested using similar procedures. The corresponding separation factor of about 3 was in accordance with the model which assumes that ethene and ethane are mainly dissolved and transported through the imbibed water. Figure 3 shows the fluxes, J, of ethane and ethene in $10^{-12}$ mole·$cm^{-2}$·$s^{-1}$ obtained through the sodium-exchanged hollow fibers and the calculated

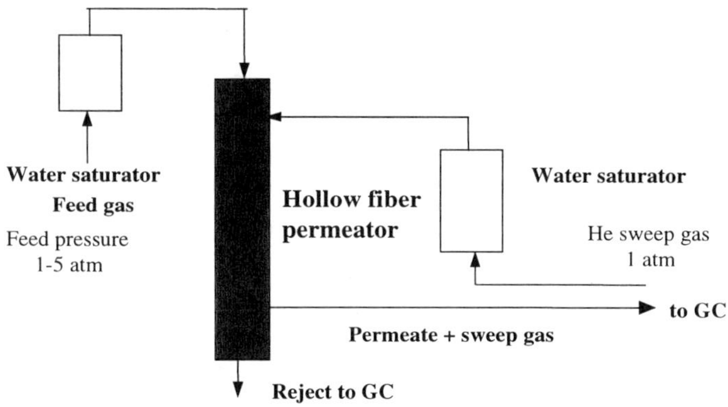


Figure 2. Test apparatus for hollow fiber modules

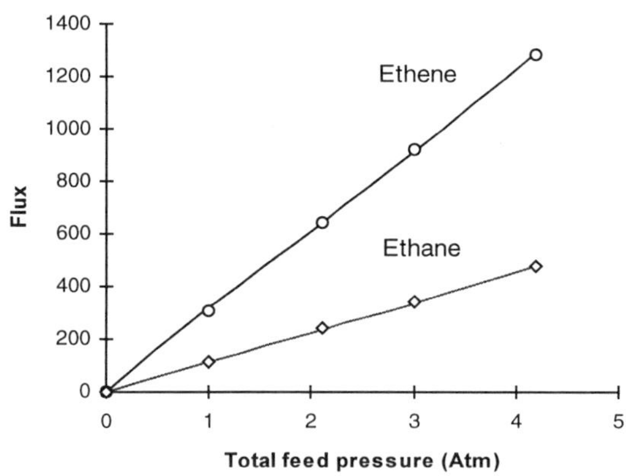


Figure 3. Fluxes J of ethane and ethene in $10^{12}$ mole·cm$^{-2}$·s$^{-1}$ through the sodium-exchanged hollow fiber module.

fluxes obtained by linear regression. The linear regression correlation coefficient is 0.99 with slopes of $(260 \pm 5)\times10^{-12}$ mole·cm$^{-2}$·s$^{-1}$·atm$^{-1}$ for ethane and $(700 \pm 24)\times10^{-12}$ mole·cm$^{-2}$·s$^{-1}$·atm$^{-1}$ for ethene at 95 % confidence limits. The low flux of ethane and ethene through the sodium-exchanged module corresponded to a recovery below 0.2 % for both feed components. The concentrations of both ethane and ethene in the purge gas leaving the permeator were below 0.04 mole %. The partial pressures of ethane and ethene in the permeate were below 0.1 % of the feed partial pressures in all tests, and the driving pressures of ethene and ethane at these conditions were set equal to the partial pressures of the feed.

The effective diffusion coefficients, $D_A$, were calculated from the observed flux, J, versus the partial feed pressure $p_F$ and the fiber wall thickness L using Equations 1 and 2 below *(2)*:

$$J \cdot L = f \cdot D_A \cdot p_F / (V_A \cdot m) \quad (1)$$

$$D_A = (dJ/dp_F) \cdot L \cdot V_A \cdot m/f \quad (2)$$

The water content, f, in the sodium-exchanged fibers was about 16 weight %. ICP analysis confirmed that the degree of proton exchange with sodium ions was near 100 %. $D_A$ for ethane and ethene calculated from equation 2 was about $1\times10^{-5}$ cm$^2$·s$^{-1}$. The effective diffusivity through the fibers is thereby of the same order as for ethene in water, for which a value of $1.8\times10^{-5}$ cm$^2$·s$^{-1}$ has been reported *(5)*.

**$Ag^+$-exchanged modules.** The $Na^+$-fibers were then ion-exchanged with $Ag^+$ and tested again with humidified feed gas. Table II summarises the flux data. The ethene fluxes through the $Na^+$-fibers were negligible compared to the fluxes through the $Ag^+$-fibers. The linearized ethene flux through the $Na^+$-fibers obtained from Figure 2 was used to calculate the facilitation factor for the $Ag^+$-fibers at a given driving pressure. The facilitation factor, F, is defined as the ratio of the ethene flux through the $Ag^+$-fibers to the ethene flux through the $Na^+$-fibers at equal driving pressures.

Table II. Ethene Fluxes J in and facilitation factors obtained for the 7 fiber $Ag^+$-module.

| Feed pressure (atm) | J (cm$^3$(STP)·cm$^{-2}$·s$^{-1}$) | δp (cm Hg) | P (Barrer) | F (Unitless) |
|---|---|---|---|---|
| 1 | $8.5\times10^{-4}$ | 33 | 2950 | 115 |
| 2 | $1.4\times10^{-3}$ | 59 | 2620 | 97 |
| 3 | $1.6\times10^{-3}$ | 87 | 2080 | 83 |
| 4 | $1.8\times10^{-3}$ | 114 | 1810 | 73 |
| 5 | $2.1\times10^{-3}$ | 137 | 1770 | 66 |

For facilitated transport Nafion® membranes in which the ethene flux is limited by diffusion through the membrane and not by mass transfer between the gas and membrane interfaces, we have previously shown that the diffusivities of the feed components and the equilibrium constant for the co-ordination of ethene to silver ions within the membrane can be calculated by a simple mathematical model *(3)*, in which it is assumed that the driving pressure equals the feed partial pressure. However, in this model the recovery of ethene during flux measurements should be as low as possible in order to neglect the back-diffusion. The content of ethene in the purge outlet for the silver ion-exchanged module was up to 24 mole %, and the model which assumes a driving pressure of ethene equal to the feed partial pressure was no longer valid. To estimate the effective diffusion coefficient for ethene, the concentrations of ethene on the shell side were considered in the mathematical model. Equation 3 gives the concentration difference, $\Delta X$, of the ethene-silver ion complex across the hollow fiber wall as a function of the equilibrium constant $K'$, the silver-ion concentration $C_T$, the partial ethene pressure on the feed side, $p_F$, and the ethene partial pressure on the permeate side of the fiber, $p_P$.

$$\Delta X = C_T \cdot K' \cdot p_F \, (1+K' \cdot p_F) - C_T \cdot K' \cdot p_P \, (1+K' \cdot p_P) \tag{3}$$

The ethene flux, J, through the fiber is assumed to be given by Equation 4 in which L is the fiber wall thickness, f is the water fraction in the fiber and $D_{AB}$ is the effective diffusion coefficient for ethene.

$$J = f \cdot D_{AB} \cdot \Delta X / L \tag{4}$$

$$J = c \cdot K' ( p_F / (1+K' \cdot p_F) - p_P / (1+K' \cdot p_P) ) \tag{5}$$

$$c = f \cdot D_{AB} \cdot C_T / L \tag{6}$$

A non-linear least-squares regression analysis was used to fit the observed ethene fluxes to Equation 5. From this analysis, $K' = 0.7 \pm 0.1$ $atm^{-1}$ and $c = (1.4 \pm 0.1) x 10^{-7}$. Based on data reported by Trueblood and Lucas *(6)*, this result is of the same order as in bulk silver nitrate aqueous solutions. A water content, f, of about 14 weight % was determined for the silver ion-exchanged fibers. Based on ICP analysis, the silver ion concentration, $C_T$, within the fibers was estimated to be about 5 M, which corresponds to an ion exchange of sodium with silver of about 80 %. The estimated effective diffusion coefficient for ethene in the facilitated mode, calculated from c in Equation 6, was on the order of $2x10^{-6}$ $cm^2 \cdot s^{-1}$. Table III shows a summary of the constants obtained for the module with an effective membrane area of 25 $cm^2$.

Table III. Summary of Parameters Calculated for Facilitated Transport Through Nafion® Hollow Fibers.

| Nafion hollow fibers | L (μm) | $D_{AB}$ Ethane ($cm^2 \cdot s^{-1}$) | $D_{AB}$ Ethene ($cm^2 \cdot s^{-1}$) | $C_T$ (M) | f (weight %) | K´ ($atm^{-1}$) |
|---|---|---|---|---|---|---|
| $Na^+$-form | 115 | $1x10^{-5}$ | $1x10^{-5}$ | -- | 16 | -- |
| $Ag^+$-form | 115 | $1x10^{-5}$ | $2x10^{-6}$ | 5 | 14 | 0.7 |

To test the stability of silver ion-exchanged Nafion® hollow fibers for olefin separation, a 50 fiber module was prepared and tested. The observed and calculated ethene flux as a function of the driving pressure is shown in Figure 4.

The permeate purity was determined as a function of feed pressure and ethene recovery (100·moles in permeate/ moles in feed) for the 50 fiber module. The recovery was varied by adjusting the feed flow to the module. The results are summarized in Figure 5. The module gave the expected behavior of decreasing purity with increasing feed pressure and with increasing recovery. This is because of saturation effects in the $Ag^+$-ethene complex transport system. The productivity of the 50 fiber module was very good.

**Long time testing.** A 107 day test was performed with the 50 fiber module at 2 bar absolute feed pressure and a flow which originally gave about 50% recovery. Note that a better recovery could be obtained by changing the feed flow rate to match the membrane capacity. The experiment was configured to run continuously and unattended except during sampling. The performance was very steady during the first 1500 hours of operation, with generally >96% purity and >40% recovery as shown in Figure 6. The three dips in recovery at times less than 1000 hours are most likely due to dehydration of the membrane. The water saturators for the feed and helium streams had to be refilled periodically. The module continued to perform well until 1500 hours. Apparently the $Ag^+$ had been deactivated because rehydration of the membrane with water did not recover the performance. At the 2300 hour mark, the membrane was regenerated with a $AgNO_3$ solution and recovery returned to 44% for experiments that are not shown in Figure 6.

## Conclusion

Facilitated transport hollow fiber modules for studying the separation of ethene from ethane can be prepared by a very simple procedure using commercial hollow fibers of Nafion® polymer. The flux data obtained through the fibers using a humidified feed at pressures up to 5 bars showed that the silver-exchanged hollow fibers have a very high selectivity to ethene. An analytical model for facilitated transport of ethene across the fibers, which assumes that the flux is proportional to the concentration gradient of the ethene-silver ion complex across the hollow fibers, was used to estimate the equilibrium constant, $K'$, for co-ordination of ethene to silver ions within the fibers. $K'$ was comparable to values reported in bulk aqueous solutions of $AgNO_3$. The transport properties from relatively thick fibers in this study can be used to evaluate a membrane separation process for recovery of ethene from ethane. In particular, the optimal feed and permeate pressures and the necessary membrane thickness can be determined. The long-term stability seems reasonable, provided the process streams are saturated with water. Further data are required to determine the reason for the performance decline. Performance may be restored by repeating the $Ag^+$ ion exchange step.

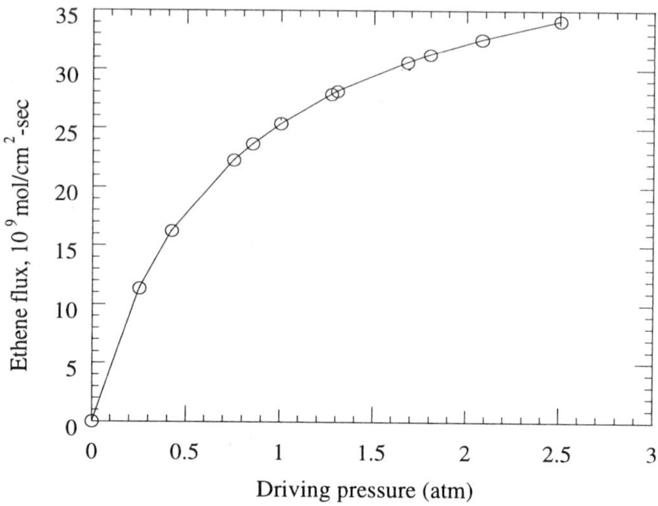


Figure 4. Ethene flux as a function of ethene pressure for the 50 fiber module in the $Ag^+$ form.

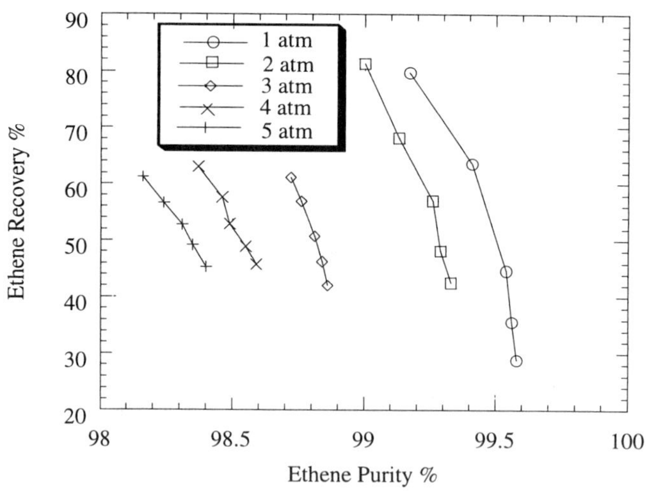


Figure 5. Ethene recovery from the feed stream for the 50 fiber module in $Ag^+$ form at various feed pressures in atm.

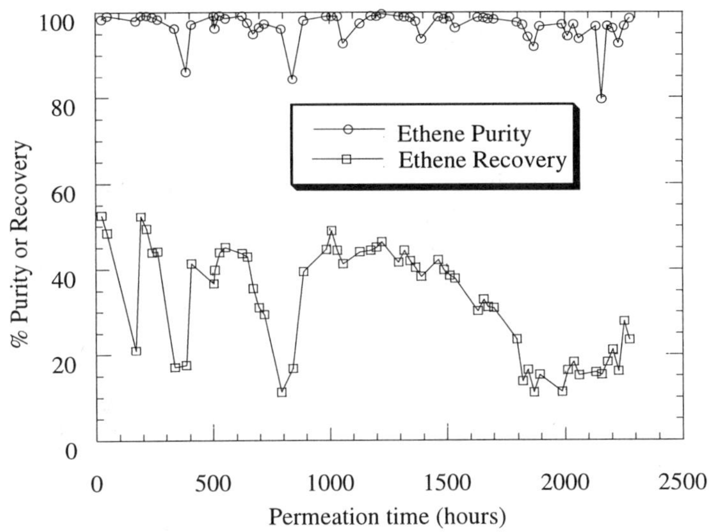


Figure 6. Long-term testing of the 50 fiber module with a constant feed and helium sweep flow rate. Feed composition was 50 vol % ethane and 50 vol % ethene at a total pressure of 2 atm.

**Acknowledgement**

We thank Phillips Petroleum Company Norway for financing this work and permitting the publication of the results.

**List of symbols**

| | |
|---|---|
| $p_F$ | feed partial pressure in atm |
| $p_P$ | permeate partial pressure in atm |
| $D_A$ | effective Fickian diffusion coefficient for ethene in $cm^2 \cdot s^{-1}$. |
| $D_{AB}$ | effective Facilitated diffusion coefficient for ethene in $cm^2 \cdot s^{-1}$. |
| J | flux in $cm^3(STP) \cdot cm^{-2} \cdot s^{-1}$ |
| L | film thickness in cm |
| $V_A$ | molar gas volume at 25 °C (24467 $cm^3 \cdot atm \cdot mole^{-1}$) |
| m | partition coefficient<br>(concentration in gas phase / concentration in liquid phase) |
| $C_T$ | carrier concentration in $mole \cdot cm^{-3}$ |
| X | concentration of silver-ethene complex in $Ag^+$-membrane in $mole \cdot cm^{-3}$ |
| K´ | equilibrium constant in $atm^{-1}$, silver-ethene coordination |
| K | equilibrium constant in $M^{-1}$, silver-ethene coordination<br>($K = K' \cdot m \cdot V_A$) |
| f | fraction of water in the membrane<br>(wet weight-dry weight)/dry weight |
| P | permeability coefficient in Barrer ($10^{-10} \cdot cm^3(STP) \cdot cm \cdot cm^{-2} \cdot s^{-1} \cdot cm\ Hg^{-1}$) |

**Literature Cited**

1. Noble, R. D., Way J. D. In *Liquid membrane, Theory and Applications;* Noble, R. D., Way J. D.; Eds. ACS Symp. Ser. No. 347, American Chemical Society; Washington, DC, **1987**, 1.
2. Eriksen, O. I., Dahl, I. M., Aksnes, E. *J. Membrane Sci.* **1993**, *85*, 89.
3. Eriksen, O. I., Dahl, I. M., Aksnes, E. *J. Membrane Sci.;* **1993**, *85*, 99.
4. Eriksen, O. I., Aksnes, E., Dahl, I. M. U.S. Patent 5,191,151 **1993**.
5. Duda, J. L., Vrentas, J. S. *AIChE J* **1968**, *14* , 286.
6. Trueblood, K. N., Lucas, H. J. *J. Am. Chem. Soc.* **1952**, *74*, 1338.

Chapter 10

# Poly(vinyl alcohol)–Silver Nitrate Facilitated Transport Membranes for the Separation of Aromatic and Aliphatic Compounds

**Carl A. Koval[1], Debra L. Bryant[1], Sarah L. Roberts[1], and Richard D. Noble[2]**

**[1]Department of Chemistry and Biochemistry, University of Colorado, Boulder, CO 80309–0215**
**[2]Department of Chemical Engineering, University of Colorado, Boulder, CO 80309–0424**

Poly(vinyl alcohol) (PVA) membranes containing silver(I) salts were prepared by several methods and found to be effective for the separation of benzene and cyclohexane using hydrated liquid feed and sweep streams. Benzene/cyclohexane selectivities exceeded 30 and benzene permeabilities exceeded 1 μm kg $m^{-2}$ $hr^{-1}$. Removal of the casting solvent by freeze-drying results in a membrane with a uniform distribution of silver. Despite the large selectivities observed in transport experiments, benzene and cyclohexane absorb equally well into the membranes and exhibit similar diffusion coefficients.

Separation of hydrocarbons by traditional distillation processes, which are energy intensive, can constitute as much as 70% of the product cost. Membranes can be used in combination with distillation in the form of membrane-distillation hybrid processes in order to lower operating costs and to produce higher throughput.(*1*) In addition, membranes can separate azeotropic and close-boiling mixtures which would be difficult to separate by distillation alone.(*2*)

Several aromatic compounds including benzene, styrene and ethylbenzene are among the top 50 chemicals produced in the United States. Separation of benzene from various saturated compounds using membranes has been addressed by several research groups.(*3-7*) Recently, we published a preliminary report on the use of poly(vinyl alcohol) (PVA) membranes containing silver(I) salts for the separation of benzene and cyclohexane.(*8*) The use of this material for the separation of butenes was originally reported by Ho et al.(*9*) In this paper, we report the results of further studies on the preparation, characterization and performance of of PVA-AgX membranes for separation of aromatic compounds.

## Preparation and Characterization of Poly(vinyl alcohol)-Silver Nitrate Membranes

Our earlier studies utilized PVA-AgX membranes that were prepared by casting from aqueous PVA solution that contained the various amounts of Ag(I) salts and formaldehyde as the cross-linking agent.(*8*) Scanning electron microscopy/energy dispersive X-ray analyses (SEM/EDXR) with a Ag probe showed the PVA-Ag(I) membranes prepared in this manner have an inhomogeneous distribution of Ag(I) throughout the cross-section of the membranes. A high concentration of Ag(I) was found in the top layer of the membrane which had been exposed to air during the solvent evaporation/curing process, while a significantly lower concentration of Ag(I) existed in the bottom membrane area which had been in direct contact with the casting dish. It was thought that use of water as the casting solvent might be responsible for the migration of Ag(I). Therefore, two alternative preparations were investigated.

As discussed in our earlier paper (*8*), PVA-$AgNO_3$ membranes containing as much as 45 wt% Ag(I) can be prepared using the above procedure. However, when these membranes are contacted with water, all but 7±2 wt% of the Ag(I) is rapidly lost. Surprisingly, the remaining 7±2 wt% of the Ag(I) remains in the PVA membrane even after months of storage in water. This suggests that the PVA-$AgNO_3$ membranes may be acting as a fixed-site carrier membrane. Based on these results, the membranes described below were prepared with 7 wt% Ag(I) either as $AgBF_4$ or $AgNO_3$. PVA-AgX membranes typically absorbed 2-3 times their dry mass when saturated with water. Therefore, the nominal concentration of Ag(I) ion in membranes prepared with 7 wt% is approximately 0.2 M.

The first alternative preparation used for preparing PVA-Ag(I) membranes was reported by Yeom et al.(*10*) The preparation used dimethylsulfoxide (DMSO) as the solvent instead of water and glutaraldehyde(GA)/para-toluenesulfonic acid(*p*-TSOH) to crosslink the polymer instead of formaldehyde. The initial PVA-Ag(I) membrane prepared using the new procedure were complicated by the phase separation of the DMSO solvent from the polymer during the curing process. As the polymer solidified, the phase separation of DMSO caused a shrinkage in the solid polymer portion of the membrane resulting in a thick membrane, approximately 1 mm, with a diameter of only approximately one quarter of the diameter of the casting dish. This problem was solved by casting the polymer solution onto a hydrophilic polyethylene (PE) support prior to evaporation of the solvent. Shrinkage was significantly reduced.

This type of PVA-AgX membrane was prepared by dissolving 10 wt% PVA in DMSO with heating. A crosslinking solution, which contained equimolar amounts of glutaraldehyde as a 50 wt% solution in water and *p*-TSOH acid dissolved in 100 mL of DMSO, was added to the polymer solution and heat was removed. $AgBF_4$, was added with stirring in dim light due to the possible reduction of Ag(I). $AgBF_4$ was used as the Ag(I) salt due to its solubility in DMSO. The solution remained clear although it became slightly yellow in color. The polymer solution was poured onto a porous PE support. Solvent evaporation was accomplished in the dark at room temperature for approximately 4 days.

Although the performance of membranes prepared using DMSO/GA were similar to those prepared using water/formaldehyde, SEM/EDXR analyses showed that the Ag(I) was also distributed inhomogeneously in these materials. Therefore, a freeze-drying procedure was developed for removing the water used as the solvent in the original preparation. PVA/AgX membranes were prepared using the initial steps detailed in our earlier report. The solution was poured into a dish which could be fitted onto a lypholizer with an air tight connection. The casting dish was immediately placed in a isopropanol/dry ice bath resulting in solidification of the polymer solution. The cast membrane was attached to the lypholizer with the ice bath still in place and vacuum was applied. The ice bath was removed after approximately 5 minutes. The cast membrane remained under vacuum, at approximately 2 x $10^{-7}$ Torr, for approximately 5 hr. The membrane appeared dry upon removal from the casting dish.

A cross-section of the PVA-Ag(I) membranes prepared by freeze-drying was analyzed by SEM(EDXR) with Ag probe. The distribution of Ag(I) in these membranes was reasonably homogeneous as shown in Figure 1. No phase separation or clear demarcation of inhomogenity was visible in the cross-section as was the case in previous PVA-Ag(I) membranes. It appears that by varying the curing process with the freeze-drying procedure, a homogeneous distribution of Ag(I) in the polymer matrix was achieved.

**Performance Characteristics**

The transport performance of PVA-Ag(I) membranes was investigated using liquid feed/liquid sweep (perstraction) transport cells described elsewhere.(*8, 11*) Membranes were equilibrated in deionized water prior to transport experiments and the feed and sweep solutions were saturated with water to insure membrane hydration was maintained throughout the length of the experiment unless otherwise indicated. The feed solution was 1:1(v:v) benzene:cyclohexane. All transport experiments had concentration change of the sweep solution monitored over time by injecting 1 μl samples into a Hewlett Packard 5890 gas chromatograph (GC) with a 25' x 0.32 mm Ultra 1 column and flame ionization detector. The column stationary phase was a 0.52 μm layer of crosslinked methyl silicone gum.

For PVA-$AgBF_4$ membranes prepared using DMSO/GA as described above with 43% crosslinking of possible sites, the benzene flux was 1.0 μm kg $m^{-2}$ $hr^{-1}$ and the benzene/cyclohexane selectivity was 53. Another membrane was prepared with 20% of possible sites crosslinked. The benzene flux was 0.9 μm kg $m^{-2}$ $hr^{-1}$ and the selectivity of 28. The lower selectivity was due to an increase in cyclohexane flux for the membrane with less crosslinking. A transport experiment using a 180 μm freeze-dried PVA-$AgNO_3$ membranes resulted in a benzene flux of 4 μm kg $m^{-2}$ $hr^{-1}$ with selectivity for benzene of 48. The fluxes and selectivities for these new formulations of PVA/AgX membranes were very similar to those previously reported for membranes cast using water and formaldehyde.(*8*) Therefore, it appears that neither the nature of the casting solvent, the cross-linking agent, or the distribution of Ag(I) has a major effect of the ability of these materials to separate benzene from cyclohexane.

One interesting and potentially useful aspect of the benzene/cyclohexane separation afforded by PVA-AgX membranes is that the benzene flux increases almost linearly with benzene feed concentration up to feed mixtures that are as high as 80% benzene by volume. This is illustrated for a PVA-$AgNO_3$ membrane in Figure 2. This result is somewhat surprising because when Ag(I) is incorporated into ion-exchange polymers such as DuPont's Nafion, the facilitation effect saturates at high concentrations of unsaturated organic compounds in the feed solution.(*12*)

An important issue regarding the eventual utility of PVA-AgX membranes for hydrocarbon separations is the hydration requirements. In transport experiments involving PVA-$AgNO_3$ membranes that were pre-equilibrated with water and using feed/sweep solutions prepared from reagent grade chemicals, the benzene flux remained constant for approximately 100 hours of continuous transport. After that time, the benzene flux diminished with time and after 200 hours of transport the benzene and cyclohexane fluxes were nearly equal. In order to test the hydration requirements in a more rigorous fashion, a PVA-$AgBF_4$ membrane cast from DMSO and 20% crosslinked with glutaraldehyde was dried thoroughly under vacuum. A transport experiment was performed with this membrane after rigorously drying the feed and sweep solutions with molecular sieves. Under these conditions, transport of benzene through the membrane was greatly diminished. Addition of small

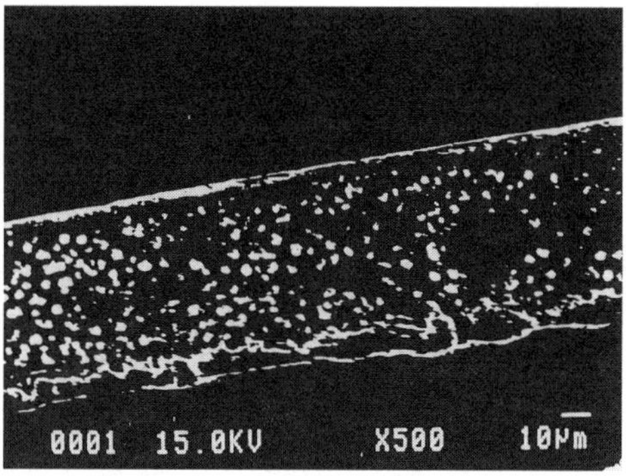


SEM of Freeze-Dried
PVA-Ag(I) Membrane

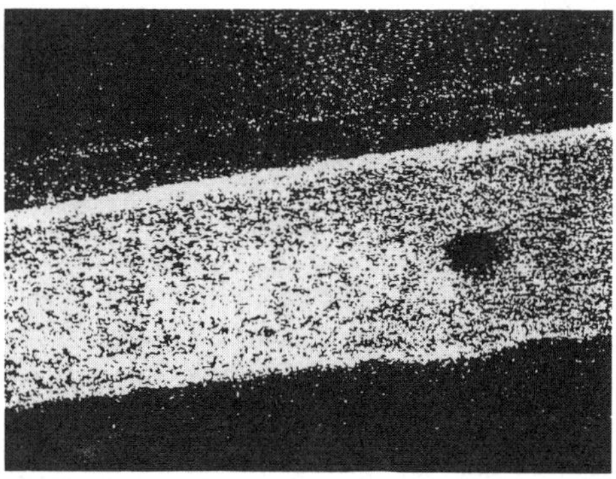

SEM(EDXR) of Freeze-Dried
PVA-Ag(I) Membrane

Figure 1. SEM (top) and SEM/EDXR demonstrating that silver is uniformly distributed in PVA-$AgNO_3$ membranes prepared using a freeze-drying procedure. The membrane contains 7% Ag by mass.

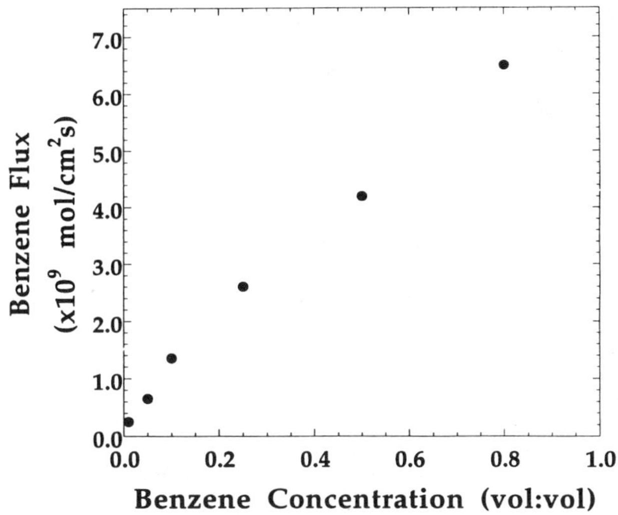


Figure 2. Plot of benzene flux vs. benzene volume fraction in the feed solution for a PVA-$AgNO_3$ membrane (7% Ag by mass) crosslinked with formaldehyde as described in reference 8. The membrane thickness was approximately 150 µm.

amounts of water to the feed and sweep solutions rapidly resulted in benzene fluxes comparable to those for membranes that had been pre-equilibrated with water.

The loss of selectivity as a result of membrane dehydration described above is similar to results reported previously for Ag(I)-facilitated transport of alkenes in perfluorosulfonic acid membranes.(*13*) For cation transport in ion-exchange materials, it is known that loss of membrane hydration results in a dramatic decrease in conductivity suggesting that the cations become much less mobile in dry membranes. Presumably, a similar phenomenon is responsible for the diminished benzene fluxes observed for dehydrated PVA-AgX membranes. Even though a Ag(I)-benzene complex can still form, hydration is necessary for the complex to be mobile and thereby provide effective facilitation of benzene.

The performance of PVA-AgX membranes was briefly examined for two other separations of aromatic compounds. The membranes achieved a styrene/ethylbenzene separation factor of 6, but showed no selectivity for the separation of *o*-xylene from *p*-xylene. Based on these results it appears that PVA-AgX membranes are more suited for separation of aromatic compounds from aliphatic compounds than for the separation of aromatics from each other.

## Unresolved Mechanistic Issues

After discovering that homogeneous PVA-$AgNO_3$ membranes could be prepared using the freeze-drying procedure described above, we attempted to investigate whether the observed benzene/cyclohexane selectivity was due to selective absorption or to selective transport. In order to measure the absorption selectivity, the dry and water saturated masses for a freeze-dried PVA-Ag(I) membrane sample were recorded. In sequential experiments, the membrane sample was placed in contacting solutions composed of benzene:cyclohexane with compositions of 4:1, 1:1, and 1:4 v:v for 4 hours. The membrane was blotted lightly to remove excess benzene/cyclohexane from the surface and placed in 25.0 ml isooctane for extraction of the organics. After approximately 4 hr the extraction solvent was tested by GC to determine benzene and cyclohexane concentrations extracted from the membrane. Further extraction with isooctane confirmed that all of the organics absorbed from the contacting solutions were removed during the 4 hour extraction period.

The results of these absorption/extraction experiments were quite surprising. For all of the contacting solutions the ratio of absorbed benzene to absorbed cyclohexane was less than 2 (values ranged from 1.2 to 1.8). Furthermore, the moles of benzene absorbed by these materials exceeded the moles of Ag(I) ion. These results suggested that most of the observed selectivity for these materials must be due to selective diffusion of benzene.

This result was tested by first absorbing benzene and cyclohexane into a freeze-dried PVA-$AgNO_3$ membrane and by measuring the rate at which the two organics were extracted into isooctane. The experimental procedure for this type of extraction experiment has been previously reported and it was shown that the data can be used to estimate the effective diffusion coefficients for the components being extracted.(*14*) The results for time-resolved extraction of benzene and cyclohexane after absorption from a 1:1 v:v contacting solution are shown in Figure 3. As in the simple extraction experiments described above, the total amount of organic removed at long extraction time indicates that the absorption selectivity is less than 2. Furthermore, the data at earlier time shows that the rates of extraction for benzene and cyclohexane, and thereby their respective effective diffusion coefficients, are nearly identical. These results are puzzling. If the membranes do not exhibit significant absorption or diffusion selectivity, than how can observed selectivities in steady-state transport experiments be routinely between 30 and 60?

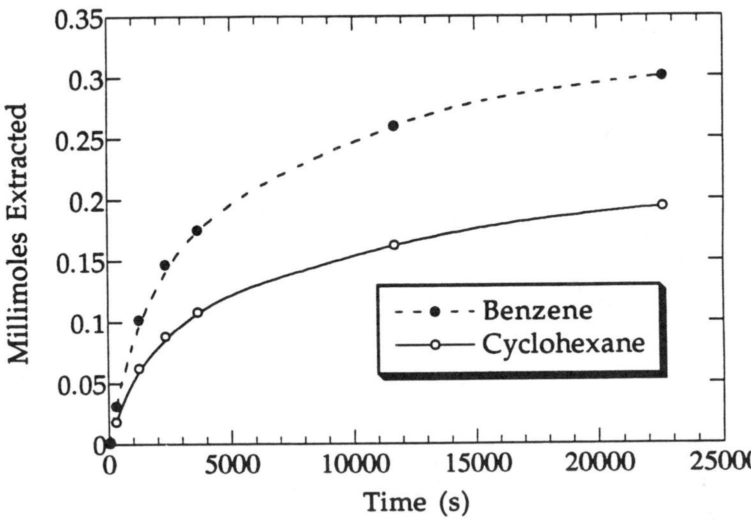


Figure 3. Time-resolved extraction of benzene and cyclohexane into isooctane from a freeze-dried PVA-$AgNO_3$ membrane (7% Ag by mass, crosslinked with formaldehyde). The membrane thickness was approximately 350 μm.

One possible explanation for these results was that the rate of transport was not being controlled by the bulk of the membrane. Possibly, transport was controlled by processes occurring only at the surface of the membranes. If this were the case, the fluxes for benzene and cyclohexane should have been relatively independent of membrane thickness rather than inversely proportional to thickness, which is usually observed for a solution-diffusion mechanism. To test this hypothesis, membranes with different thicknesses were prepared and transport fluxes were measured. Although the range of thicknesses that could be prepared was not large (factor of 2), the fluxes for benzene and cyclohexane were approximately inversely proportional to thickness.

At this point in time, the reason for the large transport selectivities remains uncertain. Possibly, the values obtained for absorption diffusion selectivities are somehow different from the values that are important in the membrane under steady-state transport conditions.

## Conclusions

PVA-AgX membranes exhibited large selectivities and benzene permeation rates for the separation of benzene and cyclohexane over a wide range of feed concentrations. Much lower selectivites were observed for styrene/ethylbenzene and *o*-xylene/*p*-xylene separations. Long term transport depends on the membranes remaining hydrated. A freeze-drying procedure was developed that provided homogeneous distribution of silver in the PVA membranes. Absorption/extraction experiments indicated that the membranes had low absorption and diffusion selectivities for benzene/cyclohexane.

## Acknowledgment

This research was supported by the NSF Center for Separations Using Thin Films.

## Literature Cited

(1) Stephan, W.; Noble, R. D.; Koval, C. A. *J.Membr. Sci.* **1995**, *99*, 259.
(2) Mulder, M. *Basic Principles of Membrane Technology*; Kluwer Academic: Dordrecht, 1991.
(3) Yamaguchi, T.; Nakao, S.; Kimura, S. *Macromol.* **1991**, *24*, 5522.
(4) Suzuki, F.; Onozato, K.; Yaegashi, H.; Masuko, T. *J. Appl. Polym. Sci.* **1987**, *34*, 2197.
(5) Enneking, L.; Stephan, W.; Heintz, A. *Ber. Bunsenges. Phys. Chem.* **1993**, *97*, 912.
(6) Acharya, H. R.; Stern, S. A.; Liu, Z. Z.; Cabasso, I. *J. Mem. Sci.* **1988**, *37*, 205.
(7) Sun, F.; Ruckenstein, E. *J. Mem. Science.* **1995**, *99*, 273.
(8) Bryant, D. L.; Noble, R. D.; Koval, C. A. *J. Membr. Sci.* **1997**, *127*, 161.
(9) Ho, W. S.; Dalrymple, D. C. *J. Membrane Sci.* **1994**, *91*, 13.
(10) Yeom, C., K,; Lee, K. H. *J. Mem. Sci.* **1996**, *109*, 257.
(11) Bryant, D. L., Ph.D. Thesis thesis, University of Colorado, Boulder, **1997**.
(12) Thoen, P. M.; Noble, R. D.; Koval, C. A. *J. Phys. Chem.* **1994**, *98*, 1262.
(13) Koval, C. A.; Bryant, D. L.; Engelhardt, H.; Manley, D.; Rabago, R.; Thoen, P.; Noble, R. D. *ACS Symposium Series.* **1996**, *642*, 286.
(14) Rabago, R.; Bryant, D. L.; Noble, R. D.; Koval, C. A. *Ind. Eng. Chem. Res.* **1996**, *35*, 1090.

Chapter 11

# Effect of Atmospheric Glow Discharge Plasma Treatment on $CO_2$ Permeability and Chemical Structure of Poly(dimethylsiloxane) Membranes

**Taisuke Maki[1], Masaaki Teramoto[1], Motomu Masuda[1], Tomohiro Morimoto[1], Qingfa Hwang[1], and Tomio Sakka[2]**

**[1]Department of Chemistry and Materials Technology, Kyoto Institute of Technology, Matsugasaki, Sakyo-ku, Kyoto 606, Japan**
**[2]Nitto Denko Company, Shimohozumi, Ibaraki 567, Japan**

The surface of poly(dimethylsiloxane) membranes was modified by atmospheric pressure glow discharge (APG) plasma treatment using He, and the $CO_2$ permeability and the selectivity of $CO_2$ over $CH_4$ were investigated in detail. Permeance of $CO_2$ ($R_{CO_2}$) decreased monotonically with an increase in the APG plasma treatment time as is the case in low pressure plasma treatment (*1*). The selectivity ($\alpha$) was improved very much by the plasma treatment and was well above the selectivity of untreated poly(dimethylsiloxane). The selectivity showed a maximum as a function of treatment time. The treatment time to achieve maximum selectivity decreased with increasing power input. At the optimum condition for $CO_2$ selectivity, $R_{CO_2}$ and $\alpha$ were $1.5 \times 10^{-5}$ mol/($m^2$•s•kPa) and 76, respectively, which indicates that the membrane has a fairly high permeability and selectivity compared with those of conventional polymer membranes. The surface of the plasma treated membrane was analyzed by XPS. The O/Si ratio of the APG treated membrane surface was about twice that of the untreated membrane. This suggests that oxygen was incorporated into the membrane surface from the external environment, probably as a result of reaction with the free radicals formed in the membrane during APG treatment. The existence of oxygen containing polar groups probably increases the affinity between the membrane and $CO_2$. The permeance of $CO_2$ and $CH_4$ decreased with time after the plasma treatment, consistent with results from low pressure plasma treatment (*1*).

Low temperature or non-equilibrium plasma processes are becoming more widely used as techniques for creating new materials which cannot be produced by conventional methods. Most of these processes are employed at low pressure (less than 10 torr), because of the stability of glow discharge plasma at low pressure. Consequently, the low pressure apparatus contains vacuum seals and an air sealing system, so it is costly. Moreover, it is not suitable for treatment of large amounts of material. However, Kanazawa et al.(*2*) demonstrated that even at atmospheric pressure, stable glow discharge occurs under the following conditions: (i) Helium is used as the diluent, (ii) an insulating plate is set on the lower electrode, and (iii) the frequency of the input power is about 3000 Hz. Kogoma et al.(*3*) demonstrated that atmospheric pressure glow discharge (APG) occurs by using gas mixtures, such as $Ar/CH_4$ and Ar/acetone as the plasma gas. The APG plasma process is superior to the low pressure process because the cost of the apparatus is lower, larger substrates may be treated, and continuous operation is possible.

The plasma process can be divided into plasma polymerization and plasma treatment (*4*). In the former, organic vapors (monomers) are subjected to electric discharge, and their polymers are formed. In the latter process, gases that are introduced into the glow discharge are nonpolymerizable in nature (*5*). The APG plasma process can be applied to both polymerization and surface treatment. Previously, surface fluorination of PET films, carbon thin film deposition, and polymerization of ethylene by APG plasma were examined (*2*, *3*, *6*). However, the preparation of separation membranes by APG plasma treatment has not been reported.

Plasma treatment is subdivided into the case where reactive gases such as $O_2$, $N_2$ and $NH_3$ are used as the plasma gases, and that where inert gases such as Ar and He are used. In the former, the direct reaction of activated gas with the polymer surface occurs, while in the latter, formation of free radicals and subsequent reactions such as cross-linking (via the recombination of free radicals and incorporation of oxygen from the extraneous environment) are the dominant processes. The APG plasma process is limited to the latter process since only He can be used as the plasma gas.

Several studies of the preparation of gas separation membranes by low pressure plasma treatment have been reported. Kawakami et al. demonstrated that the selectivity of $O_2$ over $N_2$ increased somewhat by plasma treatment of natural rubber (*7*). Yasuda et al. have found that $CO_2$ selectivity over $CH_4$ of poly(dimethylsiloxane) increased by about 20-fold after Ar plasma treatment (*4*). Matsuyama et al. investigated in detail the $CO_2$ permeability and selectivity over $CH_4$ of poly(dimethylsiloxane) membranes modified by plasma treatment using Ar, $N_2$, $O_2$ and $NH_3$ and demonstrated that the selectivity of $CO_2$ over $CH_4$ was remarkably improved by plasma treatment. They also observed that the selectivity showed a maximum with treatment time (*1*).

In this study, the surface of poly(dimethylsiloxane) membranes was modified by APG plasma. Poly(dimethylsiloxane) membranes have both the advantage of high permeability and the disadvantage of low selectivity. First, the effect of the APG plasma treatment on $CO_2$ permeability and $CO_2/CH_4$ selectivity was investigated in detail. Then the surface structure of the membrane was analyzed and the relationship between the selectivity of $CO_2$ and surface structure was studied. The results were compared with those for low pressure plasma treatment by Ar, since no data on He plasma treatment of poly(dimethylsiloxane) membranes are available.

## Experimental

**Plasma treatment.** The plasma reactor apparatus (Samco International Institute Co. Ltd.) is shown in Figure 1. It consists of a rectangular parallelepipedal reaction chamber made of stainless steel (length: 21.6 cm, width: 21.6 cm, depth: 32.0 cm), and it is equipped with a pair of parallel electrodes (distance between electrodes: 0.32 cm), and a 1-10 kHz frequency generator. A quartz plate (diameter: 12 cm, thickness: 0.5 cm), which served as an insulator for stabilizing the glow discharge, was placed on the lower electrode (diameter: 10 cm) whose temperature was controlled at 298 K by circulating temperature controlled water. The plasma gas (He) was supplied to the chamber at the desired flow rate after the system was evacuated to less than 0.05 torr. The pressure in the chamber was adjusted to 760 torr. Then, the power was turned on

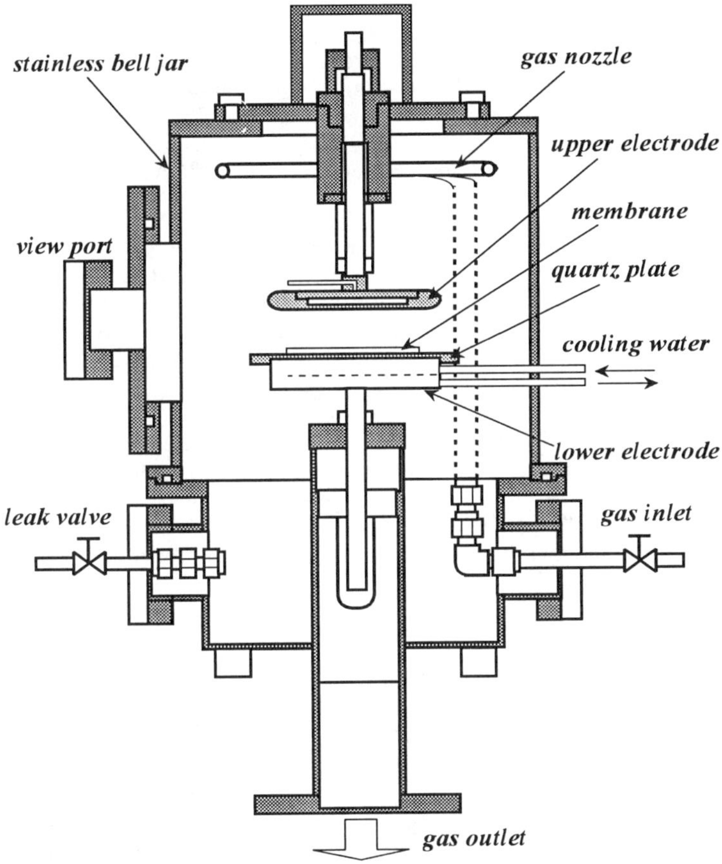


Figure 1. Schematic diagram of atmospheric pressure glow discharge plasma reactor.

and the plasma treatment was initiated. The flow of plasma gas was stopped during the plasma treatment because gas flow rate hardly influenced the properties of the treated membrane. The pressure change during the treatment was a few torr. The operating parameters such as treatment time and power input were varied. The frequency of input power was fixed at 3 kHz. The silicone membrane used as the substrate was a composite membrane (Nitto Denko Co., NTGS-2000) consisting of a top skin poly(dimethylsiloxane) layer (thickness about 1.5μm) and a porous polyimide support.

**Analysis of structure of plasma treated membrane.** To analyze the chemical structures of the membrane surface treated by plasma, X-ray photoelectron spectroscopy (XPS) was used. XPS spectra were measured with a Shimadzu electron spectrometer 750 employing MgKα excitation radiation. The $Au_{4f}$ core level at 84 eV was used for energy calibration. The structure of the inside of treated membranes was also analyzed by XPS after etching the membrane surface with Ar. The etching rate was about 50 Å/min.

**Gas permeation.** The gas permeation experiments were performed with a diffusion cell consisting of two compartments for feed and sweep gas (*1*). The plasma treated membranes were sandwiched between the two compartments. The membrane area was 9.7 $cm^2$. The feed gas (a mixture of $CO_2$ and $CH_4$, $CO_2$ mol fraction: 0.05, total flow rate: 211 $cm^3$/min) and the sweep gas (He: 50 $cm^3$/min) were supplied to the cell at atmospheric pressure. The sweep gas from the cell was analyzed by a gas chromatograph equipped with a thermal conductivity detector (Shimadzu, GC-8APT, column: activated carbon). The molar flux, N, was obtained from the flow rate of the sweep gas and its composition. The thickness of the plasma treated layer is considered to be so thin that it is difficult to measure accurately. So, we evaluated the driving-force normalized fluxes, $R_{CO_2}$ and $R_{CH_4}$ instead of the permeability coefficients (=R•L, where L is the membrane thickness). $R_{CO_2}$ and the selectivity $\alpha$ of $CO_2$ over $CH_4$ are given by

$$R_{CO_2} = N_{CO_2}/(\Delta P_{CO_2}) \quad (1)$$

$$\alpha = R_{CO_2}/R_{CH_4} \quad (2)$$

where $\Delta P_{CO_2}$ is the $CO_2$ partial pressure difference between the feed and sweep sides of the membrane. The $CO_2$ partial pressure in the feed side was estimated by accounting for the effect of He gas which couterdiffuses into the feed side from the sweep side. The diffusion cell was placed in an air bath maintained at 298 K. The APG plasma treated membrane was inserted in the diffusion cell within 5 minutes after the plasma treatment, and the permseletivities of the membrane was measured for more than 5 hr. As descried later, permeance of both gases decreased with time after plasma treatment.

## Results and discussion

**Effect of plasma treatment time and power input on membrane properties.** The effect of plasma treatment time, $t_p$, and plasma power input on $R_{CO_2}$ and $\alpha$ are shown in Figure 2. These values were measured 30 minutes after plasma treatment. $R_{CO_2}$ decreased monotonically with increasing $t_p$ as is also observed in low pressure plasma treatment (*1*). This suggests that cross-linking structures are formed via recombination of free radicals (*8*). Cross-linking by inert gas plasma is widely recognized and is

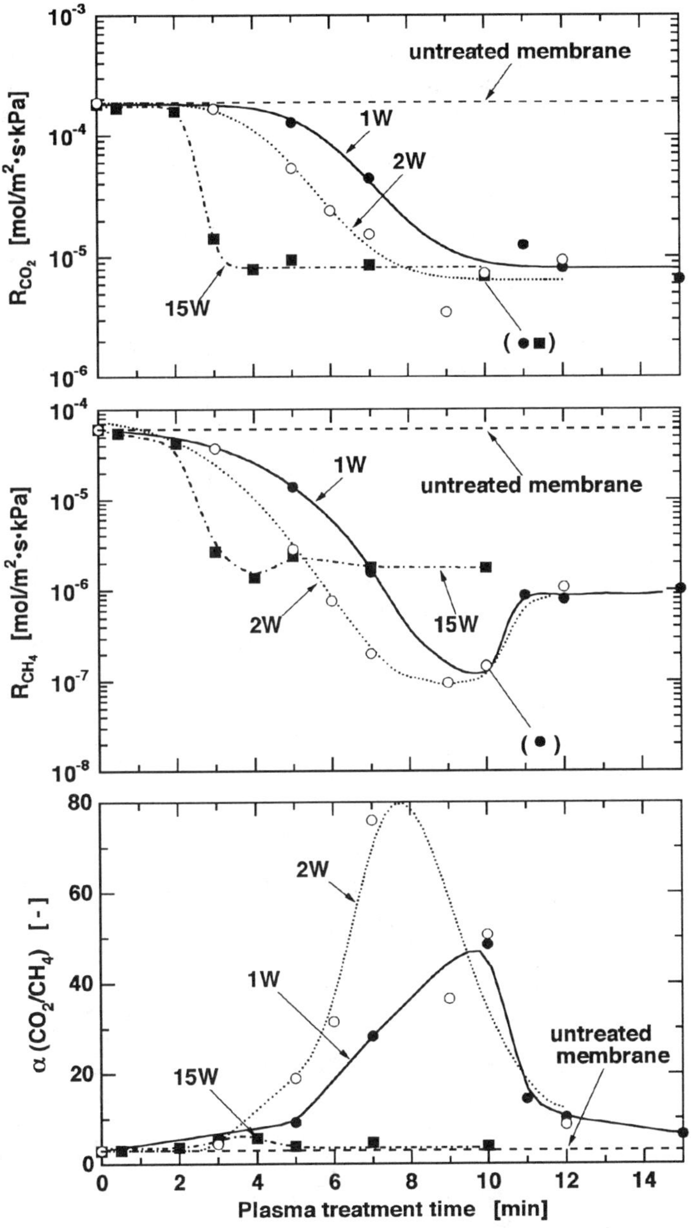


Figure 2. Effects of APG plasma treatment time and power input on $R_{CO_2}$, $R_{CH_4}$ and α of poly(dimethylsiloxane) membranes. All data are measured 30 minutes after plasma treatment. filled circles: 1W, unfilled circles: 2W, filled squares: 15W.

called CASING (*9*). The decrease of $R_{CO_2}$ was sharp when the power input was large. The selectivity, $\alpha$, was much improved by the plasma treatment, and its value was much higher than that of the untreated poly(dimethylsiioxane) membrane. This improvement in selectivity is ascribed to the permeation of small $CO_2$ molecules in the crosslinked layer being much higher than that of large $CH_4$ molecules (the kinetic sieving diameters for $CO_2$ and $CH_4$ are 0.33 and 0.38 nm, respectively). After prolonged plasma exposure, $R_{CH_4}$ increased and $\alpha$ decreased because the plasma treated layers were degraded by surface etching (ablation), and cracks or defects were formed on the surface. The treatment time at which $\alpha$ showed maximum decreased with increasing power input. When plasma power input was 2 W, a selectivity of 76, which was the maximum selectivity achieved in this study, and a $CO_2$ permeance of $1.5 \times 10^{-5}$ mol/(m$^2$•s•kPa) were obtained at $t_p$ = 7 min. However, for a power input of 15 W the selectivity increased very little after APG plasma treatment. This suggests that the APG plasma process is sensitive to plasma power and that high power input is not suitable for membrane preparation because the crosslinking and ablation occur at similar rates.

A comparison of membrane permeance/selectivity characteristics obtained in the present work with those of low pressure plasma modified poly(dimethylsiloxane) (*1*) and those of typical polymer membranes is shown in Figure 3. We evaluated the driving-force normalized fluxes, $R_{CO_2}$, for every polymer based on reported permeability coefficients and assuming that the membrane thickness was 1 $\mu$m. As shown in Figure 3, the selectivity of the APG plasma treated poly(dimethylsiloxane) membranes was significantly higher than that of conventional polymer membranes and slightly higher than that of poly(dimethylsiloxane) membranes modified by low pressure plasma. The values of $\alpha$ were obtained by considering the membrane as a single entity, and therefore, were not those of the plasma treated layer alone, $\alpha_p$. Generally, as permeability of typical polymers increases, selectivity decreases. This trend is called the trade-off rule. The performance of the plasma treated membranes deviate favorably from the general plot of permeability and selectivity of conventional polymers. Similar to poly(dimethylsiloxane) membranes treated by low pressure plasma, the APG plasma treated membranes also have higher selectivity than polyimide membranes, which have been attracting attention as a result of high selectivity of $CO_2$ over $CH_4$ (*10, 11*).

The plasma treated membrane consists of the top treated layer and the bottom silicone layer, and the resistance of each layer is connected in series. Therefore, the total permeance, $R_T$, is related to the permeance of the plasma treated layer, $R_p$, and to that of the silicone layer, $R_s$, by equation 3.

$$1/R_T = 1/R_p + 1/R_s \tag{3}$$

The thickness of the plasma treated layer is too thin to measure. We can assume that $R_s$ is equal to that of the untreated silicone membrane. The effects of the plasma treatment time on $1/R_{p,CO_2}$ and $\alpha_p$ ( $=R_{p,CO_2}/R_{p,CH_4}$) (the permeation resistance and the selectivity of the plasma treated layer alone, respectively) are shown in Figure 4. The values of $\alpha_p$ also exhibited a maximum, similar to the behavior of $\alpha$ shown in Figure 2. In the early stages of plasma treatment, $\alpha_p$ increased with treatment time. $1/R_{CO_2}$ increased not linearly but concavely with treatment time after an initial induction time. This suggests that the formation of crosslinks requires enough time to form a large amount of free radicals, and the plasma treatment proceeds gradually. It also suggests that crosslinking does not occur over the whole surface of the

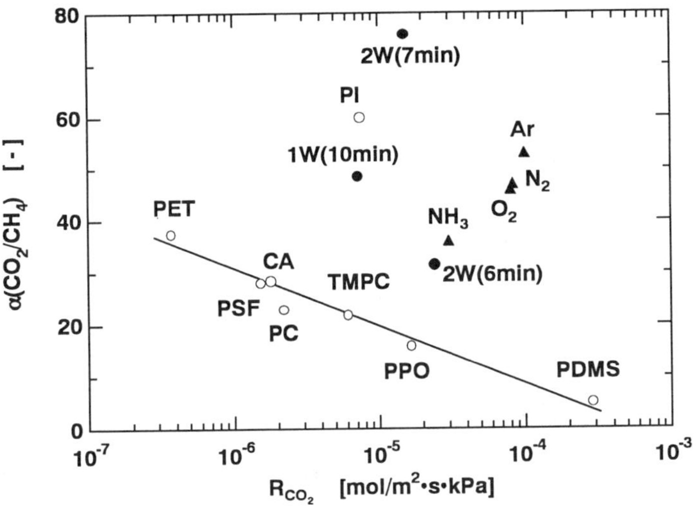


Figure 3. Comparison of membrane properties of APG plasma treated poly(dimethylsiloxane) membrane with those of low pressure plasma treated poly(dimethylsiloxane) membranes and conventional polymer membranes. filled circles: APG plasma treated membranes, filled triangles: low pressure plasma treated membrane from (1), unfilled circles: conventional polymer membrane data from (10, 11). In all cases, a membrane thickness of 1 μm was used to calculate permeance, $R_{CO_2}$, from permeability data.

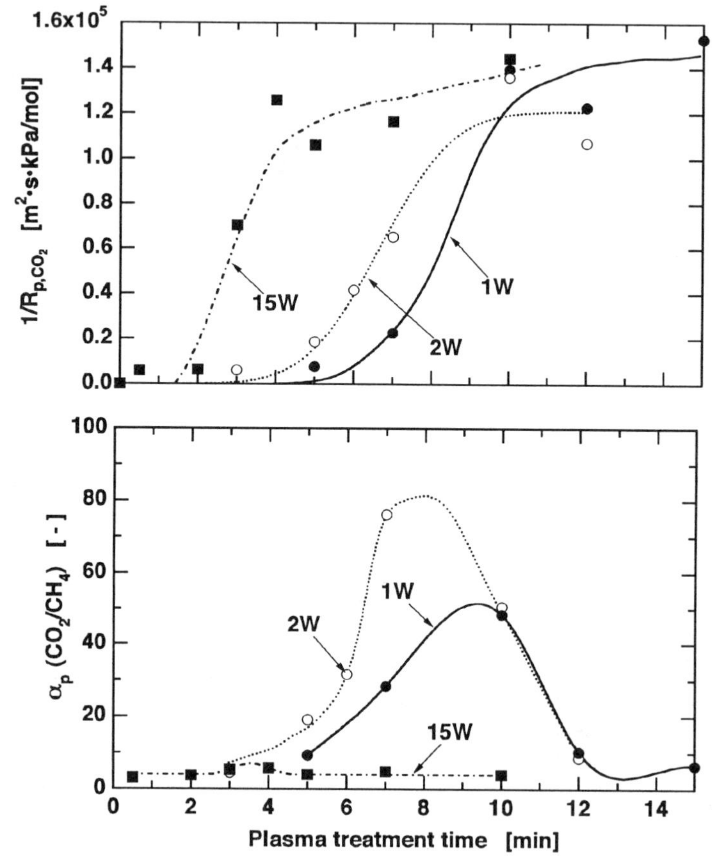


Figure 4. Effect of plasma treatment time on $1/R_{p,CO_2}$ and $\alpha_p$. filled circles: 1W, unfilled circles: 2W, filled squares: 15W.

membrane, and that some untreated areas remain. After longer plasma treatment, $\alpha_p$ decreases and the slopes of the $1/R_{CO_2}$ vs. $t_p$ curves become lower due to degradation of the treated layer by surface etching. These results are similar to those obtained via low pressure plasma treatment (*1*).

The behavior of APG plasma treatment is summarized as follows. At short treatment times, large areas of the membrane surface remain effectively untreated, the degree of crosslinking is low, and the membrane selectivity is low. With an increase in treatment time, the plasma treated layer, *i.e.* the crosslinked layer, covers the whole membrane surface, and at the same time crosslinking proceeds inside the membrane. In this case, the highest selectivity is obtained. With further plasma exposure, the selectivity decreases due to the formation of cracks from surface etching. When power input is high, the crosslinking and surface etching occur at the same time, and high selectivity is never obtained. Consequently, relatively low power input is desirable for the APG plasma treatment process.

**Membrane structure.** The effect of plasma treatment time on the $Si_{2p}$ and $C_{1s}$ spectra by XPS measurement are shown in Figure 5. For the untreated membrane, the $Si_{2p}$ and $C_{1s}$ spectra showed a single peak at 102.2 eV, assigned to Si-O, and a peak at 284.9 eV, assigned to aliphatic carbons and C-Si bonds (*12*). These results show good agreement with those expected from the structure of poly(dimethylsiloxane). The peak of the $Si_{2p}$ spectrum of the plasma treated membrane shifted to higher binding energy. This indicates that the oxidation number of the Si atom increased (*13*). The $C_{1s}$ spectra of the treated membrane increased at 287 eV(C-O) and at 289 eV (C = O) (*14*). Figure 6 shows the effect of plasma treatment time on the ratios of oxygen and carbon to silicon at the membrane surface. The O/Si ratio of the treated membrane was about twice that of the untreated membrane while the C/Si ratio decreased with an increase in plasma treatment time. Therefore, oxygen was incorporated into the membrane surface from the contiguous environment, probably by reaction with free radicals formed during plasma processing, and hydrocarbon fragments were released from the membrane surface due to cleavage of the Si-C bond. The formation of polar groups containing oxygen at the membrane surface is another possible explanation for the improvement of $CO_2/CH_4$ selectivity. Additionally, the formation of crosslinked structures improves the separation by increasing the size-sieving ability of the polymer matrix. The affinity between $CO_2$ and the membrane surface was probably increased by the existence of polar groups such as C-OH. Similar results were also observed for low pressure Ar plasma treatment (*1*).

As explained earlier, the plasma treated membranes were etched with Ar to remove the surfaces treated layer and examined the structure of the underlying membrane. The etching rate was 50 Å/min. The effect of etching time on the atomic ratios is shown in Figure 7. While the O/Si and C/Si values of the plasma treated membrane surface at $t_e$ = 0 were remarkably different from those of the untreated membrane, these values approached those of the untreated membrane with an increase in etching time. For membranes plasma treated for 3 minutes and 7 minutes, the atomic ratios at etching times of 1 and 3 minutes, respectively, were equal to those of the untreated membrane. For the membrane plasma treated for 11 minutes, atomic ratios varied little with etching time and were different from those of the untreated membrane at $t_e$=6 min. In fact, even at $t_e$ = 20 min, atomic ratios were different from those of the untreated membrane. (These data are not shown in Figure 7.) This suggests that when plasma treatment time increases, plasma modification extends deep inside the membrane with increasing plasma treatment time, and the structure of membrane becomes more and more homogeneous.

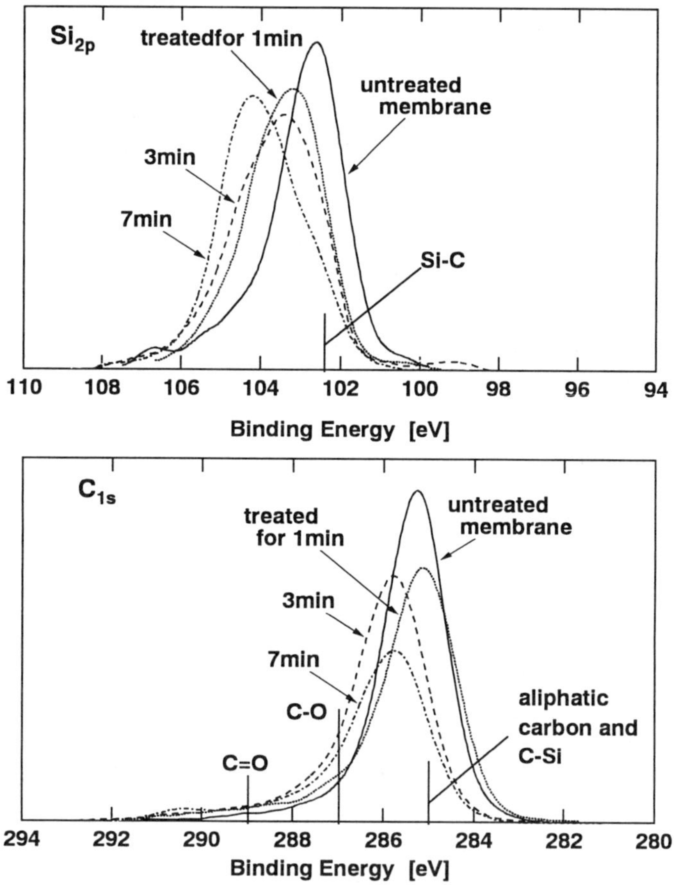


Figure 5. $Si_{2p}$ and $C_{1s}$ spectra of APG plasma treated and untreated membrane surface. power input: 2W.

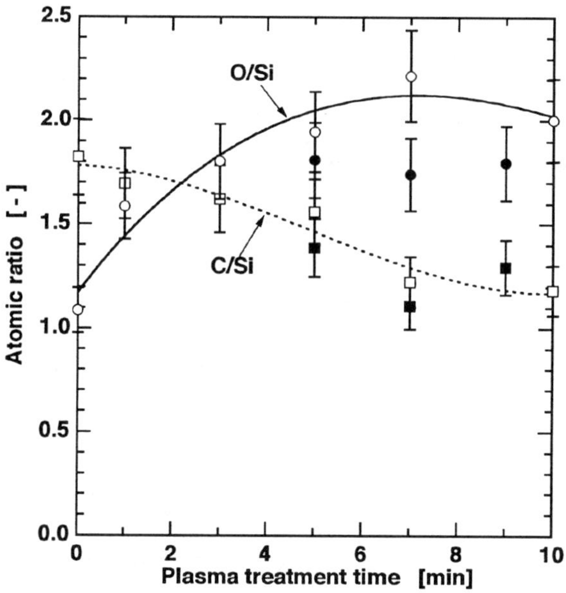


Figure 6. Effect of plasma treatment time on atomic ratios at the membrane surface. power input: 2W. unfilled symbols: measured within 1 hour after APG plasma treatment, filled symbols: measured about 1 week after APG plasma treatment.

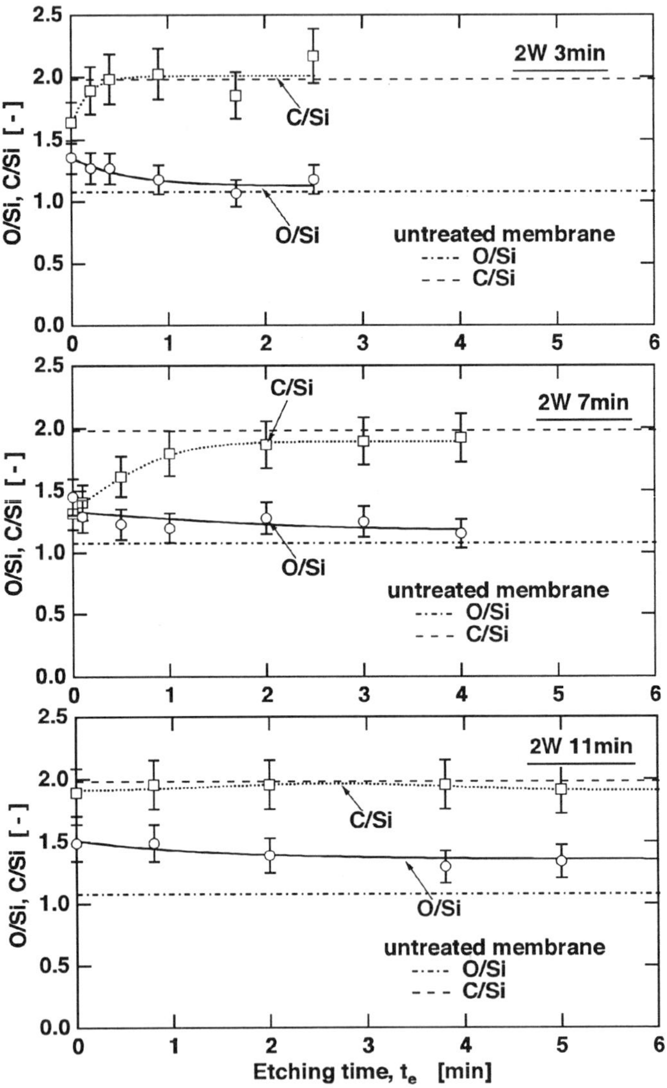


Figure 7. Effect of etching time and plasma treatment time on atomic ratio s of membrane. power input: 2W.

**Time dependence of membrane properties.** Figure 8 shows the time dependence of the permeation properties of APG plasma treated membranes. Both $R_{CO_2}$ and $R_{CH_4}$ decreased monotonically with time after plasma treatment. The values of $\alpha$ also decreased monotonically because $R_{CH_4}$ decreased slightly more than $R_{CO_2}$. One possible cause of the increase in the mass transfer resistance is continuous crosslinking via the recombination of the residual free radicals after the plasma treatment. In the case of treatment for 10 minutes, the values of $R_{CH_4}$ were nearly constant, while $R_{CO_2}$ decreased, which resulted in a marked decrease in $\alpha$. Under these plasma treatment conditions, cracks or defects are believed to form on the membrane surface by etching because $\alpha$ at these conditions started to decrease from its maximum value, as shown in Figure 4. $CH_4$, having a low permeance through the treated layer, passes mainly through the cracks with no selectivity. On the other hand, $CO_2$, which has a high permeance, is transported through the cracks and the uncracked regions and is, therefore, sensitive to post-treatment crosslinking. Therefore, the permeance of $CH_4$ is hardly influenced by the continuous formation of crosslinks in the treated layer after plasma treatment.

Another possible cause is a decrease in the affinity between $CO_2$ and the membrane surface. For membranes treated for 5 and 7 minutes, $R_{CO_2}$ decreased significantly relative to $R_{CH_4}$. At these plasma conditions, cracks or defects are not believed to form on the membrane surface by etching. If continuous crosslinking is the only reason for degradation, the decrease of $R_{CH_4}$ should be larger than that of $R_{CO_2}$ because $CH_4$ is larger than $CO_2$. This result is different from that of low pressure plasma treatment, where $R_{CO_2}$ and $R_{CH_4}$ decrease. The filled symbols in Figure 6 show the C/Si and O/Si values of the treated membrane measured about one week after plasma treatment. While the C/Si value is nearly constant, the O/Si value decreases with time. This shows that the amount of presumably polar oxygen moieties decrease with time. As a result, one may hypothesize that the affinity between $CO_2$ and the membrane decreases with time. One of the reasons for the decrease in the O/Si ratio at the membrane surface may be that poly(dimethylsiloxane) can migrate from the bulk to the surface of the plasma treated composite material because poly(dimethylsiloxane) possesses a relatively low surface free energy. Figure 9 shows the effect of the O/C ratio of the membrane surface on $CO_2$ selectivity of the treated membrane. The unfilled circles and filled circles show data measured less than one hour and about one week after treatment, respectively. Irrespective of the elapsed time, $CO_2$ selectivity is well correlated with the O/C values. Therefore, the main reason for the decrease of $R_{CO_2}$ and $\alpha$ with time after APG plasma treatment is the decrease in the affinity between $CO_2$ and the membrane caused by the decrease in the surface oxygen concentration.

## Conclusions

The effect of atmospheric pressure glow discharge (APG) plasma treatment on $CO_2$ permeability and selectivity of poly(dimethylsiloxane) membranes was investigated in detail. The selectivity of $CO_2$ over $CH_4$ was remarkably improved at low power input conditions (i.e. a few watts), and showed a maximum as a function of treatment time. The maximum $CO_2/CH_4$ selectivity obtained was 76, and the permeance of $CO_2$ obtained was $1.5 \times 10^{-5}$ mol/($m^2$•s•kPa) in this case. The selectivity was significantly higher than that of conventional polymer membranes and was slightly higher than poly(dimethylsiloxane) membranes modified by low pressure plasma.

The surface chemical structure of the treated membrane was analyzed by XPS. The O/Si ratio on the treated membrane surface and inside the treated membrane

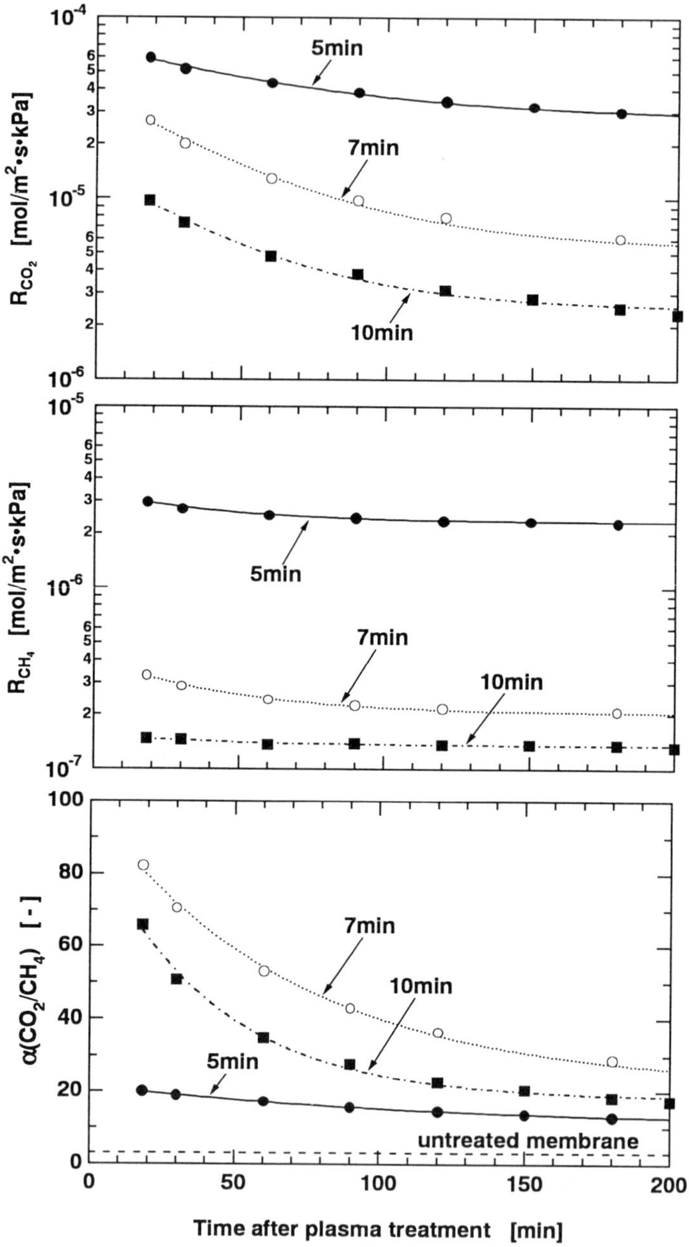


Figure 8. Time dependence of membrane properties after APG plasma treatment. power input: 2W.

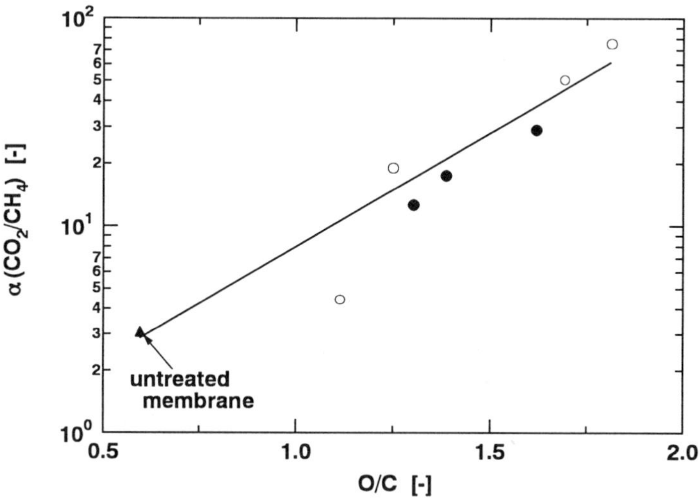


Figure 9. Correlation between surface structure and $CO_2/CH_4$ selectivity of APG plasma treated membranes. unfilled circles: measured within 1 hour after APG plasma treatment. filled circles: measured about 1 week after APG plasma treatment.

increased with an increase in plasma treatment time due to the incorporation of oxygen moieties into the surface, while the C/Si ratio decreased. The $CO_2$ selectivity increased due to an increase in the affinity between $CO_2$ and the membrane surface resulting from an increase in the surface concentration of polar groups containing oxygen.

The permeance of both $CO_2$ and $CH_4$ decreased with time after plasma treatment. This decrease was due to a decrease in oxygen content of the membrane surface.

## Literature Cited

1. Matsuyama, H.; Teramoto M.; Hirai K. *J. Membrane Sci.* **1995**, *99*, 147.
2. Kanazawa, S.; Kogoma, M.; Moriwaki, T.; Okazaki, S. *J. Phys. D: Appl. Phys.* **1988**, *21*, 838.
3. Kogoma, M.; Okazaki, S.; Kanda, N.; Uchiyama, H.; Jinno, H. *Proc. Jpn. Symp. Plasma Chem.* **1991**, *4*, 345.
4. Yasuda, H. *Plasma Polymerization*; Academic Press: Orlando, FL, 1985, pp10-16
5. Kramer, P.W.; Yeh, Y.S.; Yasuda, H. *J. Membrane Sci.* **1989**, *46*, 1.
6. Yokoyama, T.; Kogoma, M.; Kanazawa, S.; Moriwaki, T.; Okazaki, S. J. *Phys. D: Appl. Phys.* **1990**, *23*, 374.
7. Kawakami, M.; Yamashita, Y.; Iwamoto, M.; Kagawa, S.; *J. Membrane Sci.* **1984**, *19*, 249.
8. Chang, F.Y.; Shen, M.; Bell, A.T. *J. Appl. Polym. Sci.* **1973**, *17*, 2915.
9. Hansen, R. H.; Schonhorn, H. *J. Polym. Sci., Polym. Lett.* **1966**, *4*, 203
10. Kim, T.H.; Koros, W.J.; Husk, G. R.; O'Brien, K.C.; *J. Membrane Sci.* **1988**, *37*, 45.
11. Tanaka, K.; Kita, H.; Okamoto, M. *Polymer* **1992**, *33*, 585.
12. Inagaki, N.; Kishi, A. *J. Polym. Sci., Polym. Chem. Ed.* **1983**, *21,* 1847.
13. Nordberg, R.; Brecht, H.; Albridge, R. G. *Inorg. Chem.* **1970**, *92*, 469.
14. Inagaki, N.; Yamazaki, H. *J. Appl. Polym. Sci.* **1984**, *29*, 1369.

Chapter 12

# Group Contribution Modeling of Gas Transport in Polymeric Membranes

**D. V. Laciak, L. M. Robeson, and C. D. Smith**

**Corporate Science and Technology Center, Air Products and Chemicals, Inc., Allentown, PA 18195–1501**

While gas separation via polymeric membranes is a commercially practiced technology, there remains an ongoing search for new and improved polymers. Clearly, this search would be most efficient when guided by a structure-property relationship model. We have developed a relatively simple group contribution method which accurately predicts the gas permeation properties of linear, amorphous polymers by segmenting the polymer into a set of volume elements. The volumes of these elements, or groups, were calculated using the Quanta molecular modeling program. Multiple linear regression techniques were used to determine the permeability contribution of each volume element. In this manner, we have shown that the gas permeability of over 87 polymers including polysulfones, polyethers, polyarylates, polycarbonates, and polyimides can be predicted from a set of just 35 unique groups. In this paper we describe the model and compare its predictions to the experimental values for the gases $CO_2$, $CH_4$. Moreover, the model identifies those groups which enhance permselectivity and so is an excellent tool for guiding polymer synthesis efforts.

There is an ongoing and widely recognized need for gas separation polymers with improved permeability and permselectivity. Over the last 15 years a large database of experimentally determined permeability appearing in the membrane literature has demonstrated that the search for improved gas permeation properties has largely been empirical one and the relationship of polymer structure to permeability is still not well understood. Several researchers have shown that increased polymer chain rigidity and decreased chain packing density lead to improved permselectivity (*1,2*). Nonetheless, predicting such effects for all bus a small family of polymers is not straightforward.

It would clearly be beneficial to have the ability to predict polymer permeability based only on polymer structure. The use of molecular dynamics has proven useful in modeling gas permeation through rubbery polymers, particularly silicone rubber (*3-5*). The application of molecular dynamics to glassy polymers however has proven problematic owing to the nonequilibrium nature of the glassy state. Group contribution methods express polymer properties as the sum of properties of unique chemical fragments (*6*). Several group contribution approaches to predicting gas and vapor permeability have been published. Those by Salame (*7*) and Bicerano (*8*) involve complex connectivity indices to generate the required property data while the methods of Jia and Xu (*9*) Park and Paul, (*10*) correlate gas permeability with fractional free volume. The latter proved to be an improvement of Bondi's methods for predicting gas permeability (*12*). In the fractional free volume approach the free volume was calculated from experimental density measurements. The method presented in this paper utilizes only one experimental parameter, gas permeability. The other parameter of interest, subunit molar volume, was calculated using commercial computer software. The application of this Qualitative Structure-Property Relationship, or QSPR, or model to oxygen, nitrogen and helium permeation has been described previously (*12*). In this paper we extended the database from 65 to 87 polymers and the number of subunits from 24 to 35 while demonstrating the general applicability of the method by applying the QSPR model to the separation of carbon dioxide from methane, a critical unit operation in natural gas processing.

## Datasets and Calculation Methods

The journal and patent literature was searched to identify the 87 polymers used in this work. They include members from the polymers, their structure, and gas transport properties are provided in the Appendix. Carbon dioxide is reported to permeate many polymers by a dual mode transport mechanism (*13*). It has also been noted that $CO_2$ has a plasticizing effect on some polymers leading to decreased permselectivity at high pressure (*14*). For these reasons we have preferentially used pure gas permeability data obtained at 10 atm $CO_2$ and 35°C.

The QSPR group contribution model involves a least squares fit of the equation

$$\ln P = \sum \Phi_i \ln P_i \qquad (1)$$

where P is the polymer permeability; $\Phi_i$ is the volume fraction of group i; and $P_i$ is the permeability of group i.

The groups (i's) consists of chemical bonds, or parts of chemical bonds, and associated structural units. For example, polysulfone

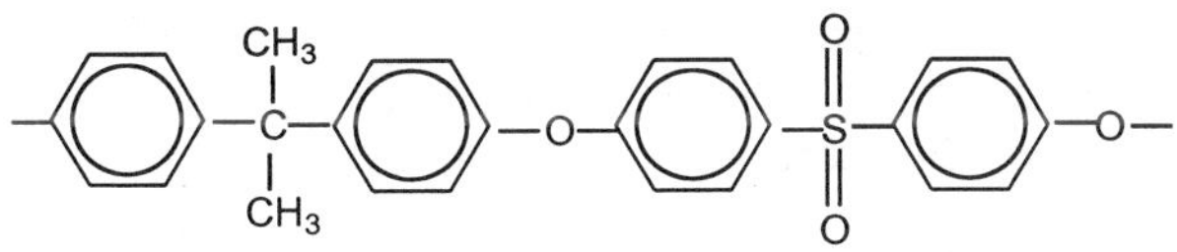

can be represented by the equation 4P2 + 2P3 + 2P4, where P2, P3 and P4 are defined by the structural units

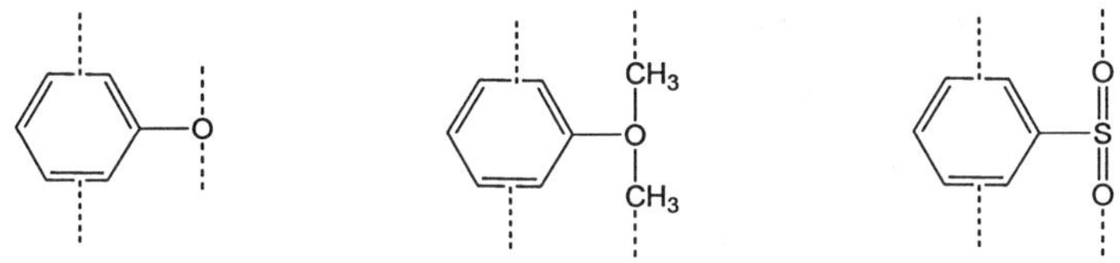

**P2** $\Phi = 39.00A^3$ **P3** $\Phi = 54.94A^3$ **P4** $\Phi = 49.40A^3$

The groups are defined by the molecular volume contained between the dotted lines and not necessarily defined along chemical bonds or complete functional units. Indeed, in many cases a group may contain only one-half of an atom. Correlation factors can be strongly influenced by the choice of structural and bonding units comprising the group. In particular, incorporation of nonsymmetry considerations into the model was critical. Aromatic substitution patterns *outside* the subunit volume element, while having minimal impact on the subunit volume, allows for different permeability contributions of structurally similar groups. Likewise, para vs. meta catenation patterns have minimal effect on subunit molar volume but account for some of the largest differences in group permeability contributions. For example, there are two phenyl ether subunits defined by their catenation within the polymer as shown below which differ significantly in the permeability contributions.

p phenyl ether m-phenyl ether

Taking these feature into account, we found that the 87 polymers in the dataset could be represented by a linear combination of 35 unique subunits or 'groups'. The molar volume of each group was calculated in $A^3$ using the mainframe-based computer modeling package Quanta (Molecular Simulations, Inc., San Diego, CA). The details of calculating the van der Waals molar volume by this method have been given a previous publication (*12*). Briefly, it involves generating the structure of a model compound is minimized and the molar volume calculated. To extract the molar volume of the subunit, the volume of structural fragments extraneous to the subunit are subtracted. This process is illustrated below for the 'phenyl ether' subunit. First, the model compound diphenyl ether is constructed. It can be seen the volume of the subunit (defined by the volume between the dashed lines) is equal to one half the volume of diphenyl ether minus the volume of benzene. This method was applied to calculate the molar volumes listed in Table I.

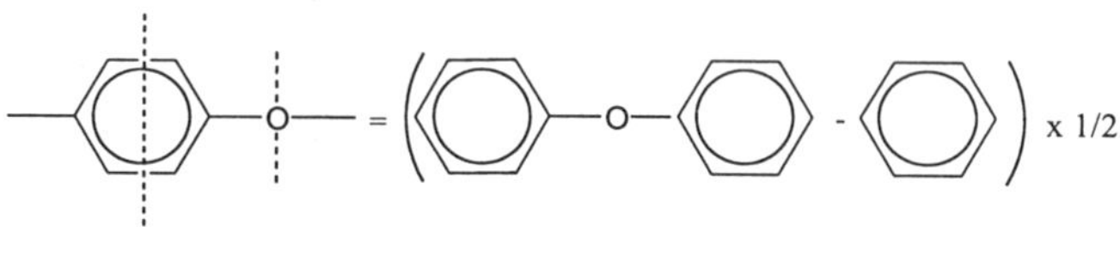

Application of equation 1 using the volumes in Table I to the polymer database (Appendix) results in a matrix of 87 equations and 35 unknowns. The solution was obtained by a least squares fit of the data using multiple linear regression techniques.

## Results

The subunits defined in the model are listed in Table I. Also shown are the group molar volume and the group contribution to the $CO_2$ permeability and the $CO_2/CH_4$ permselectivity; the best fit solution to the matrix of linear equations. The gas permeability of each polymer in the dataset was calculated from the resultant subunit permeability indicated in Table I and the normalized structural equation for each polymer. The $CO_2$ and $CH_4$ permeability (in Barrers) predicted by the group contribution model is compared to experimental values in Figure 1. An excellent correlation is evident for both gases. The correlation between model predicted and experimental $CO_2/CH_4$ selectivity is shown in Figure 2.

Goodness of fit statistics are represented as the sum of the residuals form the lest squares fit subroutine. Agreement between experimental and model predicted permeability is quire good for both gases. A second measure of fit is given by the average percent error which is 17% for $CO_2$ and 20% for $CH_4$.

| Gas | $\Sigma$ residuals | APE |
|---|---|---|
| $CO_2$ | 2.03 | 17.3% |
| $CH_4$ | 2.29 | 19.9% |

$$APE = \sum_{i=1}^{n} \left| \left( P_{exp} - P_{calc} \right) \right| * 100 / n$$

Figure 3 shows the permselectivity contributions of each of the 35 unique structural groups plotted against the $CO_2/CH_4$ upper bound.(7) Particularly attractive groups identified include P1, P5, P12, P15, P18 and P35.

**Table I. Predicted Permeability and Permselectivity Data for Structural Units Comprising the Group Contribution Approach**

| | Structural Unit | Volume $V_i(\mathrm{\AA}^3)$ | $CO_2$ Permeability (Barrers) | Permselectivity $\alpha(CO_2/CH_4)$ |
|---|---|---|---|---|
| **P1** | 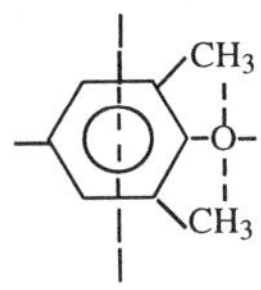 | 65.56 | 309.0 | 19.3 |
| **P2** | 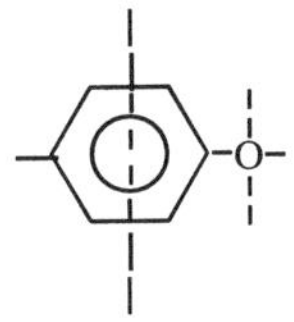 | 39.0 | 3.34 | 17.4 |
| **P3** | 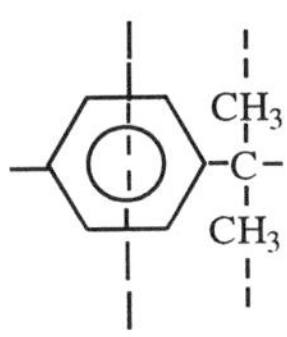 | 54.94 | 16.0 | 16.2 |
| **P4** | 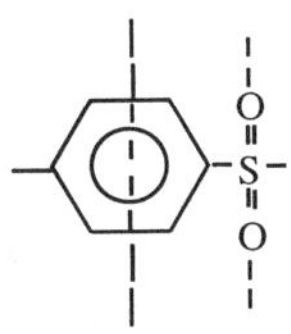 | 49.40 | 2.83 | 44.5 |
| **P5** | 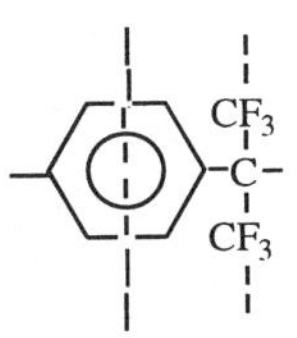 | 773.19 | 278.7 | 24.4 |
| **P6** | 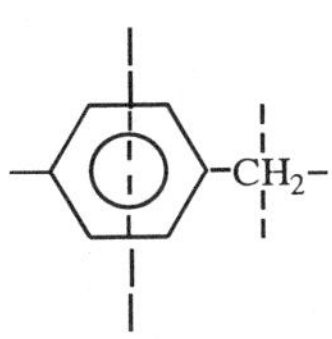 | 41.25 | 7.59 | 29.4 |

*Continued on next page.*

**Table I. Predicted Permeability and Permselectivity Data for Structural Units Comprising the Group Contribution Approach (Cont'd)**

| | Structural Unit | Volume $V_i(\text{Å}^3)$ | $CO_2$ Permeability (Barrers) | Permselectivity $\alpha(CO_2/CH_4)$ |
|---|---|---|---|---|
| **P7** | 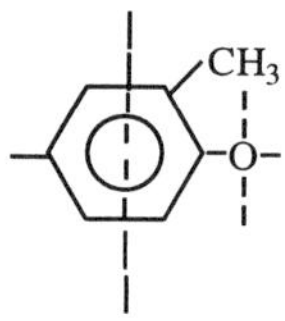 | 51.07 | 0.123 | 63.1 |
| **P8** | 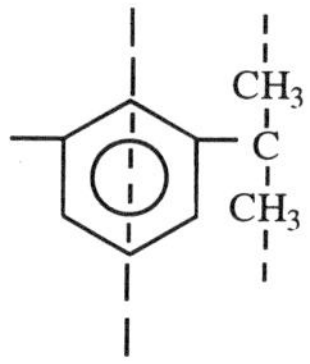 | 54.82 | 0.220 | 41.9 |
| **P9** | 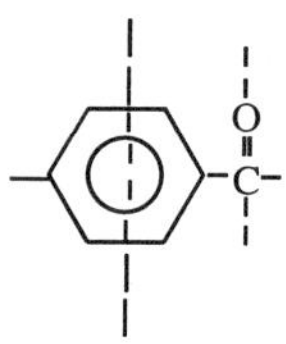 | 44.32 | 0.648 | 34.4 |
| **P10** | 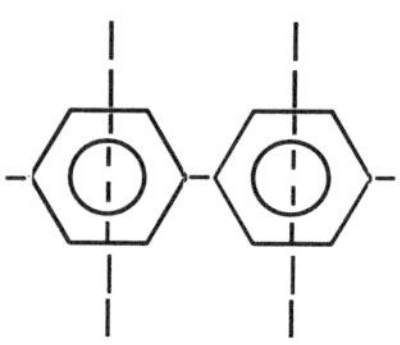 | 69.0 | 76.4 | 20.2 |
| **P11** | 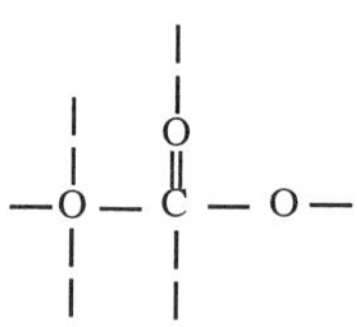 | 14.88 | 0.049 | 45.9 |

**Table I. Predicted Permeability and Permselectivity Data for Structural Units Comprising the Group Contribution Approach (Cont'd)**

| | Structural Unit | Volume $V_i(\AA^3)$ | $CO_2$ Permeability (Barrers) | Permselectivity $\alpha(CO_2/CH_4)$ |
|---|---|---|---|---|
| **P12** | 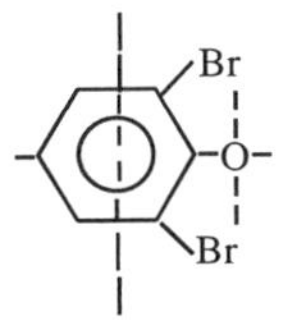 | 74.5 | 10.66 | 62.7 |
| **P13** | 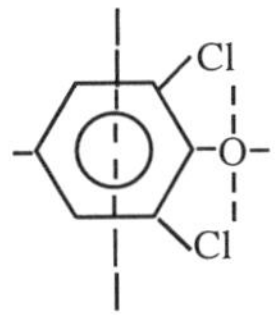 | 68.0 | 34.18 | 41.7 |
| **P14** | 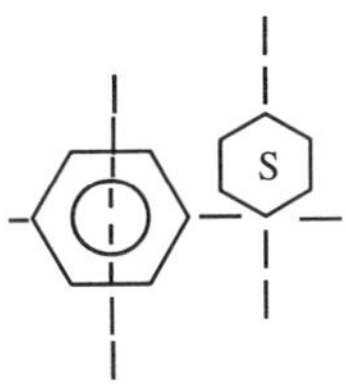 | 71.75 | 3.07 | 25.5 |
| **P15** | 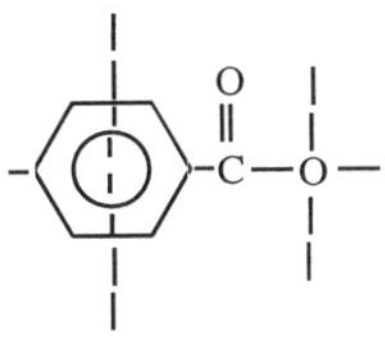 | 59.0 | 43.97 | 45.9 |
| **P16** | 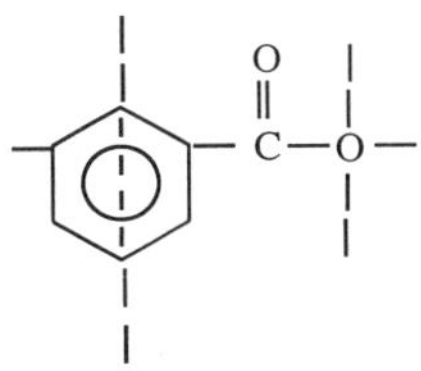 | 58.13 | 0.871 | 41.9 |

*Continued on next page.*

**Table I. Predicted Permeability and Permselectivity Data for Structural Units Comprising the Group Contribution Approach (Cont'd)**

| | Structural Unit | Volume $V_i(\AA^3)$ | $CO_2$ Permeability (Barrers) | Permselectivity $\alpha(CO_2/CH_4)$ |
|---|---|---|---|---|
| **P17** | 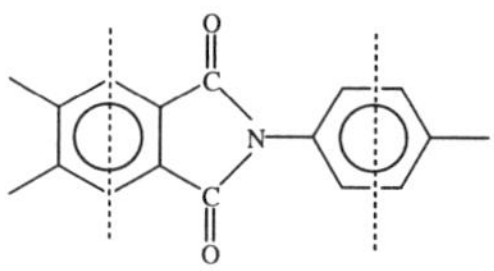 | 110.94 | 3.514 | 54.1 |
| **P18** | 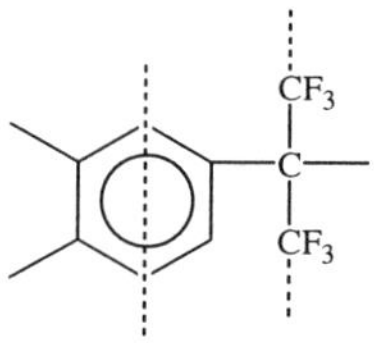 | 73.19 | 319.2 | 47.0 |
| **P19** | 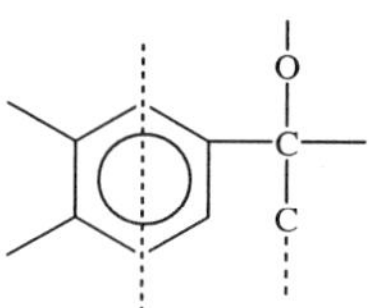 | 45.38 | 0.011 | 180.5 |
| **P20** | 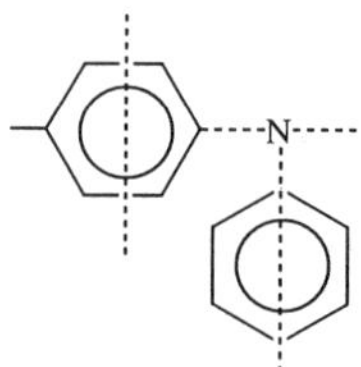 | 74.75 | 47.7 | 8.31 |
| **P21** | 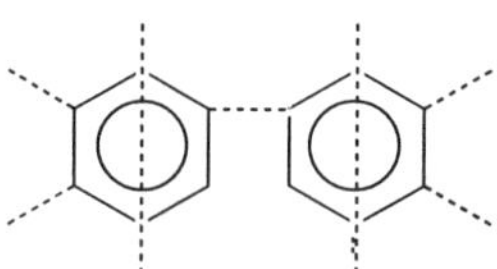 | 68.12 | 0.128 | 70.7 |

**Table I. Predicted Permeability and Permselectivity Data for Structural Units Comprising the Group Contribution Approach (Cont'd)**

| | Structural Unit | Volume $V_i(\text{Å}^3)$ | $CO_2$ Permeability (Barrers) | Permselectivity $\alpha(CO_2/CH_4)$ |
|---|---|---|---|---|
| **P22** | 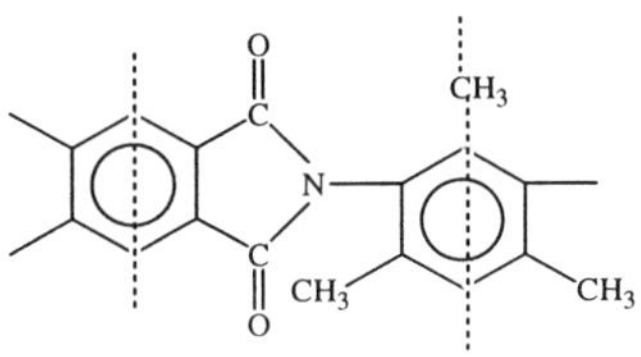 | 131.07 | 607.9 | 11.3 |
| **P23** | 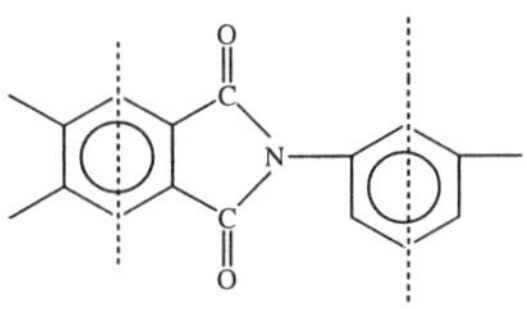 | 111.76 | 0.92 | 63.6 |
| **P24** | 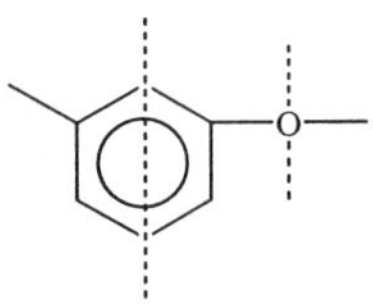 | 40.75 | 0.021 | 42.4 |
| **P25** | 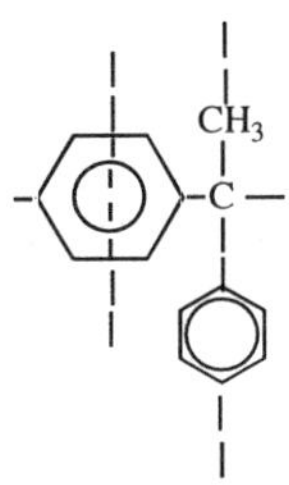 | 83.56 | 37.9 | 23.6 |
| **P26** | 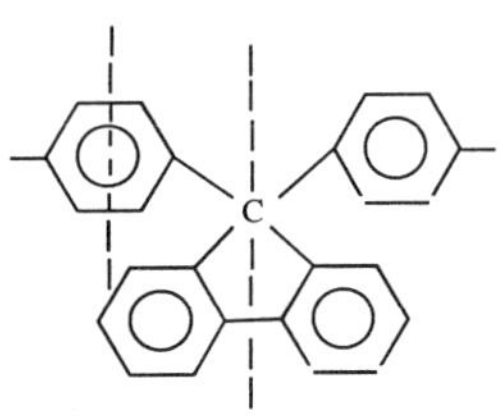 | 107.63 | 82.1 | 20.3 |

*Continued on next page.*

**Table I. Predicted Permeability and Permselectivity Data for Structural Units Comprising the Group Contribution Approach (Cont'd)**

| | Structural Unit | Volume $V_i(\text{Å}^3)$ | $CO_2$ Permeability (Barrers) | Permselectivity $\alpha(CO_2/CH_4)$ |
|---|---|---|---|---|
| **P27** | 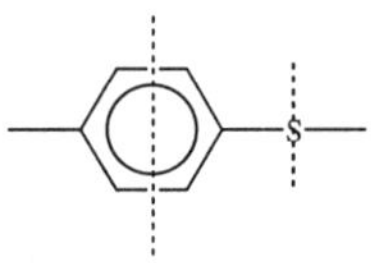 | 43.38 | 1.91 | 19.5 |
| **P28** | 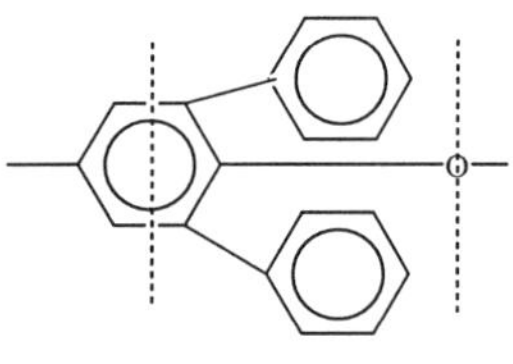 | 179.28 | 68.3 | 14.3 |
| **P29** | 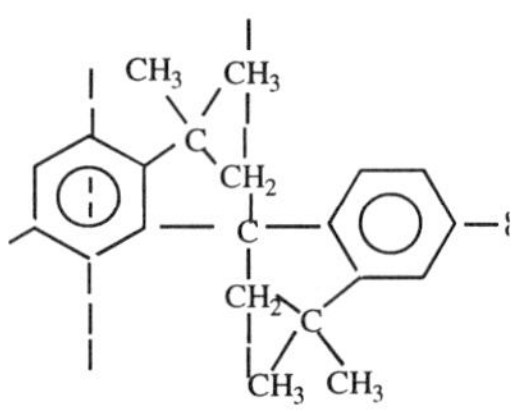 | 89.25 | 522.7 | 8.6 |
| **P30** | 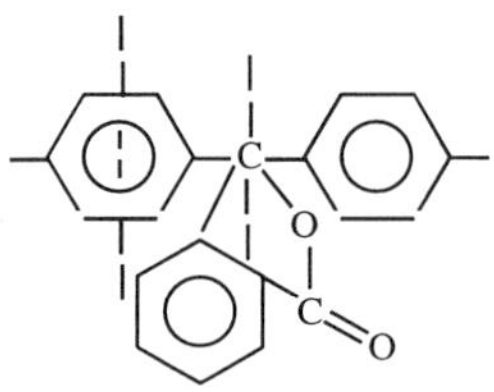 | 86.0 | 33.2 | 39.1 |
| **P31** | 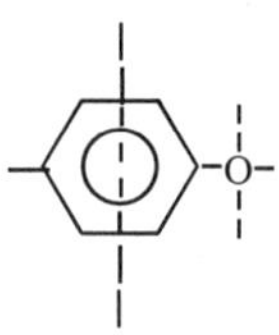 | 40.75 | 0.010 | 124.7 |

**Table I. Predicted Permeability and Permselectivity Data for Structural Units Comprising the Group Contribution Approach (Cont'd)**

| | Structural Unit | Volume $V_i$(Å$^3$) | $CO_2$ Permeability (Barrers) | Permselectivity $\alpha(CO_2/CH_4)$ |
|---|---|---|---|---|
| **P32** | 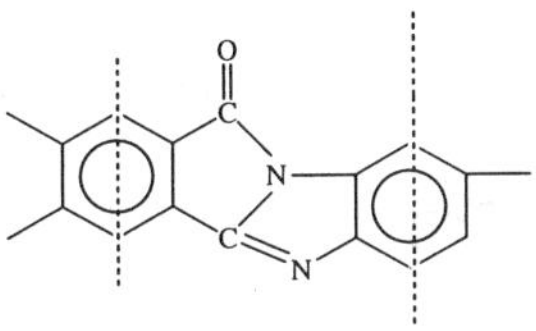 | 103.0 | 23.52 | 45.4 |
| **P33** | 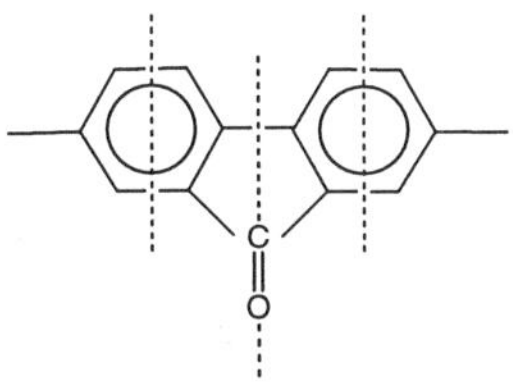 | 98.57 | 7.33 | 46.9 |
| **P34** | 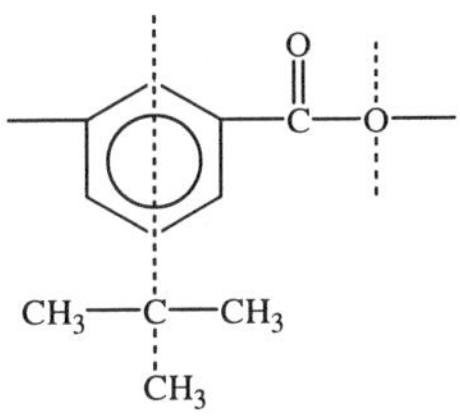 | 86.07 | 65.9 | 13.3 |
| **P35** | 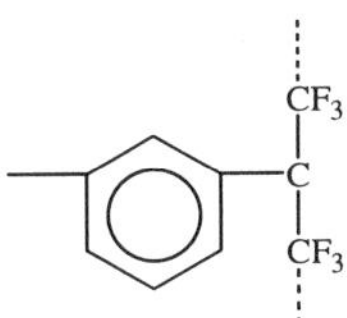 | 73.19 | 5.73 | 72.14 |

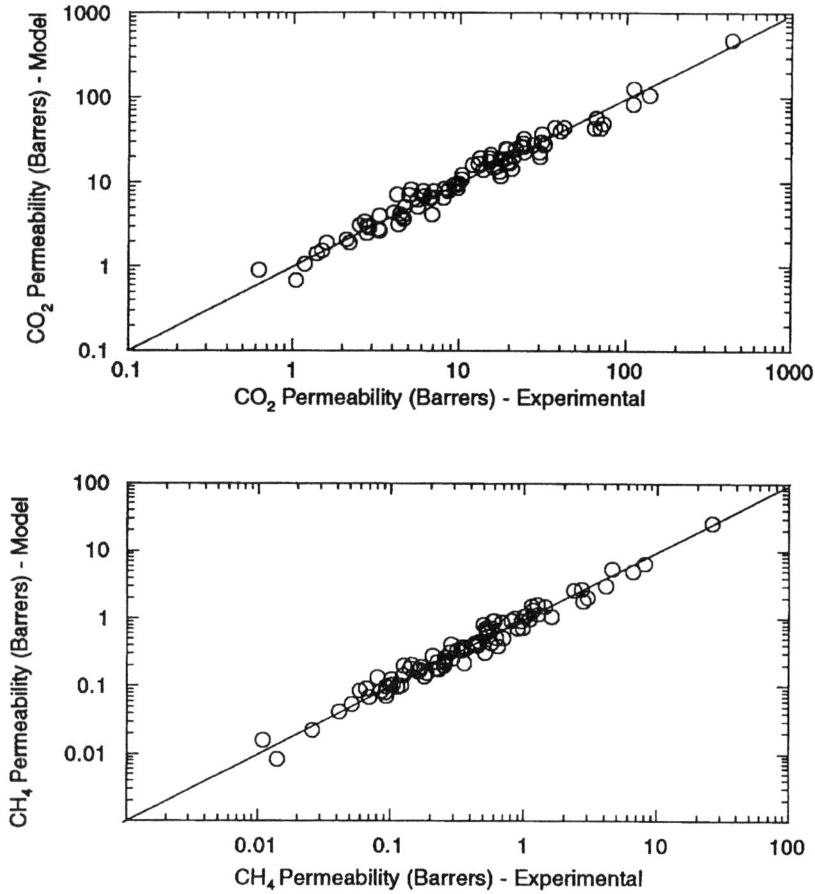


Figure 1. Comparison of Experimental and Model Predicted Permeability.

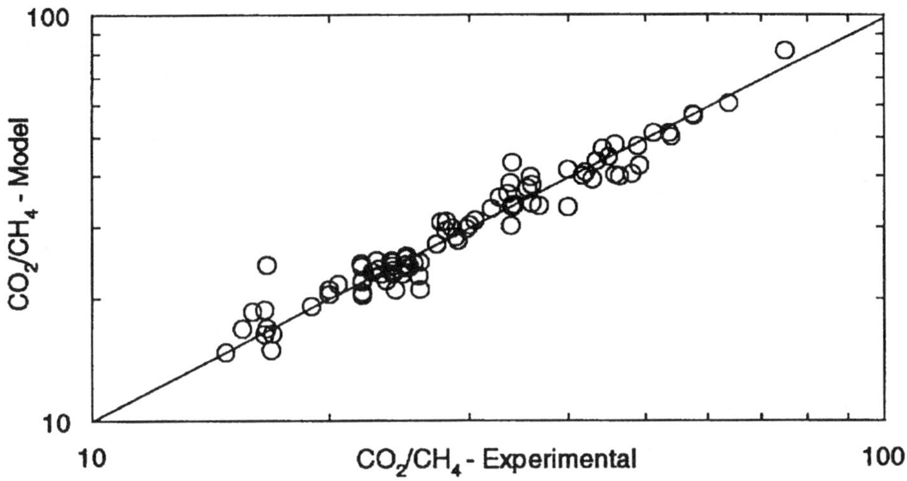


Figure 2. Comparison of Model and Experimental $CO_2/CH_4$ Selectivity.

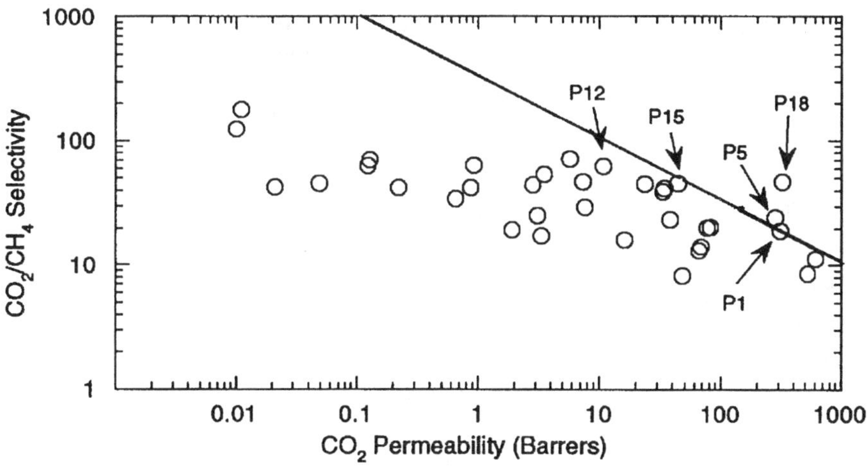


Figure 3. Comparison of Subunit Permeability Contribution to the $CO_2$ - $CH_4$ Upper Bound (15).

It is interesting to note the effect of neighboring groups; that is, the effect of structural elements adjacent to the group. This effect is illustrated below. The group P12, found in a PMDA-based polyimide has significantly different permeability contributions than the group P35 even though the structural elements defined within the group are the same.

| Structural Unit | Group | $PoCO_2$ (barrer) | $\alpha CO_2/CH_4$ |
|---|---|---|---|
| 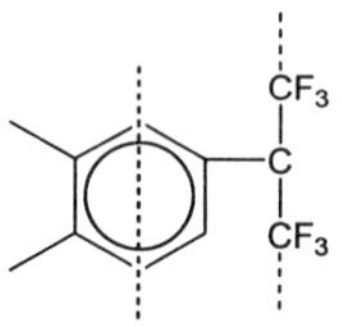 | P12 | 319.2 | 47.0 |
| 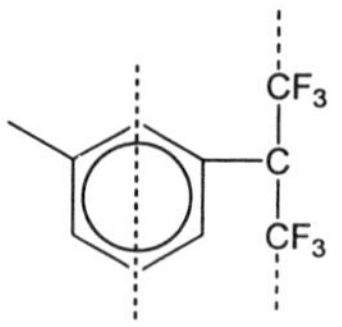 | P35 | 5.73 | 72.1 |

This QSPR model also provides a quantitative measure for the structure-property relationships observed by other researchers. For example, it is well known that the *para* isomer is almost always more permeable and less selective than is the *meta*. This effect is illustrated below for the isopropylidene and hexafluoro isopropylidene units. The *para* isomers are 50-100x more permeable and with only about 30-40% of the selectivity of the *meta* isomers.

| | | |
|---|---|---|
| | 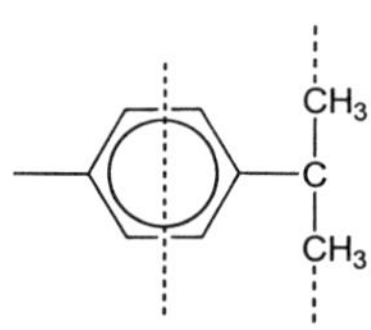 | 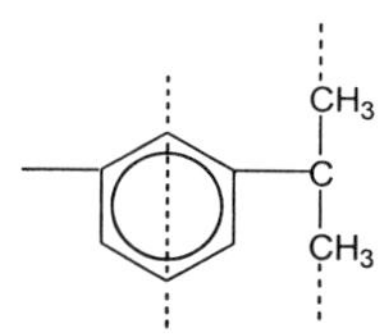 |
| | ***para*** | ***meta*** |
| $PoCO_2$(barrer): | 16.0 | 0.22 |
| $\alpha CO_2/CH_4$: | 16.2 | 41.9 |

| | | |
|---|---|---|
| | ***para*** | ***meta*** |
| | 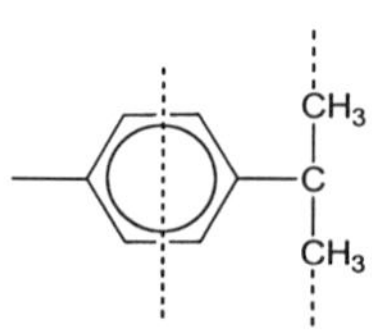 | 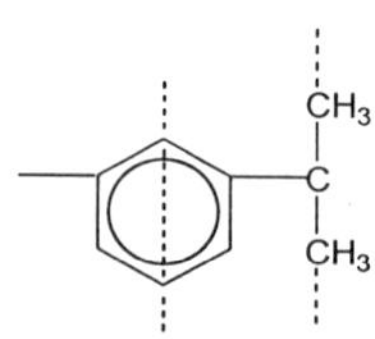 |
| $PoCO_2$(barrer): | 278.7 | 5.73 |
| $\alpha CO_2/CH_4$: | 24.4 | 72.1 |

Another example is comparison of tetramethyl bisphenol-A polysulfone with dimethyl bisphenol-A polysulfone. The $CO_2$ permeability and $CO_2/CH_4$ selectivity are 21.0 barrer and 22.0 and 2.1 barrer and 30, respectively. These polymers differ only in the extent of methyl substitution on the bisphenol- rings. It has been noted that asymmetrical, dimethyl substitution allows for a 'tighter' packing of polymer chains and therefore lower gas permeability and higher selectivity (*16*). The new group contribution model quantifies this effect. The differences in polymer permselectivity are manifested in the permeability contributions of these two groups as shown below. The symmetrically substituted group P1 is 1000x more permeable but with only one third the selectivity of the asymmetrically substituted group P7.

## Conclusions

A new group contribution model has been developed which accurately describes the $CO_2$ and $CH_4$ permeability and selectivity of 87 polymers. Unlike previous group contribution models, it is not based on fractional free volume estimated from density measurements but rather, in this approach, the volume fractions of the structural elements comprising the polymer serve as the basis for correlation. The volume of these groups were calculated from commercially available computer software. This new model quantifies a variety of structure-property relationships which have been reported in the literature.

## Acknowledgements

The authors wish to acknowledge Ken Anselmo of Air Products and Chemicals, Inc., MIS-Research Engineering Services for assistance in solving the regression matrix and Christine Vermeulen for drawing the polymer structures.

## Literature Cited

1. Mohr, J.M., Paul, D.R., Tullos, G.L. and Cassidy, P.E. *Polymer* ***1991****, 32,* 2387.
2. McHattie, J.S., Koros, W.J. and Pau, D.R. *Polymer* **1992**, *33* 1701.
3. Muller-Plathe, F., *J. Chem. Phys*. **1991** 94, 319.
4. Muller-Plathe, F., *J. Chem. Phys*. **1992** 96, 3200.
5. Sonnenburg, J., Gao, J. and Weiner, H. *Macromolecules*, **1990** *23*, 4653.
6. Van Krevelen, D. W. "Properties of Polymers", Elsevier; Amsterdam 1990.
7. Salame, M. *Poly. Eng. Sci*. **1986,** *26*, 321.
8. Bicerano, J. J. Macromol. Sci. **1996**, 36, 161.
9. Jia, L. and Xu, *J. Polym. J*. **1991,** *23* 417.
10. Park, Y. and Paul, D. R. *J. Mem. Sci*. **1997**, *125* 23.
11. Bondi, A. Properties of Molecular Liquids, Crystals and Glasses; John Wiley and Sons; NY, **1968**.
12. Robeson, L.M., Smith, C.D., and Langsam, M. *J. Memb. Sci*. **1997**, *132*, 33.

13. Chern, R.T. *J. Mem. Sci.,* **1983** *15*, 157.
14. Koros, W.J., and Fleming, G.K. *J. Mem. Sci.* **1993**, *83*, 1.
15. Robeson, L.M. *J. Mem. Sci.* **1991**, *62*, 165.
16. Aiken, C.L. *J. App. Poly. Sci.* **1982**, *27*, 2997.
17. McHattie, J.S., Koros, W.J., and Paul, D.R. *Polymer* **1992**, *32*, 585.
18. Aiken, C.L., Koros, W.J. and Pau, D.R. *Macromolecules* **1992**, *25*, 3424.
19. Toi, K., Morel, G., and Paul, D.R. *J. Appl. Poly. Sci.* **1982,** *27*, 2997.
20. Aitken, C.L., Koros, W.J. and Paul, D.R. *Macromolecules* **1992**, *25*, 3651.
21. Hellums, M.W., Koros, W.J., G.R. Husk, and Paul, D.R. *J. Appl. Polym. Phys.* **1991**, *29*, 731.
22. McHattie, J.S., Koros, W.J., and Paul, D. R. *J. Polym. Sci.*, **Part B:** *Polym Phys.* **1991**, *29*, 731.
23. Hellums, M.W., Koros, W.J., G.R. Husk, and Paul, D.R. *J. Mem. Sci.* **1989**, *46,* 93.
24. Maruruganandum, N., Koros, W.J., and Paul, D.R. *J. Polym. Sci.*, **Part B:** *Polym. Phys.* **1987**, *25* 1999.
25. Barbari, T.A., Koros, W.J., and Paul, D.R. *J. Mem. Sci.* **1989**, *42*, 69.
26. Kawakami, J. Mauruganandum, N., and Brode, G.L. U.S. Patent No. 5,055,114 (1981).
27. Kim, T.H., Koros, W.J., G.R. Husk and O'Brien, K.C. *J. Mem. Sci.* **1988**, *37*, 45.
28. O'Brien, K. C., Koros, W.J., and Husk, G.R. *J. Mem. Sci.* **1988**, *35*, 217.
29. Tanaka, K., Kita, H., Okano, M., and Okamoto, K. *Polymer* **1992**, *33*, 585.
30. Okamoto, K. *Polymer Journal* **1992**, *24*, 451.
31. Kim, T.H., Koros, W.J., and Husk, G.R. *Sep. Sci. & Tech.* **1988**, *23*, 1611.
32. Tanaka, K., Okano, M., Toshino, H., Kita, H., and Okamoto, K. *J. Polymer. Sci.,* Part B: *Polym Phys.* **1992**, *30*, 907.
33. Auguilar-Vega, M. and Paul, D.R. *J. Polym. Sci.* **1993**, *31*, 1599.
34. Glatz, F. P., Mulhaupt, R., Shultze, J.D., and Springer, *J. J. Mem. Sci.* **1994**, *90*, 151.
35. Krizan, T. U.S. Patent No. 5,080,698 **1992**.
36. Pessan, L.A., Koros, W.J., Schmidhauser, J.C., and Richards, W.D. *J. Poly. Sci.:* **Part B**: *Poly. Phys.* **1995**, *33*, 487.
37. Pixton, M. and Paul, D.R. *J. Poly. Sci.:* **Part B:** *Poly. Phys.* **1995**, *33,* 1135.
38. Pixton, M. and Paul, D.R. *J. Poly. Sci.:* **Part B:** *Poly. Phys.* **1995**, *33*, 1353.
39. Pessan, L.A. PhD Dissertation, University of Texas, Austin **1993**.
40. Paul, D.R. unpublished results.

## Appendix A. Permeability Data

| | Polymer Structure | $P(CO_2)$ Barrer | $P(CH_4)$ Barrer | Ref. |
|---|---|---|---|---|
| 1. | 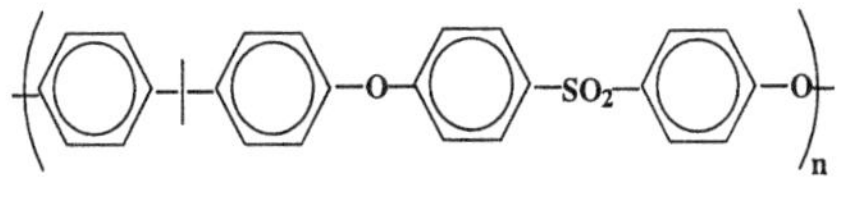 | 5.6 | 0.25 | 17 |
| 2. | 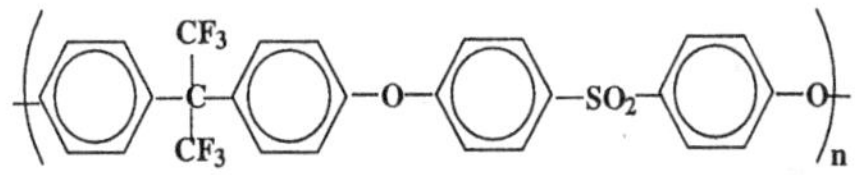 | 12.0 | 0.524 | 17 |
| 3. | 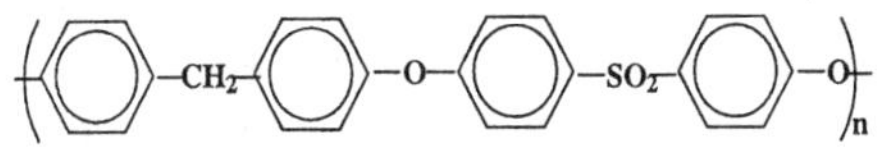 | 4.5 | 0.188 | 17 |
| 4. | 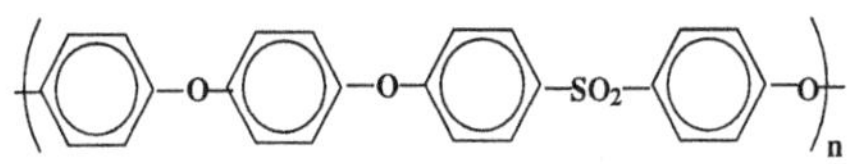 | 4.3 | 0.188 | 17 |
| 5. | 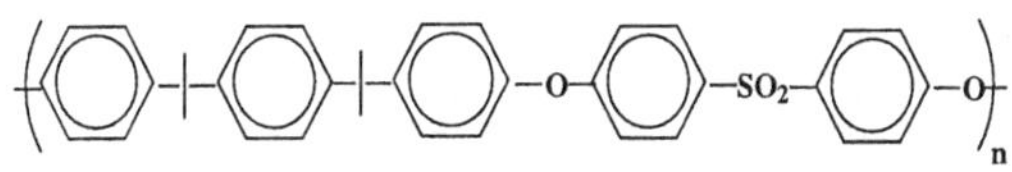 | 6.8 | 0.340 | 18 |
| 6. | 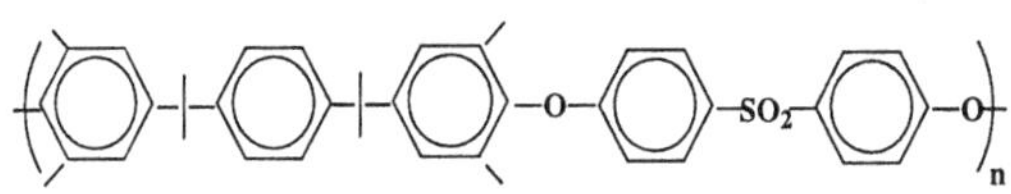 | 13.2 | 0.600 | 18 |
| 7. | 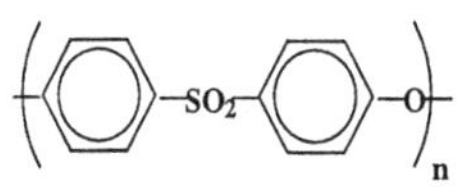 | 2.8 | 0.100 | 17 |
| 8. | 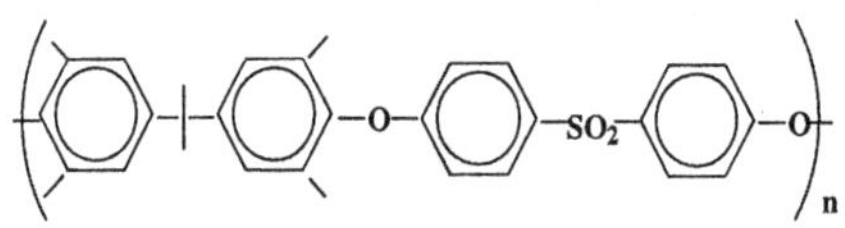 | 21.0 | 0.955 | 18 |
| 9. | 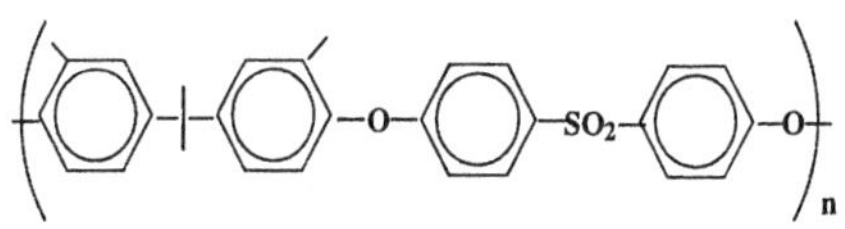 | 2.1 | 0.070 | 18 |
| 10. | 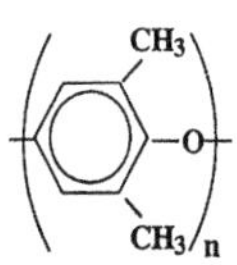 | 65.5 | 4.1 | 19 |

*Continued on next page.*

## Appendix A. Permeability Data (Cont'd)

| | Polymer Structure | $P(CO_2)$ Barrer | $P(CH_4)$ Barrer | Ref. |
|---|---|---|---|---|
| 11. | 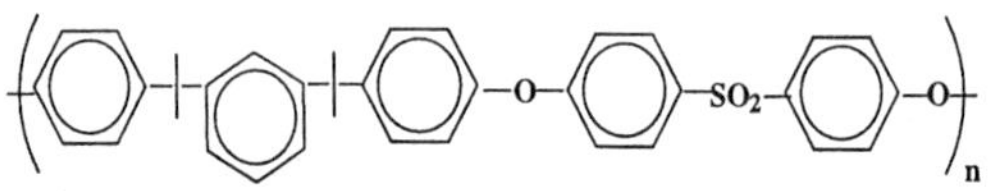 | 2.8 | 0.112 | 18 |
| 12. | 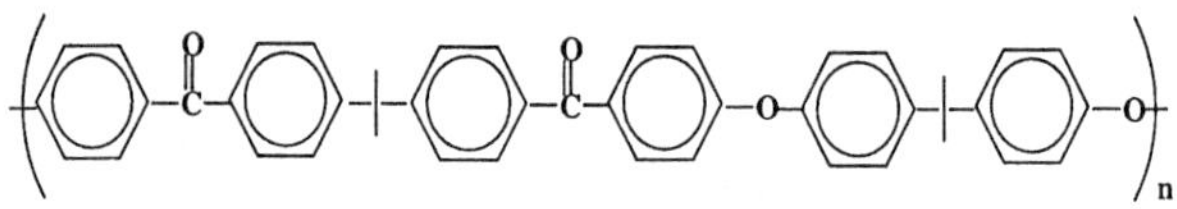 | 4.6 | 0.23 | 1 |
| 13. | 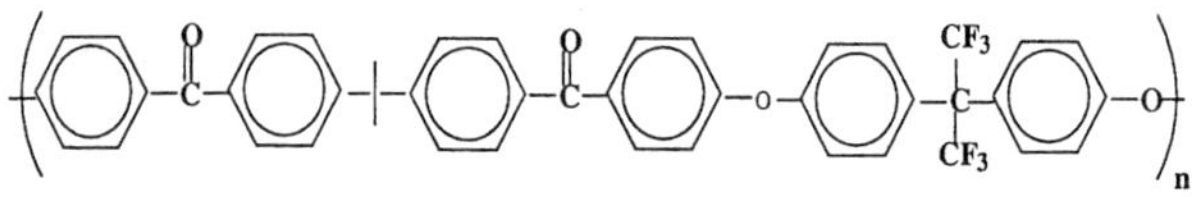 | 8.6 | 0.37 | 1 |
| 14. | 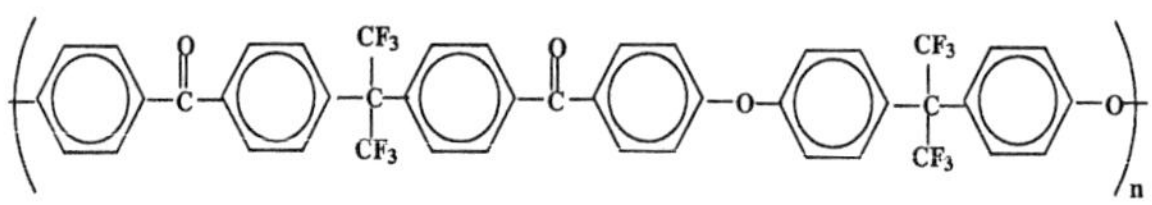 | 12.9 | 0.54 | 1 |
| 15. | 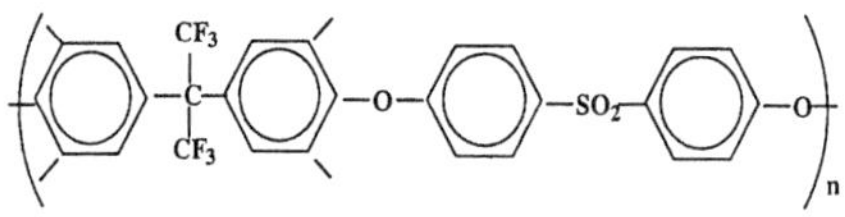 | 72.0 | 3.0 | 2 |
| 16. | 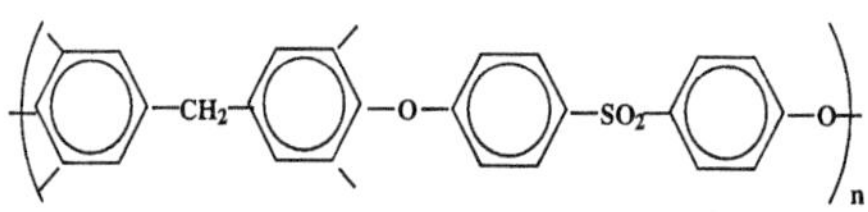 | 15.0 | 0.54 | 2 |
| 17. | 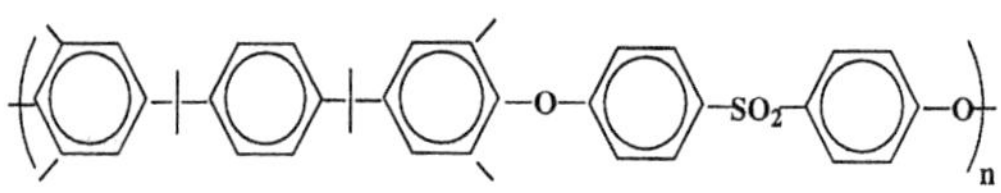 | 7.0 | 0.280 | 18 |
| 18. | 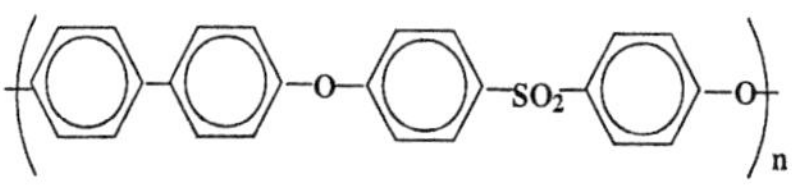 | 5.6 | 0.255 | 20 |
| 19. | 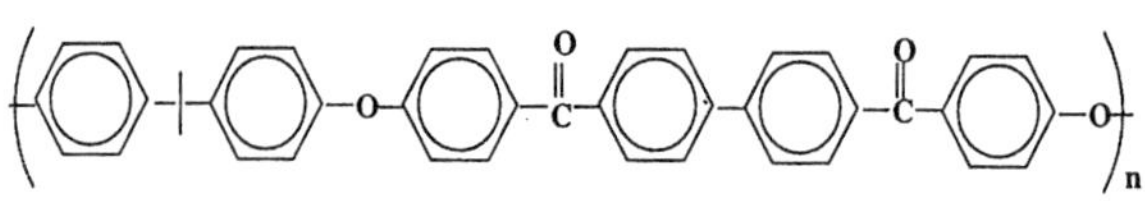 | 3.3 | 0.14 | 20 |
| 20. | 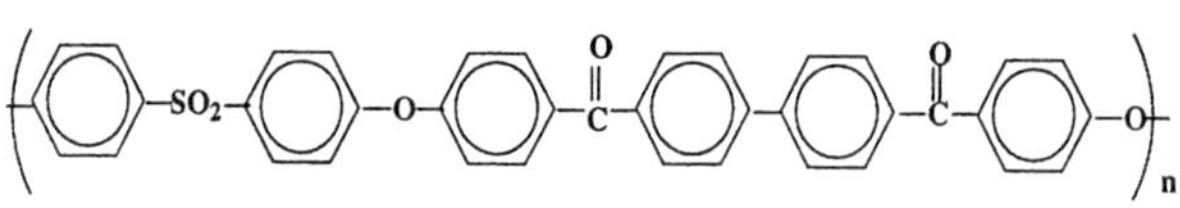 | 3.27 | 0.12 | 10 |

## Appendix A. Permeability Data (Cont'd)

| | Polymer Structure | $P(CO_2)$ Barrer | $P(CH_4)$ Barrer | Ref. |
|---|---|---|---|---|
| 21. | 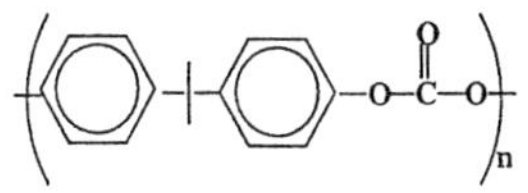 | 6.8 | 0.358 | 21 |
| 22. | 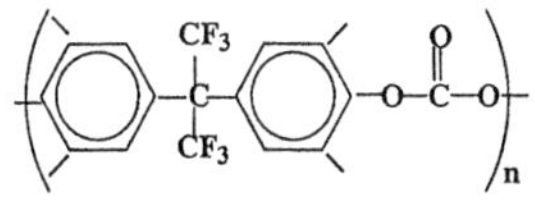 | | | |
| 23. | 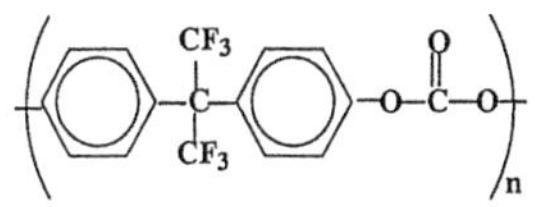 | 110.0 | 4.58 | 23 |
| 24. | 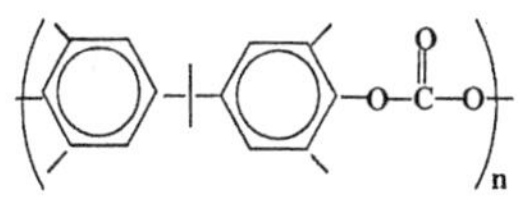 | 17.6 | 0.800 | 24 |
| 25. | 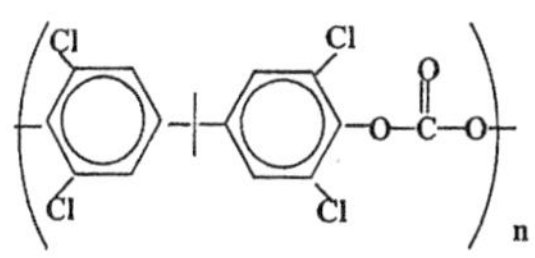 | 6.66 | 0.224 | 24 |
| 26. | 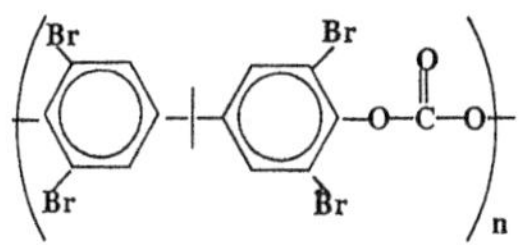 | 4.23 | 0.126 | 24 |
| 27. | 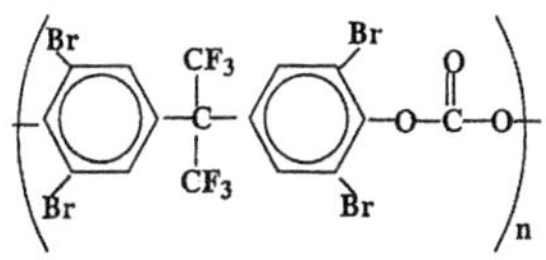 | 32.0 | 0.893 | 21 |
| 28. | 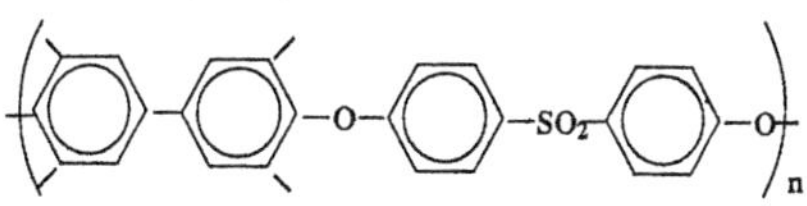 | 31.8 | 1.27 | 20 |
| 29. | 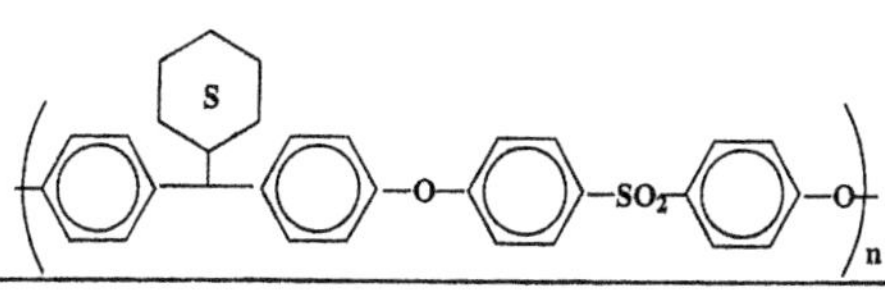 | 2.54 | 0.102 | 18 |

*Continued on next page.*

## Appendix A. Permeability Data (Cont'd)

| | Polymer Structure | $P(CO_2)$ Barrer | $P(CH_4)$ Barrer | Ref. |
|---|---|---|---|---|
| 30. | 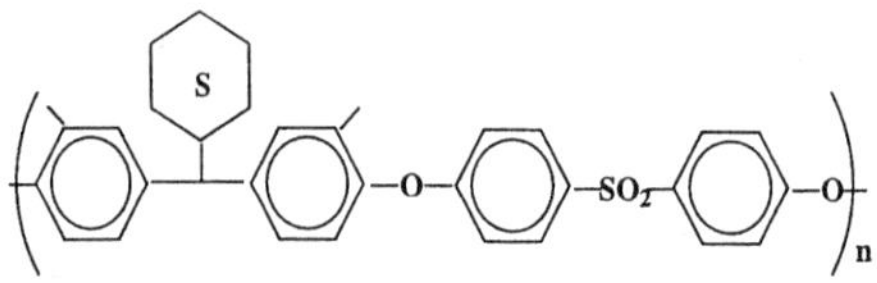 | 1.4 | 0.042 | 18 |
| 31. | 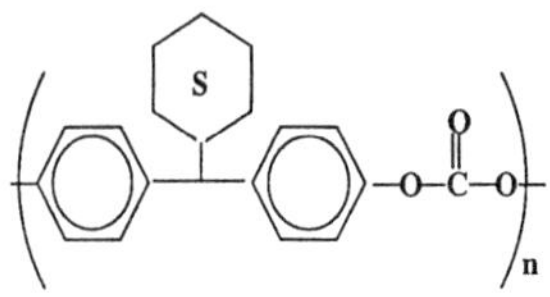 | 2.2 | 0.126 | 22 |
| 32. | 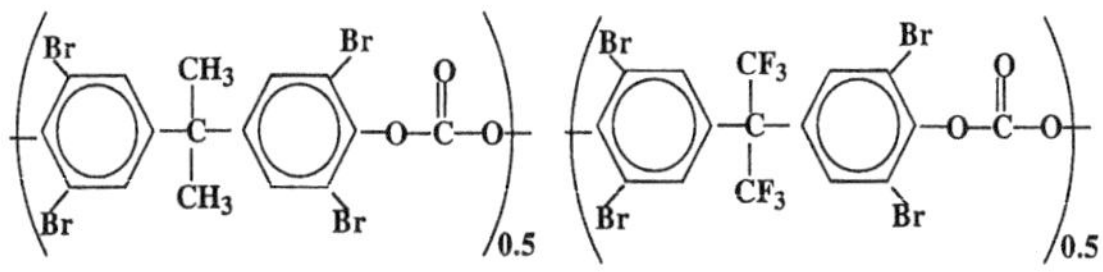 | 16.0 | 0.445 | 21 |
| 33. | 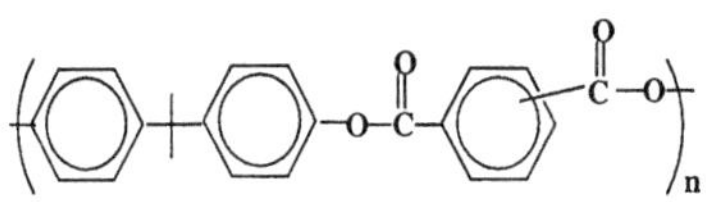 50/50 iso/tere | 10.2 | 0.465 | 25 |
| 34. | 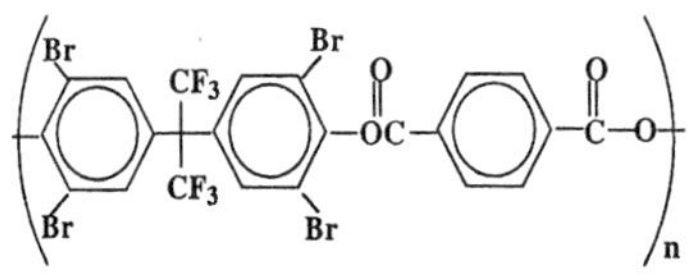 | 19.9 | 0.429 | 26 |
| 35. | 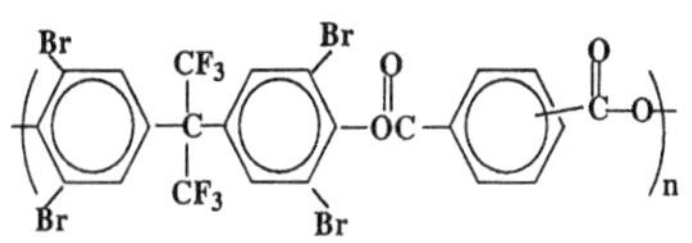 80/20 iso/tere | 24.0 | 0.5 | 26 |
| 36. | 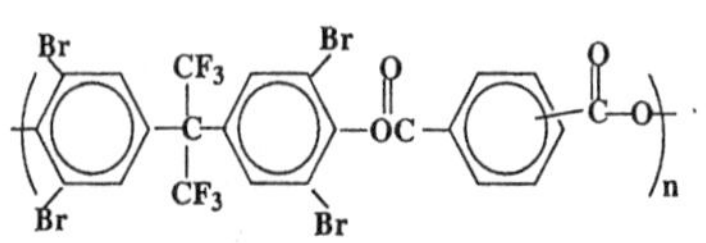 25/75 iso/tere | 42.0 | 0.959 | 26 |

## Appendix A. Permeability Data (Cont'd)

| | Polymer Structure | $P(CO_2)$ Barrer | $P(CH_4)$ Barrer | Ref. |
|---|---|---|---|---|
| 37. | 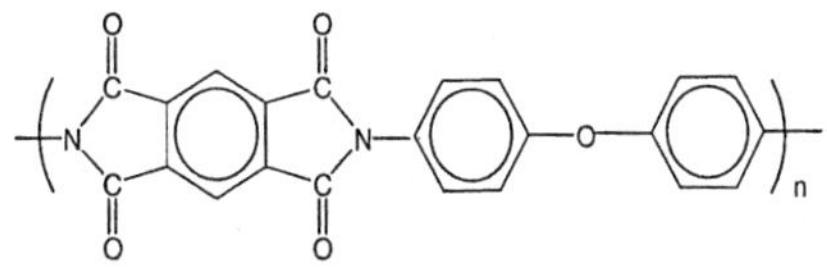 | 7.05 | 1.214 | 27 |
| 38. | 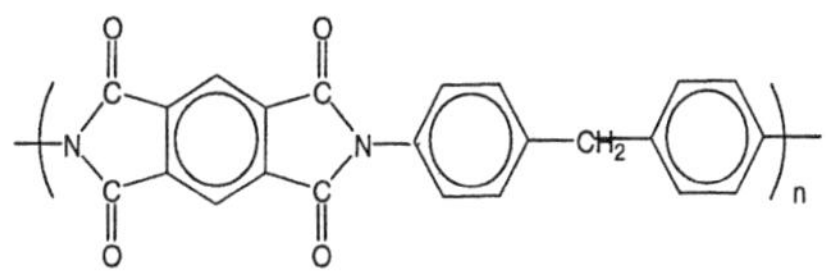 | 4.03 | 0.094 | 28 |
| 39. | 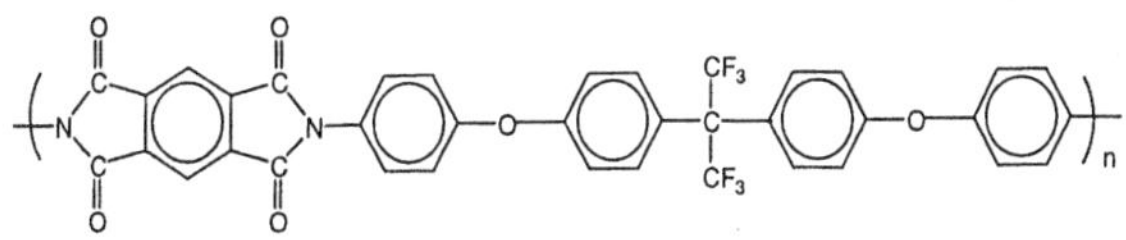 | 17.6 | 0.638 | 29 |
| 40. | 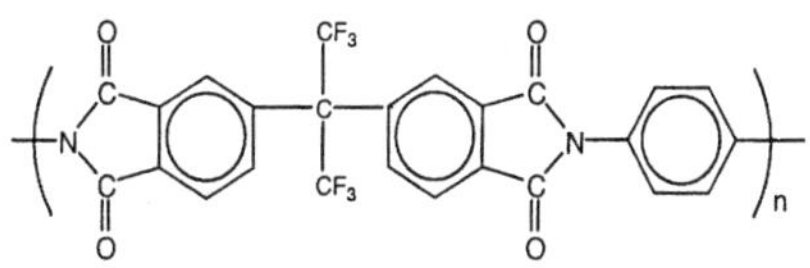 | 15.3 | 0.286 | 30 |
| 41. | 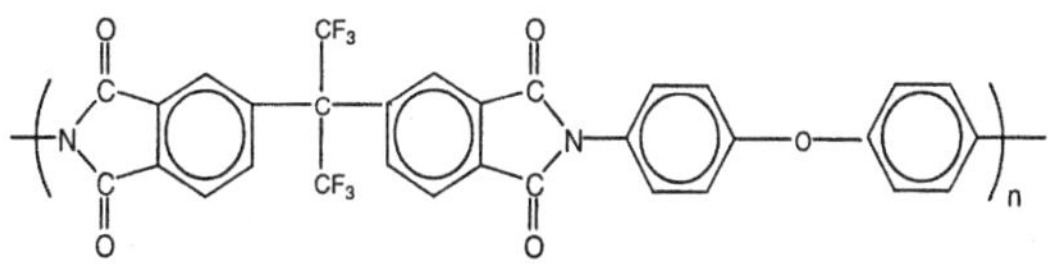 | 16.7 | 0.34 | 31 |
| 42. | 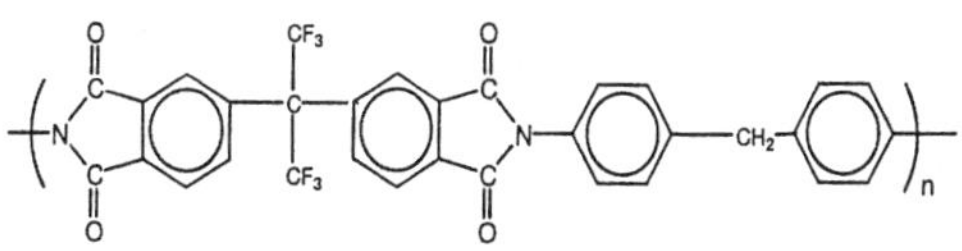 | 19.3 | 0.430 | 21 |
| 43. | 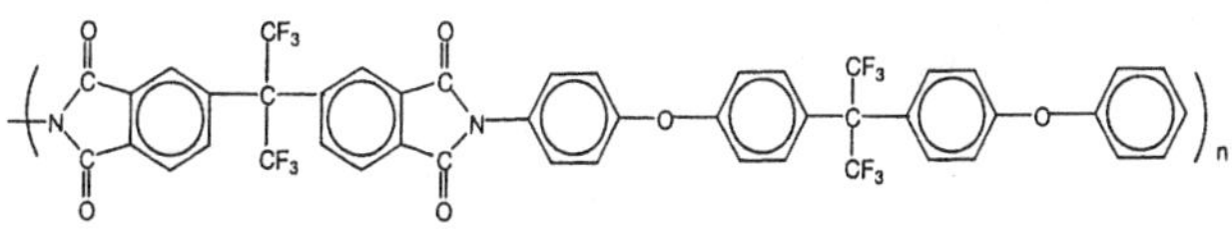 | 19.1 | 0.520 | 29 |
| 44. | 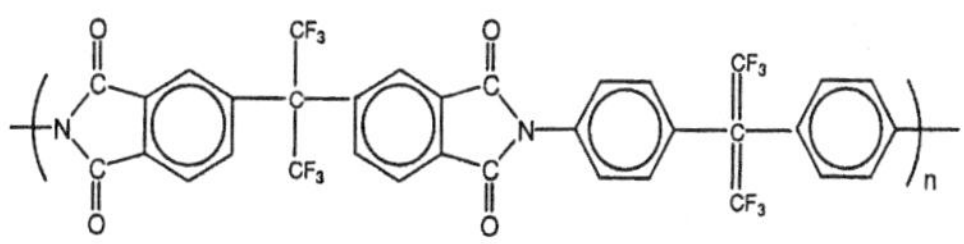 | 63.9 | 1.602 | |

*Continued on next page.*

**Appendix A. Permeability Data (Cont'd)**

| | Polymer Structure | $P(CO_2)$ Barrer | $P(CH_4)$ Barrer | Ref. |
|---|---|---|---|---|
| 45. | 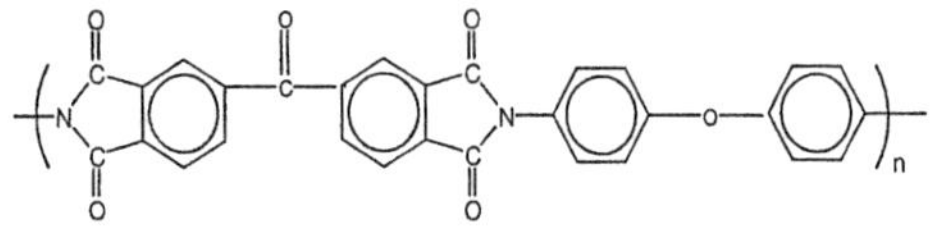 | 0.625 | 0.109 | 29 |
| 46. | 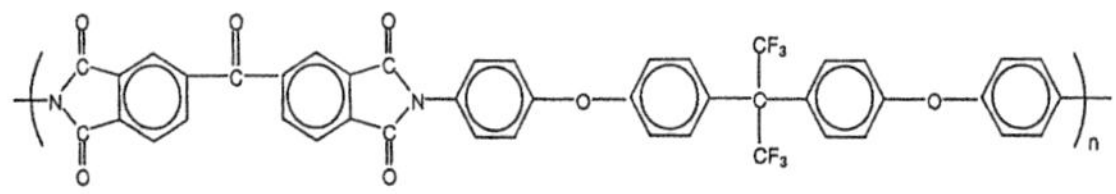 | 4.37 | 0.105 | 29 |
| 47. | 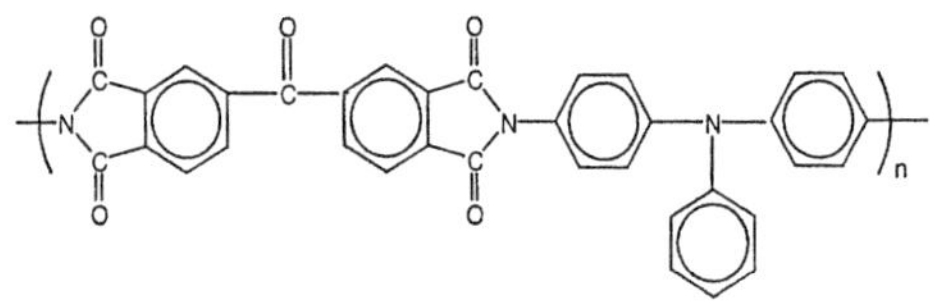 | 3.3 | 0.093 | 30 |
| 48. | 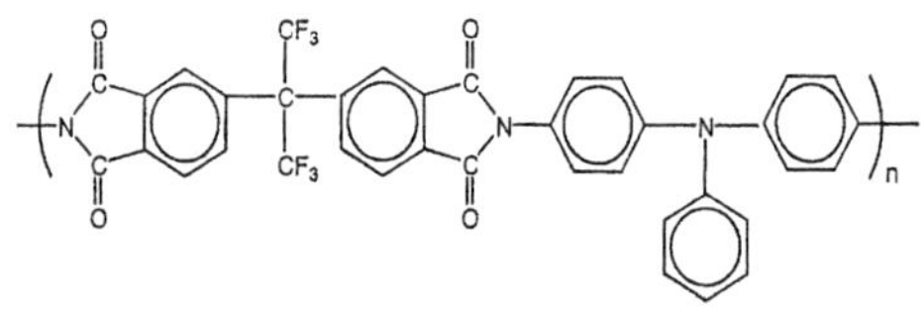 | 23.0 | 0.68 | 30 |
| 49. | 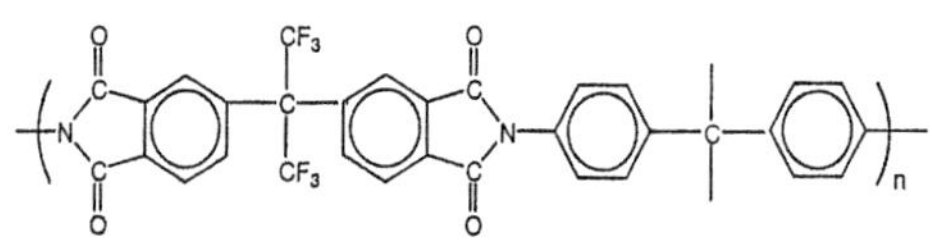 | 30.0 | 0.70 | 31 |
| 50. | 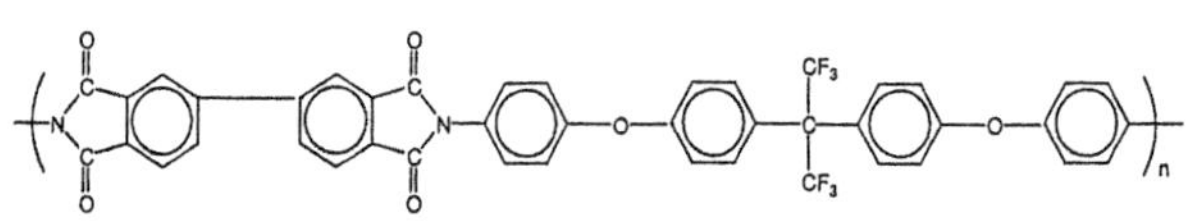 | 4.96 | 0.145 | 29 |
| 51. | 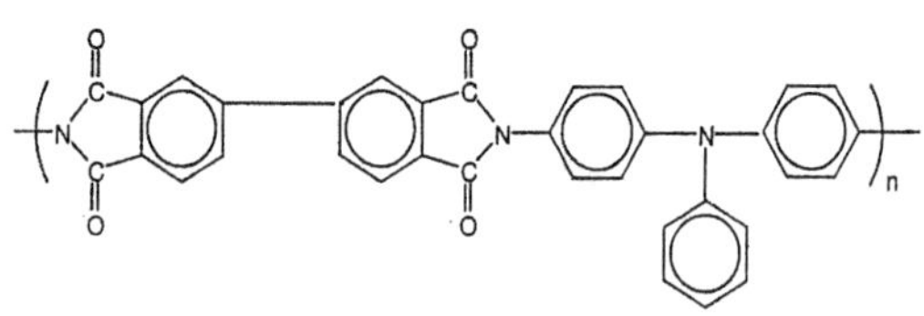 | 4.7 | 0.165 | 30 |
| 52. | 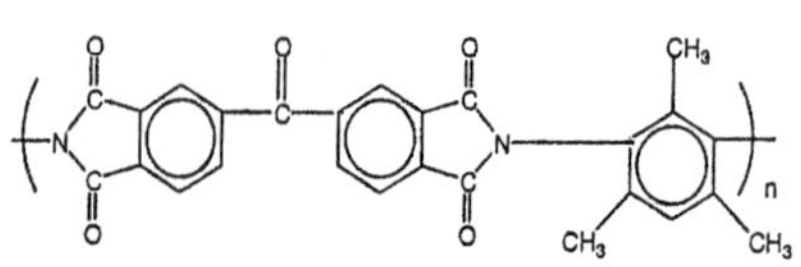 | 30.9 | 1.25 | 32 |

**Appendix A. Permeability Data (Cont'd)**

| | Polymer Structure | $P(CO_2)$ Barrer | $P(CH_4)$ Barrer | Ref. |
|---|---|---|---|---|
| 53. | 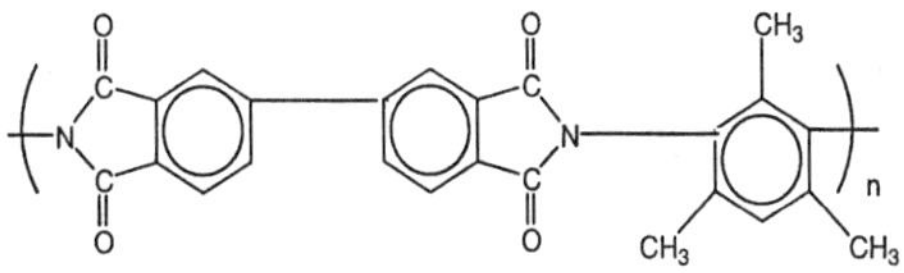 | 137 | 8.08 | 32 |
| 54. | 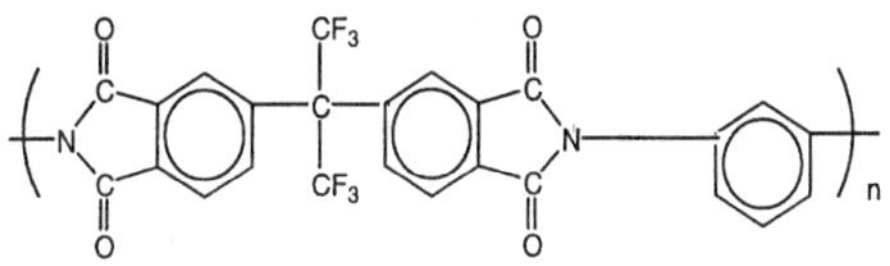 | 9.2 | 0.160 | 32 |
| 55. | 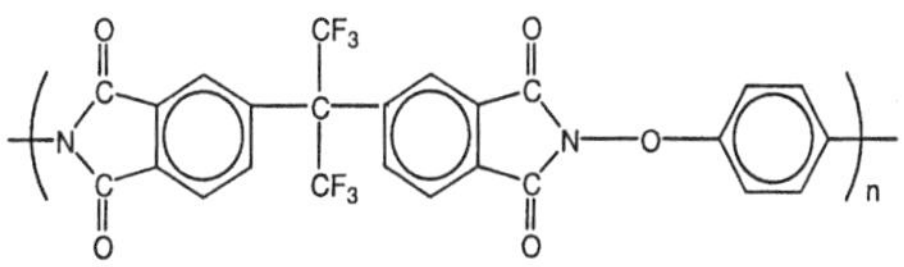 | 6.11 | 0.125 | 29 |
| 56. | 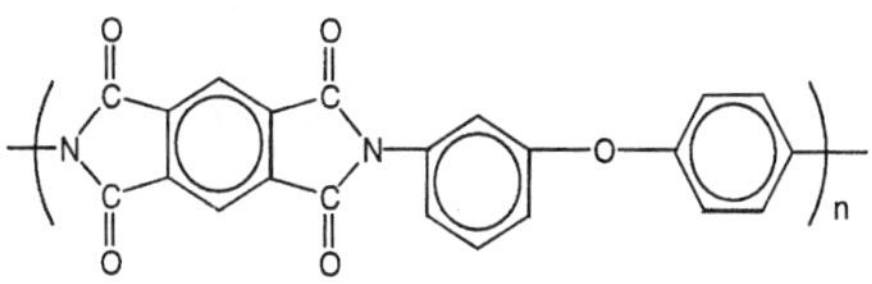 | 1.18 | 0.023 | 29 |
| 57. | 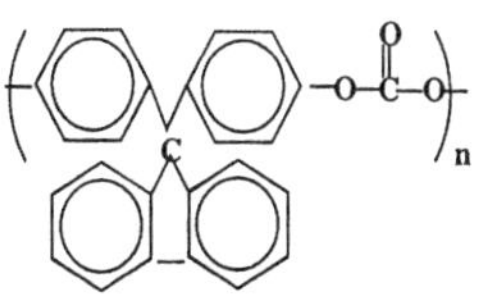 | 15.1 | 0.581 | 33 |
| 58. | 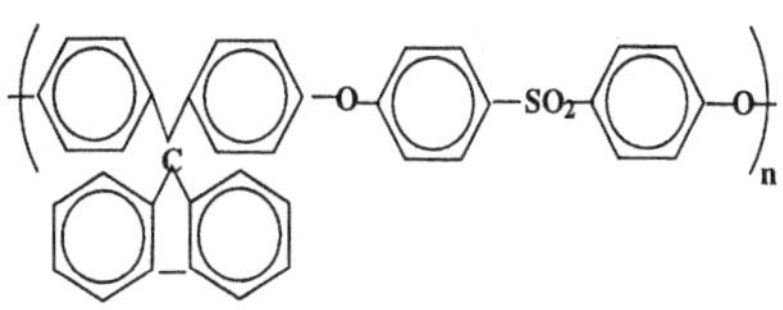 | 13.8 | 0.533 | 33 |
| 59. | 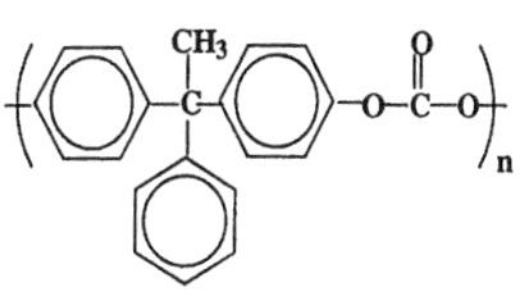 | 9.48 | 0.419 | 33 |

*Continued on next page.*

## Appendix A. Permeability Data (Cont'd)

| | Polymer Structure | $P(CO_2)$ Barrer | $P(CH_4)$ Barrer | Ref. |
|---|---|---|---|---|
| 60. | 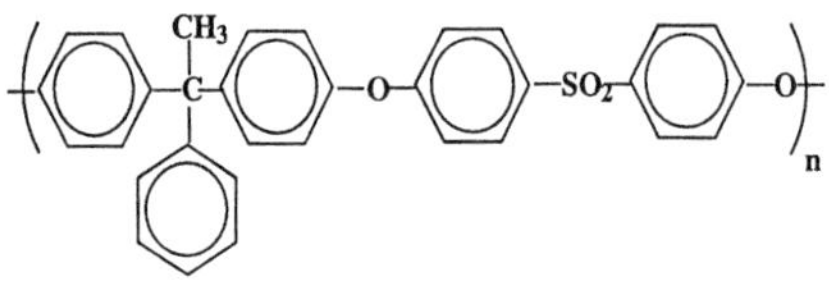 | 8.12 | 0.318 | 33 |
| 61. | 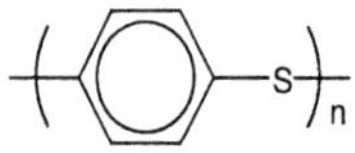 | 1.60 | 0.066 | 40 |
| 62. | 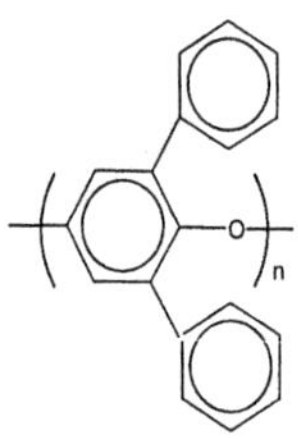 | 39.9 | 2.7 | 40 |
| 63. | 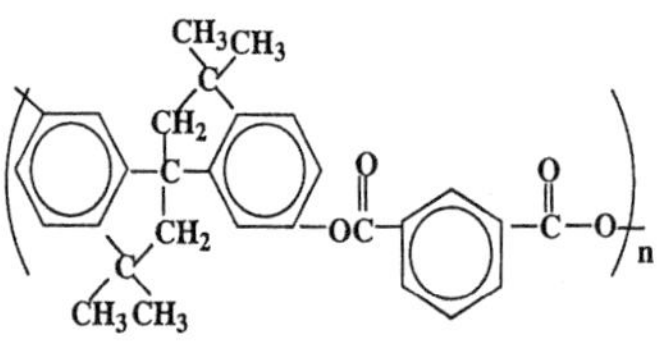 | 109.7 | 6.57 | 36 |
| 64. | 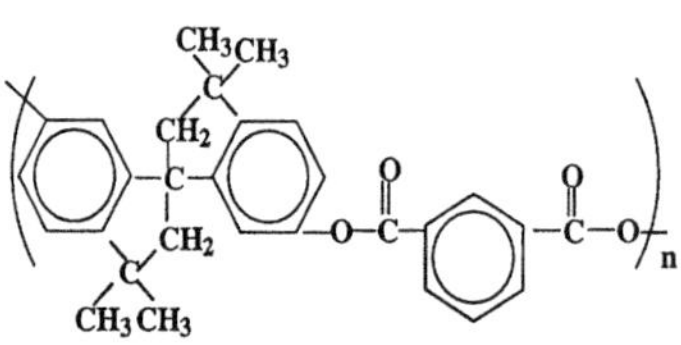 | 18.0 | 1.139 | 36 |
| 65. | 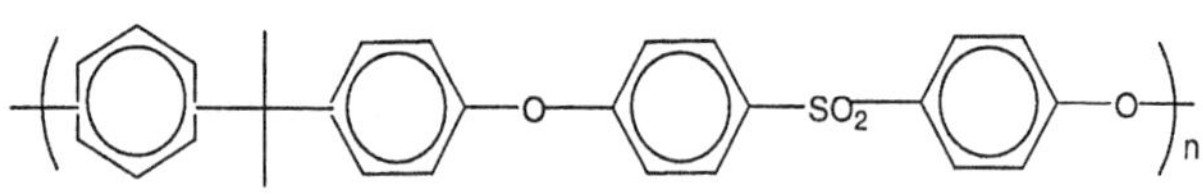 | 1.5 | 0.052 | 18 |
| 66. | 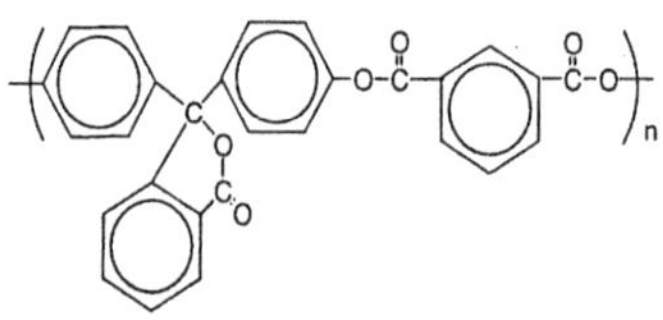 | 6.7 | 0.168 | 37 |

**Appendix A. Permeability Data (Cont'd)**

| | Polymer Structure | $P(CO_2)$ Barrer | $P(CH_4)$ Barrer | Ref. |
|---|---|---|---|---|
| 67. | 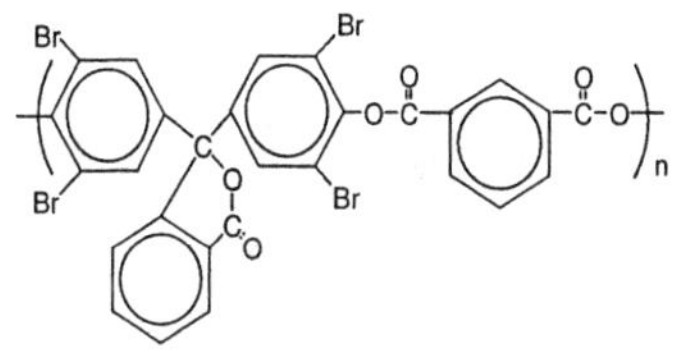 | 9.7 | 0.220 | 38 |
| 68. | 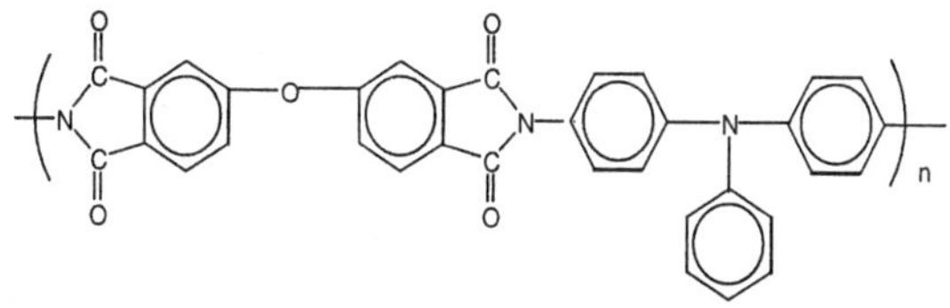 | 2.88 | 0.085 | 30 |
| 69. | 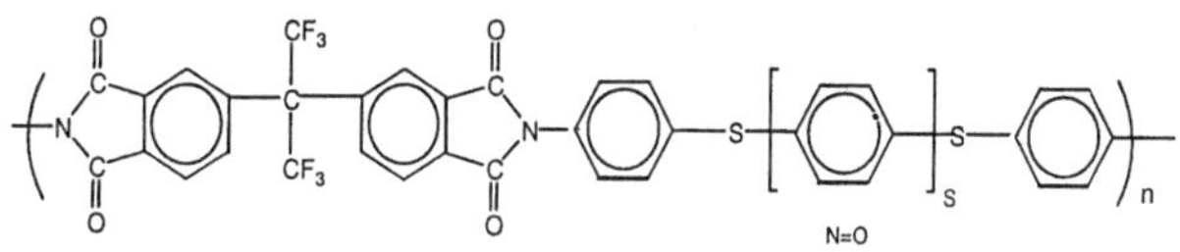 | 17.3 | 0.51 | 34 |
| 70. | 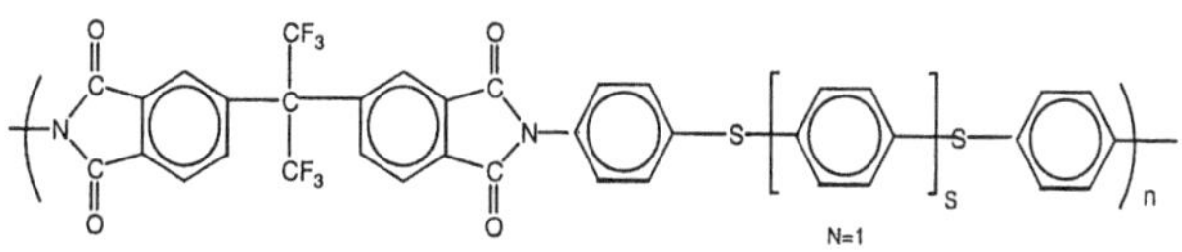 | 9.8 | 0.29 | 34 |
| 71. | 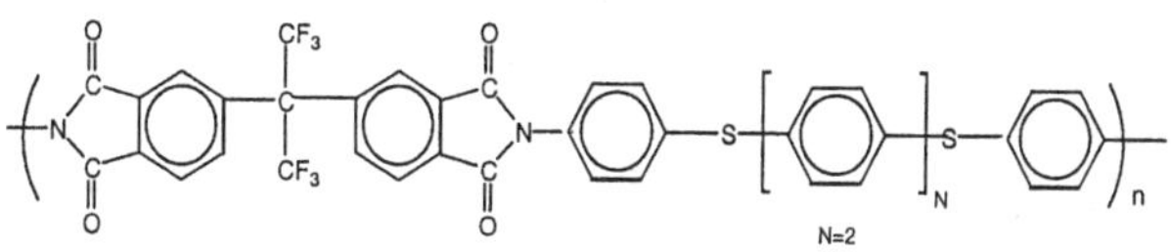 | 8.5 | 0.26 | 34 |
| 72. | 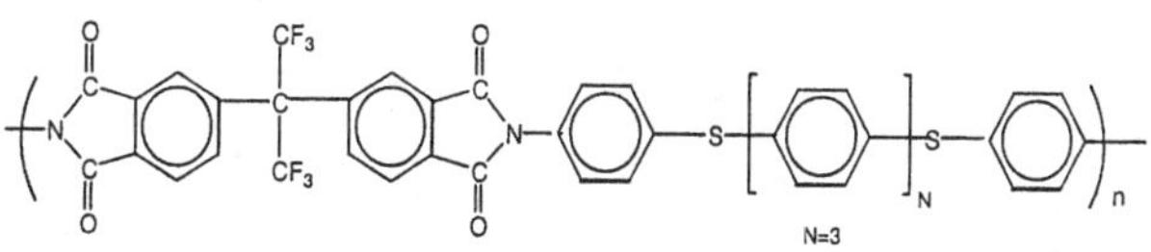 | 8.0 | 0.25 | 34 |
| 73. | 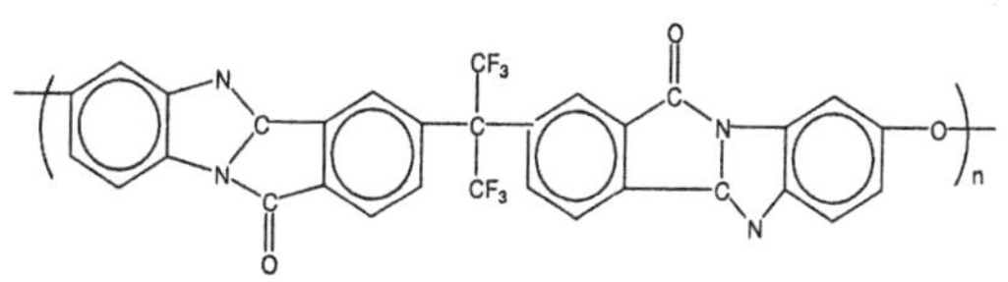 | 27.6 | 0.540 | 10 |

*Continued on next page.*

**Appendix A. Permeability Data (Cont'd)**

| | Polymer Structure | $P(CO_2)$ Barrer | $P(CH_4)$ Barrer | Ref. |
|---|---|---|---|---|
| 74. | 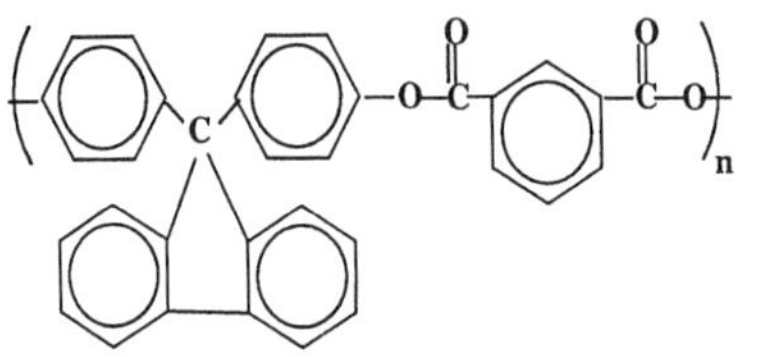 | 10.3 | 0.616 | 37 |
| 75. | 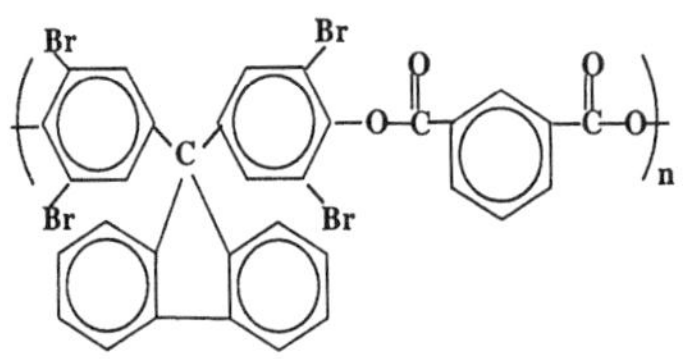 | 20.4 | 0.567 | 38 |
| 76. | 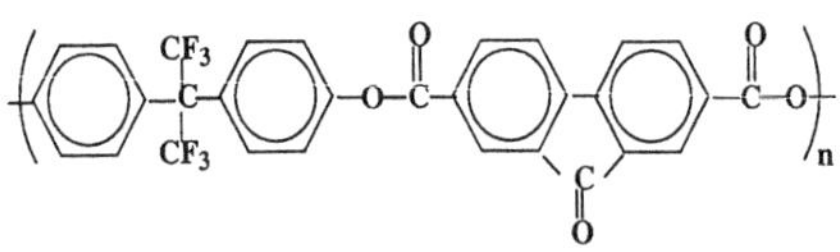 | 6.0 | 0.208 | 39 |
| 77. | 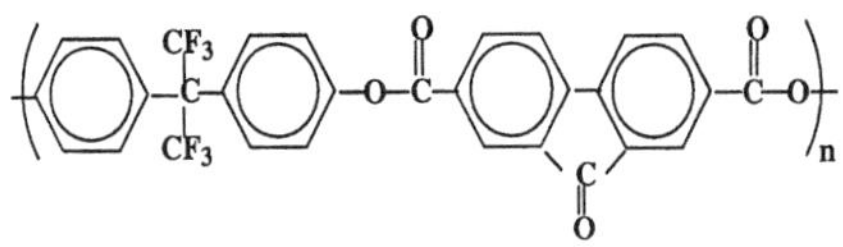 | 29.9 | 0.980 | 39 |
| 78. | 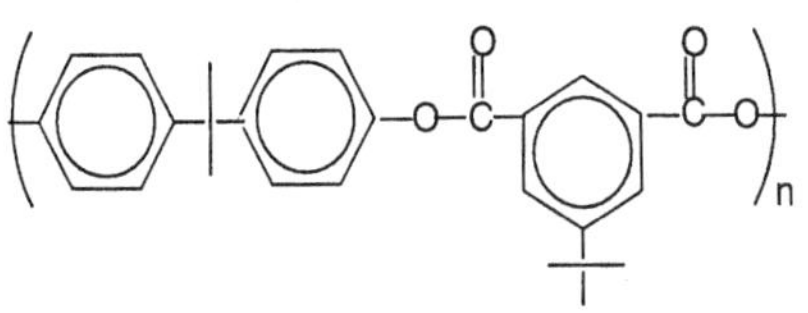 | 24.2 | 1.432 | 37 |
| 79. | 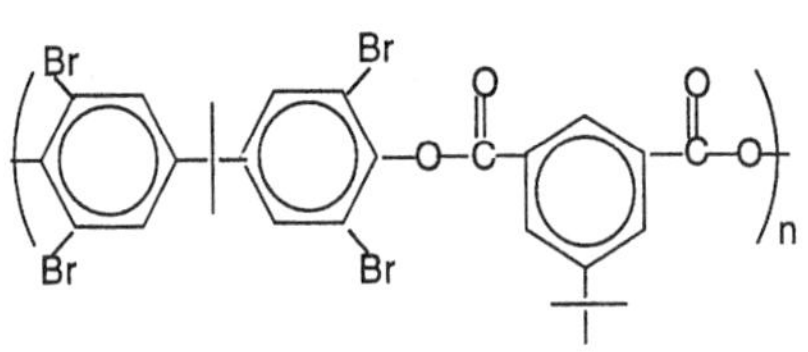 | 21.5 | 0.853 | 38 |
| 80. | 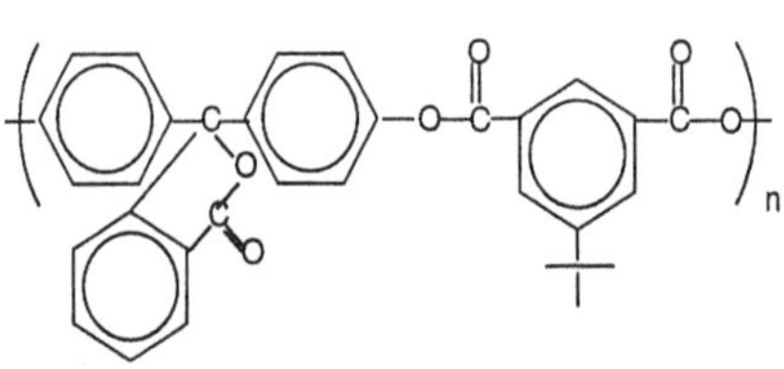 | 23.8 | 1.16 | 37 |

**Appendix A. Permeability Data (Cont'd)**

| | Polymer Structure | $P(CO_2)$ Barrer | $P(CH_4)$ Barrer | Ref. |
|---|---|---|---|---|
| 81. | 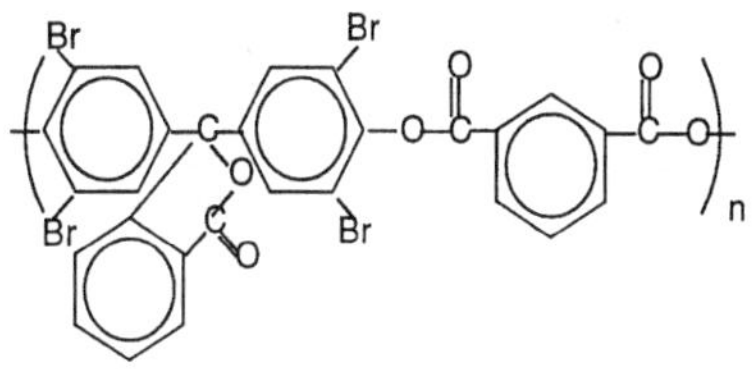 | 30.6 | 1.09 | 17 |
| 82. | 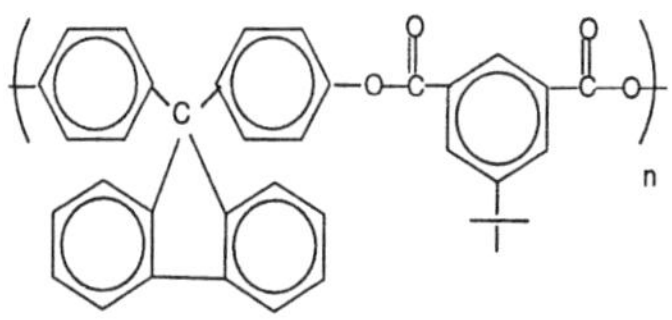 | 36.8 | 2.37 | 36 |
| 83. | 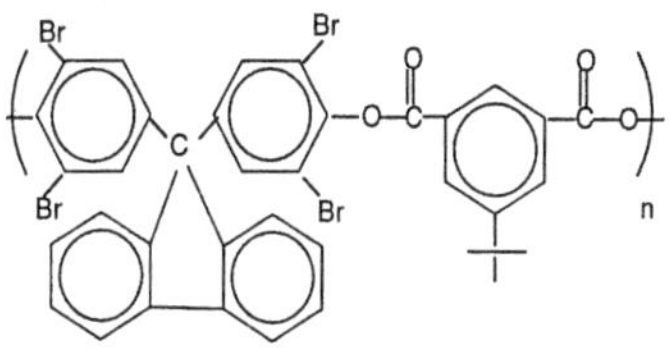 | 69.5 | 2.78 | 17 |
| 84. | 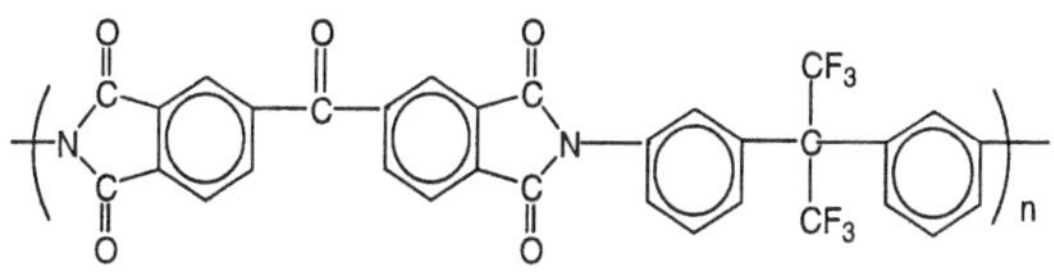 | 1.05 | 0.014 | 35 |
| 85. | 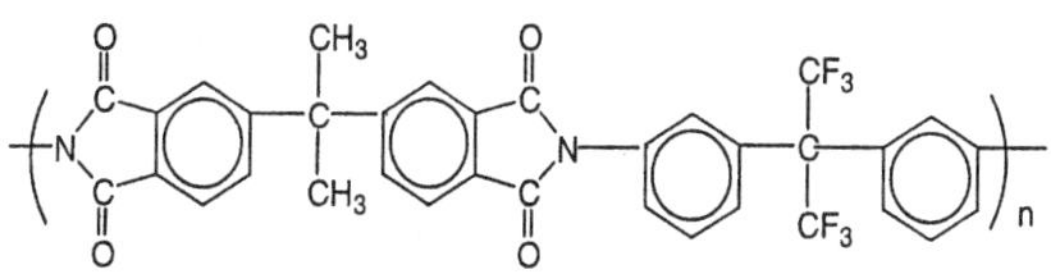 | 5.1 | 0.08 | 35 |
| 86. | 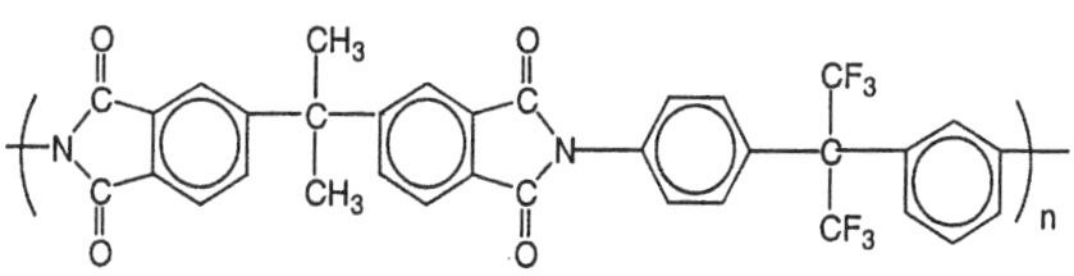 | 18.6 | 0.345 | 35 |
| 87. | 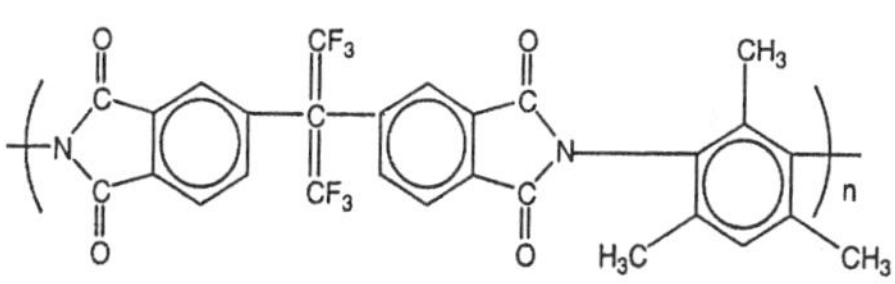 | 4.31 | 26.0 | 32 |

Chapter 13

# Solubility of Gaseous Mixtures in Glassy Polymers: NELF Predictions

**Ferruccio Doghieri, Michele Canova, and Giulio C. Sarti**

**Dipartimento di Ingegneria Chimica, Mineraria e delle Tecnologie Ambientali, Università di Bologna, Viale Risorgimento 2, 40136 Bologna, Italy**

A thermodynamic model for the chemical potential of low molecular weight species in glassy polymers has been presented recently and has proven to be a reliable tool for the prediction of pure component solubility in amorphous polymers below the glass transition temperature (*1*). The model allows evaluation of the solubility coefficient through a purely predictive procedure when the volumetric data for pure penetrant and polymeric species are known. In this work the model is extended to polymeric mixtures with several low molecular weight solutes, and a procedure for the prediction of solubility of gaseous mixtures in glassy polymers is proposed. A comparison between the model predictions and experimental data for the solubility of $CO_2/C_2H_4$ mixtures in PMMA is presented, and the results confirm the reliability of the model.

Evaluating the solubility of low molecular weight species in glassy polymers is essential in applications such as packaging or membrane separations, and reliable experimental procedures have been established to measure the solubility of pure gases or vapours in a polymeric material. On the other hand, the experimental evaluation of the solubility of the components of a gaseous mixture in solid polymers requires much more complicated and time consuming procedures, and few examples of such data, indeed, are known in the technical literature. In that sense, it is of great interest to develop mathematical models which permit the calculation of the solubility of low molecular weight species in polymeric materials on the basis of the pure components properties and a limited number of binary data. Additionally, a model which provides a purely predictive procedure for the evaluation of solubility would allow a priori estimates of the performance of new polymers designed as barrier materials or separation membranes.

In fact, several thermodynamic models have been developed for evaluating the solubility of gases and gaseous mixtures in rubbery polymers. These models are based on the activity coefficient approach or proper equations of state (*2*, *3*). On the other hand, the correlation of solubility data in glassy polymers is still confined to empirical models which do not allow for the prediction of the data. Indeed, the solute content in a glassy material depends not only on temperature and pressure, as is the case for equilibrium rubbery amorphous polymers, but it also changes according to the history of the solid sample. A thermodynamic description of the properties of glassy mixtures must, therefore, take into account the non-equilibrium state of the system.

A thermodynamic model was recently proposed to calculate the solubility of small molecules in glassy polymers. This model is based on the assumption that the density of the polymer matrix can be considered as a proper order parameter for the non-equilibrium state of the system (*1*). In this chapter, the fundamental principles of the model are reviewed and the relation of the model to the rheological properties of the polymeric matrix is developed. In particular, a unique relation between the equilibrium and non-equilibrium properties of the polymer-penetrant mixture can be obtained on the basis of a simple model for the stress-strain relationship.

With respect to our previous work, the model is extended to the case of multicomponent mixtures and a procedure is described for the prediction of gas mixture solubility in glassy polymers based only on pure component equilibrium properties and on the knowledge of the pure polymer density in the glassy state. The reliability of the model is then tested by comparing model predictions with experimental data for the solubility of $CO_2/C_2H_4$ mixtures in PMMA (*4*).

## Generalities and Rheological Model

It is well known that the mechanical deformation of a solid polymer affects its thermodynamic properties and ultimately influences gas and vapour solubility. Indeed, at the same temperature, sorption driven by a fixed activity jump in a solid polymer sample follows different sorption kinetics and results in different equilibrium penetrant contents depending on the mechanical constraints applied to the sample.

Several models have been proposed to describe the effect of deformation on the thermodynamic properties of elastic solids, both for the general case (*5*, *6*, *7*) and specifically for solid polymers (*8*, *9*). The case of crosslinked rubbery polymers is of special interest since they typically support large deformations and the elastic energy is significantly affected by the presence of a penetrant species. Expressions for the free energy of such elastic polymeric mixtures were obtained as a function of temperature, composition and principal elongations (*10*). These models permit a reliable estimate of the effect of non-isotropic deformations on the chemical potential of the solvent and polymeric species. Conversely, the thermodynamic properties of viscoelastic materials depend on the history of temperature, composition and deformation. However, for this case, reliable expressions for the free energy are not available, to the best of our knowledge.

Expressions for the free energy as a function of the state variables have been obtained for rubbery polymers on the basis of thermodynamic arguments for equilibrium systems. Nonetheless, a different approach must be used for the analysis of the thermodynamic properties of solid polymers below the glass transition temperature because non-equilibrium states need to be considered for such glassy materials.

In this article, simple thermodynamic arguments are used to derive expressions for the free energy of glassy polymeric mixtures. The case of an isotropic material will be considered experiencing only isotropic deformations. In this case, all the thermodynamic properties in a general non equilibrium state are determined by assigning the values of temperature $T$, volume $V$, composition $Y_i$ and pressure $p$. The variables $T$, $V$, $Y_i$ and $p$ can be chosen independently of each other under non-equilibrium conditions; a viscoelastic rheological equation holds true for the evolution in time of pressure (or volume). Vice versa, in an equilibrium state the value of pressure is uniquely determined by the known values of $T$, $V$ and $Y_i$. In particular, under non-equilibrium conditions, the Helmholtz free energy $A$ is given by a function, $a^{NE}$, of the current value of temperature $T$, pressure $p$, volume $V$ and composition:

$$A^p = a^{NE}\left(T, p, Y_i, V^p\right) \tag{1}$$

In the above equation the superscript $p$ labels the property value per unit mass of polymer, while $Y_i$ is the number of moles of low molecular weight penetrant in the mixture per unit mass of polymer. As already said, the free energy of the system is assumed to depend only on the present value of temperature, pressure and volume at given mixture composition. No further dependence on the past histories of the independent variables is included beyond that required by the rheological constitutive equations.

Equation 1 for the non-equilibrium Helmholtz free energy may be compared with the corresponding equation for the same system under equilibrium conditions, when the free energy has a unique equilibrium value at any given temperature, volume and composition:

$$A^p = a^{EQ}\left(T, Y_i, V^p\right) \tag{2}$$

In this work we assume that the rate of variation of the specific volume depends only on temperature, pressure, volume and composition. In particular, the time derivatives of temperature, pressure or composition do not influence the time rate of change of the specific volume $V^p$.

$$\frac{dV^p}{dt} = f\left(T, p, Y_i, V^p\right) \tag{3}$$

The simplest rheological model which describes stress-strain relationships in the form of equation 3 is Voigt's model, which refers to the case of an elastic and a viscous element arranged in parallel.

We now consider the non-equilibrium states of the system which result in pseudo-equilibrium conditions. For the case of glassy polymers, which are characterised by extremely high viscosity, the function $f$ in equation 3 may have negligibly small values even when the volume $V^p$ is far from its equilibrium value at T, p, and $Y_i$. In this situation, the swelling of the polymer is inhibited for kinetic reasons, and a pseudo-equilibrium condition is reached for the specific volume $V^p$:

$$f\left(T,p,Y_i,V^p\right) \approx 0 \qquad \textit{at pseudo-equilibrium for } V^p \qquad (4)$$

## Thermodynamic analysis

A specific thermodynamic analysis is needed to derive expressions for the non-equilibrium properties of polymeric mixtures below the glass transition temperature. On the basis of the viscoleastic stress-strain relationship which has been considered in equation 3, specific relations may be obtained between equilibrium and non-equilibrium properties of the mixture. Indeed, when a mixture of one polymeric species and $N_p$ penetrant components is considered, the following balance equations for mass, energy and entropy may be written in local form, with respect to a polymer-fixed frame (*11*).

$$\frac{dY_i}{dt} + \nabla\cdot\left(V^p\,\mathbf{n}_i\right) = 0 \qquad (5)$$

$$\frac{dU^p}{dt} + \nabla\cdot\mathbf{q} + p\frac{dV^p}{dt} + \sum_{i=1}^{i=Np}\nabla\cdot\left(\bar{H}_i\,V^p\,\mathbf{n}_i\right) = 0 \qquad (6)$$

$$\frac{dS^p}{dt} + \nabla\cdot\left(\frac{\mathbf{q}}{T}\right) + \sum_{i=1}^{i=Np}\nabla\cdot\left(\bar{S}_i\,V^p\,\mathbf{n}_i\right) \geq 0 \qquad (7)$$

where $U^p$ and $S^p$ are the internal energy and entropy of the mixture per unit mass of polymer, respectively, $\frac{d}{dt}$ represents the material time derivative with respect to the polymer component. The scalars $\bar{H}_i$ and $\bar{S}_i$, as yet unspecified, must be determined by the thermodynamic analysis. $\mathbf{n}_i$ represents the molar flux density of component $i$ relative to the polymeric species, and $\mathbf{q}$ is the heat flux density.

Equations 5-7 together with equation 3 hold true for any process, that is for any value of the time derivative of those state variables which can be arbitrarily changed. In this regard, the time derivatives of temperature, pressure and composition must be considered arbitrary quantities as long as heat and mass fluxes may assume any value. The latter assumption is consistent with the idea that energy, momentum and mass flux directly depend on the gradients of the state variables. On the contrary, the time derivative of the volume $V^p$ is not an arbitrary quantity in view of equation 3.

For the sake of simplicity, we confine our attention to isothermal processes in which the local time derivative and spatial gradient of temperature are zero. In this case the entropy balance, equation 7, can be written in the following form:

$$\frac{d\left(TS^p\right)}{dt} + \nabla\cdot\mathbf{q} + \sum_{i=1}^{i=Np}\nabla\cdot\left(T\bar{S}_i\,V^p\,\mathbf{n}_i\right) \geq 0 \qquad (8)$$

Combining equations 5 through 8, we obtain

$$\frac{d\left(U^P - TS^P\right)}{dt} + p\frac{dV^P}{dt} + \sum_{i=1}^{i=Np}\left[\left(\bar{H}_i - T\bar{S}_i\right)\underline{\nabla}\cdot\left(V^P\,\mathbf{n}_i\right)\right] + \sum_{i=1}^{i=Np}\left[V^P\,\mathbf{n}_i\cdot\underline{\nabla}\left(\bar{H}_i - T\bar{S}_i\right)\right] \le 0 \tag{9}$$

When the term $\bar{H}_i - T\bar{S}_i$ is written as $\mu_i$ and the mass balance in equation 6 is used, one can write the inequality in equation 9 in the following form:

$$\frac{dA^P}{dt} + p\frac{dV^P}{dt} + \sum_{i=1}^{i=Np}\mu_i\frac{dY_i}{dt} + \sum_{i=1}^{i=Np}V^P\,\mathbf{n}_i\cdot\underline{\nabla}\mu_i \le 0 \tag{10}$$

By explicitly considering the dependence of $A^P$ as indicated in equation 1, the above balance may be written as

$$\frac{\partial a^{NE}}{\partial p}\frac{dp}{dt} + \left(\frac{\partial a^{NE}}{\partial V^P} + p\right)\frac{dV^P}{dt} + \sum_{i=1}^{i=Np}\left[\left(\frac{\partial a^{NE}}{\partial Y_i} - \mu_i\right)\frac{dY_i}{dt}\right] + \sum_{i=1}^{i=Np}\left(V^P\,\mathbf{n}_i\cdot\underline{\nabla}\mu_i\right) \le 0 \tag{11}$$

Since equation 11 must hold for any value of the time derivatives $dp/dt$ and $dY_i/dt$, we conclude that

$$\mu_i = \frac{\partial a^{NE}}{\partial Y_i} = \left.\frac{\partial A^P}{\partial Y_i}\right|_{T,p,V^P,Yj\neq i} \tag{12}$$

$$\frac{\partial a^{NE}}{\partial p} = \left.\frac{\partial A^P}{\partial p}\right|_{T,V^P,Y_i} = 0 \tag{13}$$

That is, at any pressure the non-equilibrium Helmholtz free energy per unit mass of polymer is independent of pressure and is allowed to vary only with $T$, $V^P$ and $Y_i$, just as the equilibrium Helmholtz free energy function; since the equilibrium conditions are obtained when p is the equilibrium pressure $p^{EQ}$, based on equation 13 one has that at any pressure $p$

$$a^{NE}\left(T,p,Y_i,V^P\right) = a^{NE}\left(T,p^{EQ},Y_i,V^P\right) = a^{EQ}\left(T,Y_i,V^P\right) \tag{14}$$

Equations 13 and 14 are of crucial importance for the later developments, since they enable us to know explicitly the Helmholtz free energy in any non equilibrium state, once its corresponding equilibrium function is given in terms of $T$, $V^P$ and $Y_i$. Clearly, when the equilibrium properties are given directly in terms of $T$, $V^P$ and $Y_i$, equation 14 is rather straightforward to apply and it may appear trivial. However, its great importance and novelty must not be overlooked, since equation 14 permits the evaluation of all the thermodynamic properties but pressure under any non equilibrium conditions. In any case, its use is much less trivial when the equilibrium properties are expressed in terms of temperature, pressure and composition. Finally we remark that the above result is based on the assumption that the stress strain relationship for the polymer matrix is represented by equation 3. After equation 14 we can easily recognise that, at any pressure $p$, the following relation holds

$$\frac{\partial a^{NE}}{\partial V^P}\left(T,p,Y_i,V^P\right) = \frac{\partial a^{EQ}}{\partial V^P}\left(T,Y_i,V^P\right) \tag{15}$$

As otherwise usual the latter term in equation 15 represents the opposite of the equilibrium pressure $p^{EQ}$ at given values of temperature, specific volume and composition

$$\frac{\partial a^{EQ}}{\partial V^P}\left(T,Y_i,V^P\right) = -p^{EQ}\left(T,Y_i,V^P\right) \tag{16}$$

Based on equations 15 and 16 the residual inequality equation 11 is finally rewritten as

$$\left(p - p^{EQ}\right)\frac{d\,V^P}{d\,t} + \sum_{i=1}^{i=Nc} V^P\, \underline{n}_i \cdot \underline{\nabla}\mu_i \leq 0 \tag{17}$$

and it is demanded to hold for any process.

In particular, when all chemical potential gradients are zero and use is made of equation 3, we find that

$$\left(p - p^{EQ}\right) f \leq 0 \tag{18}$$

The latter inequality was derived for the case of zero chemical potential gradients but, since $f$, $p$ and $a^{NE}$ are insensitive to the gradients of the state variables, it holds true also in general. Inequality 18 represents a constraint for the constitutive equation function $f$ in equation 3, insofar as it requires that $f$ be positive when the actual pressure is lower that the equilibrium value and that $f$ be negative in the opposite case; while both $f$ and the difference $p$ - $p^{EQ}$ are both demanded to be zero in equilibrium conditions.

Conversely, the pseudo equilibrium condition considered in equation 4 just refers to negligible values of the time rate of change of the specific volume and this situation may occur in glassy systems for states in which the actual pressure $p$ is significantly different from the equilibrium pressure $p^{EQ}$ corresponding to the given values of temperature, volume and composition.

It is useful to summarise the assumptions considered for the thermodynamic analysis discussed above, and the main results obtained. Based on the stress strain relationship for the glassy system in the form of equation 3, the chemical potential of a solute component in the polymeric mixture may be calculated as the derivative of the specific non-equilibrium Helmholtz free energy $A^P$ with respect to the moles of solute per polymer mass $Y_i$ at constant temperature, pressure and specific volume, as expressed in eqaution 12. On the other side, under the same assumption, the non-equilibrium Helmholtz free energy has a unique value at given temperature, specific volume and composition, whatever is the pressure of the system, as stated in equation 13.

In this case, equation 14 provides a rational framework to extend existing expressions of the Helmholtz free energy from the equilibrium curve to the entire space of the non-equilibrium states of the system. Any mathematical model for the equilibrium thermodynamic properties of polymeric mixtures could be used to derive an expression for the non-equilibrium Helmholtz free energy according to the procedure described above. On the basis of statistical thermodynamic arguments, lattice fluid theory allows for the direct calculation of the equilibrium internal energy and entropy of polymeric mixtures at given temperature, volume and concentration. Thus, lattice fluid theory provides simple and physically reasonable expressions for the equilibrium Helmholtz free energy, $a^{EQ}(\bullet)$, in the form of equation 2. In this article, $a^{EQ}$ is taken from the Sanchez and Lacombe version of the lattice fluid theory.

## Sanchez-Lacombe Lattice Fluid Model

On the basis of statistical thermodynamic arguments and the mean field approximation, the lattice fluid theory provides the following expression for the equilibrium Helmholtz free energy per polymer mass $a^{EQ}$, in the limit of infinite coordination number (*12*):

$$\frac{a^{EQ}}{RT}=\left[\frac{r_{Np+1}}{M_{Np+1}}+\sum_{i=1}^{Np}\left(r_i Y_i\right)\right]\left[-\frac{1}{\tilde{v}\tilde{T}}+\left(\tilde{v}-1\right)\ln\left(\frac{\tilde{v}-1}{\tilde{v}}\right)+\sum_{i=1}^{Np+1}\left(\frac{\phi_i}{r_i}\ln\left(\frac{\phi_i}{\tilde{v}}\frac{v_i^*}{v^*}\right)\right)\right] \tag{19}$$

Equation 19 is written for the mixture a polymer and $N_p$ low molecular weight components. In the lattice fluid theory of Sanchez and Lacombe, each species $i$ is characterised by three parameters which refer to the given component in the pure phase: the number of cells occupied by a molecules of the species, $r_i$, the characteristic volume of a lattice cell, $v_i^*$, and the characteristic energy, $\varepsilon_i^*$. In equation 19 the subscript $N_p+1$ has been used to label polymeric properties so that $r_{Np+1}$ and $M_{Np+1}$ refer to the number of lattice cells occupied by a polymer molecule, and to the polymer molecular weight respectively.

In equation 19 the free energy of the mixture is expressed as function of the dimensionless temperature $\tilde{T}$, the dimensionless specific volume $\tilde{v}$ and the volume fraction of each component in the mixture, $\phi_j$ $\{j\in[1,N_p]\}$. These variables are defined as follows:

$$\tilde{T}=\frac{T}{\varepsilon^*/R} \tag{20}$$

$$\tilde{v}=\frac{V^P}{v^*\left[\frac{r_{Np+1}}{M_{Np+1}}+\sum_{i=1}^{Np}\left(r_i Y_i\right)\right]} \tag{21}$$

$$\phi_j=\frac{r_j n_j}{\left[\frac{r_{Np+1}}{M_{Np+1}}+\sum_{i=1}^{Np}\left(r_i Y_i\right)\right]} \tag{22}$$

In the above relations $T$ is the absolute temperature, $V$ is the total volume of the mixture and $\varepsilon^*$ and $v^*$ are the characteristic energy and cell volume of the lattice, respectively. In the relation developed by Sanchez and Lacombe, $\varepsilon^*$ and $v^*$ depend on the composition of the mixture according to the following mixing rules:

$$v^* = \sum_{i=1}^{Np+1} \left(\phi_i v_i^*\right) \qquad (23)$$

$$\varepsilon^* = \sum_{i=1}^{Np+1} \left(\phi_i \varepsilon_i^*\right) - 0.5 \sum_{i=1}^{Np+1} \left[ \phi_i \frac{v_i^*}{v^*} \sum_{j=1}^{Np+1} \left( \phi_j v_j^* \Delta P_{ij}^* \right) \right] \qquad (24)$$

In equation 24 $\Delta P_{ik}^*$ is a binary parameter characterising the interaction energy between elements of species $i$ and $j$. Sanchez and Lacombe (*12*) suggested the following expression as an estimation of the binary interaction parameter:

$$\Delta P_{ij}^* = \left( \sqrt{\frac{\varepsilon_i^*}{v_i^*}} - \sqrt{\frac{\varepsilon_j^*}{v_j^*}} \right)^2 \qquad (25)$$

From equation 19, the equilibrium Helmholtz free energy for a mixture can be calculated as a function of temperature, volume and composition, when the pure component parameters, $\varepsilon_i^*$, $v_i^*$ and $r_i$, together with the binary parameters, $\Delta P_{ik}^*$, are known. Alternatively, the Hemholtz free energy of the polymeric mixture may be calculated at a given temperature, pressure and composition from equation 19 when the corresponding equilibrium density is calculated form the Sanchez Lacombe equation of state:

$$\frac{1}{\tilde{v}^2} + \tilde{p} + \tilde{T} \left[ \ln\left( \frac{\tilde{v} - 1}{\tilde{v}} \right) + \left( 1 - \sum_{i=1}^{Np+1} \frac{\phi_i}{r_i} \right) \frac{1}{\tilde{v}} \right] = 0 \qquad (26)$$

If equation 25 is used to estimate the binary lattice fluid parameters, then the free energy, $A$, as a function of $T$, $V$ and $\phi_\iota$ may be calculated from equation 19 using only pure component lattice fluid parameters. These parameters may be determined from pure component equilibrium properties such as $P$-$V$-$T$ data or vapour pressure (*12*).

Following the procedure described in the previous section, an expression for the non-equilibrium chemical potential, $\mu_i$, of the penetrant species $i$ in the mixture may be obtained by extending the expression of the free energy in equation 19 to non-equilibrium states according to the relation offered in equation 14 and using equation 12. We then obtain the following expression for the chemical potential of the penetrant species $i$:

$$\mu_i^{NE} = 1 + \ln\left(\frac{\phi_i}{\tilde{v}}\right) - \left[ r_i + r_i \left( \frac{v_i^*}{v^*} - 1 \right) \right] \ln\left( \frac{\tilde{v} - 1}{\tilde{v}} \right) - r_i \frac{v_i^*}{v^*} + \frac{r_i}{RT} \frac{1}{\tilde{v}} \left[ \varepsilon_i^* + \sum_{j=1}^{Np+1} \left( \frac{v_j^*}{v^*} \phi_j \left( \varepsilon_j^* - v_j^* \Delta P_{ij}^* \right) \right) \right] \qquad (27)$$

Equation 27 represents the basic equation for the NELF model based on the Sanchez and Lacombe lattice fluid theory; it provides the explicit dependence of the chemical potential of each penetrant species of a multicomponent mixture on temperature, volume and composition. In view of equation 12 and equation 14 at given temperature, volume and composition this equation is valid for any pressure applied on the system (either at equilibrium or non-equilibrium pressure) and thus holds true in any non-equilibrium state.

## Solubility Calculations based on the NELF Model

The results obtained in the previous section permit calculation of penetrant solubility in a glassy polymer as a function of temperature, penetrant fugacity and polymer density. Contrary to the situation for the equilibrium solubility calculation, the polymer density under pseudo-equilibrium conditions must be known in order to calculate the corresponding penetrant content. For pure penetrants, the comparison of model predictions with experimental isotherms is rather satisfactory for the cases in which volume dilation data are also available (*1*, *13*).

The requirement to specify the polymer density may represent a serious limitation for the practical application of the NELF model. Indeed, the dilation of the polymer matrix at high penetrant pressure could be significant and difficult to estimate without specific experimental data. On the other hand, the use of the non-equilibrium polymer density is actually a powerful tool to represent complex non-equilibrium phenomena. It has been shown, for example, that it allows a description of sorption-desorption hysteresis (*1*) as well as the influence of pretreatments on the solubility isotherms of gases in glassy polymers. For such cases, the different pseudo-equilibrium solubility values at the same prevailing temperature and penetrant fugacity are satisfactorily accounted for by considering the different pseudo-equilibrium polymer densities.

It is also interesting to observe that the solubility coefficient of a penetrant in a glassy polymer, in the limit of low pressure, is independent of the partial molar volume of the penetrant component in the mixture and can be reliably calculated from the NELF model using only the pure polymer density (*14*). Thus, also in the absence of volume dilation data the NELF model can be reliably used, albeit in a limited pressure range often in the order of few bars at least (*14*).

In the calculations considered in the following sections, the dilation of the polymer matrix is ignored since no direct experimental evidence is available for the volume dilation associated to the sorption of the mixture considered. The solubility of penetrants in the solid polymer will then be calculated by assuming that the polymer density at pseudo-equilibrium conditions is equal to the pure polymer density at the same temperature.

**Solubility of Gaseous Mixtures: Comparison with Experimental Data.** The experimental determination of the solubility of gas mixture components in a solid polymer is a very complicated task relative to the measurement of solubility of a pure gas or vapour. Indeed, very few mixed gas sorption data are available in the technical literature. To compare with predictions of the NELF model, only the experimental data for the solubility of $CO_2/C_2H_4$ mixtures in PMMA at 35°C (*4*, *15*, *16*) will be considered in this article. Since the volume dilation produced by the gas mixture is not available, the analysis of the data is confined to the low pressure range, where the swelling of the glassy polymer may be neglected. Indeed, in the calculation which follows the non-equilibrium solubility is estimated from the NELF model assuming that the polymer density at pseudo-equilibrium conditions may be approximated by the density of the pure polymer at the same temperature.

The pure component lattice fluid parameters for PMMA, $CO_2$ and $C_2H_4$ have been obtained using the *P-V-T* data reported in (*17*) for the polymer and in (*18*) for the penetrants. The parameters values obtained in this work for the Sanchez-Lacombe EOS are recorded in Table I. They are in good agreement with the values reported in the literature for the same components (*19*).

The polymer used by Koros and co-workers (*4*, *15*, *16*) is a PMMA with a rather high molecular weight ($M_w \approx 600000$ kg/kmol), and with a relatively high glass transition temperature ($T_g \approx 120$°C). Its density at 35°C was not reported and thus has been estimated by considering the experimental specific volume of the rubbery phase at T=120°C, after Olabisi and Simha (*20*), and the cubic expansion coefficient for the glassy phase as indicated in (*21*). At 35°C the density of the glassy PMMA is estimated to be 1.176 kg/m$^3$.

The comparison of NELF predictions with the experimental data for the solubility of pure $CO_2$ and $C_2H_4$ in PMMA is reported in Figure1. The sorption isotherms obtained from the NELF model for $C_2H_4$ in PMMA present a slight overestimation relative to the experimental values in the low pressure range, while the prediction is more than satisfactory for the solubility of $CO_2$. Following the general trend observed in (*13*, *14*), the underestimation of $CO_2$ solubility observed at higher pressures is a consequence of neglecting the swelling of the polymer matrix in the NELF model calculation. A more detailed discussion on the volume dilation effect on the solubility calculation may be found in (*1*, *13*, *14*).

All three binary lattice fluid interaction parameters, $\Delta P_{ij}^*$ , which are involved in the free energy for the three-component-mixtures of $CO_2$ and $C_2H_4$ in PMMA, were obtained from the pure component parameters using the simple relation in equation 25. For the $CO_2$-$C_2H_4$ pair, a better estimation of the binary interaction parameter may be obtained from the analysis of VLE data of the binary mixture. Indeed, by using the data of vapour and liquid saturation pressure for the $CO_2$-$C_2H_4$ mixtures reported by Mollerup (*22*), the interaction parameter $\Delta P^*_{CO_2\ C_2H_4}$ was directly calculated. The fitting procedure, based on the SL EOS with the pure component parameter reported in Table I, leads to $\Delta P^*_{CO_2\ C_2H_4} = 65$ MPa while its first order approximation calculated through equation 25 gives 43 MPa. However, when the binary parameter $\Delta P^*_{CO_2\ C_2H_4}$ changes from 43 to 65 MPa the solubility of $CO_2/C_2H_4$ mixtures, predicted through the NELF model, undergoes a variation always smaller than 0.1% in the conditions here considered. Clearly, the influence of the $CO_2/C_2H_4$ binary interaction parameters has a minor relevance in the solubility calculation as long as very dilute solutions of $CO_2$ and $C_2H_4$ in PMMA are obtained.

In Figures 2 and 3, two examples of the solubility isotherms predicted by the NELF model are presented and compared with the experimental data for $CO_2/C_2H_4$ mixtures in PMMA. Such examples show the main features observed in the comparison of the model prediction with all of the experimental data reported in refs (*4*, *15*, *16*).

In Figure 2 the solubility of $CO_2$ in PMMA is considered as a function of $CO_2$ partial pressure, and the partial pressure of $C_2H_4$ is held constant at 2.06±0.08 atm. The NELF model correctly predicts the solubility coefficient in the low pressure limit

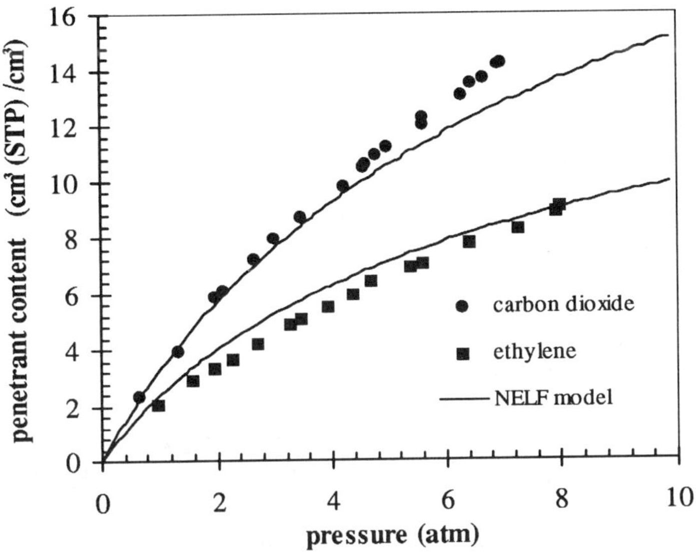


Fig. 1. Solubility isotherms of pure $CO_2$ and $C_2H_4$ in PMMA at 35°C. Experimental data by [16].

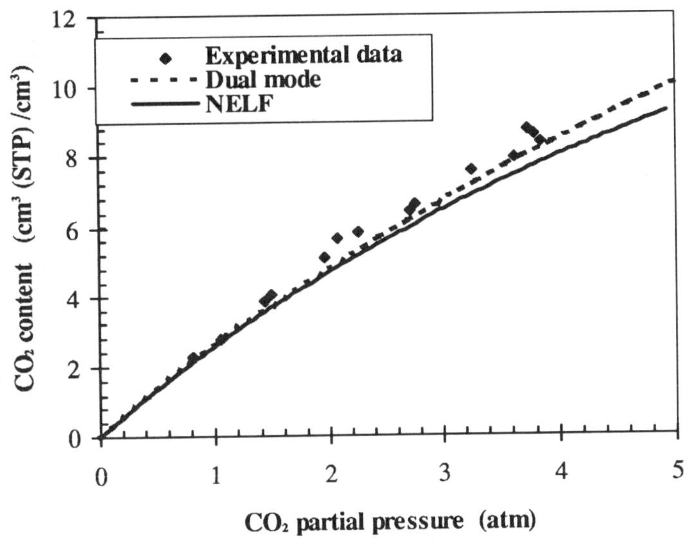


Fig. 2. Solubility of $CO_2$ in PMMA at 35°C; dependence on $CO_2$ partial pressure at a fixed $C_2H_4$ partial pressure of 2.06 atm. Experimental data by [16].

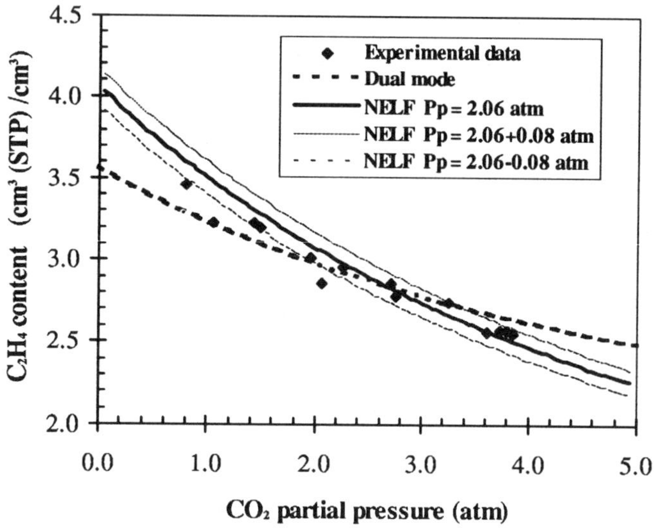


Fig. 3. Solubility of $C_2H_4$ in PMMA at 35°C; dependence on $CO_2$ partial pressure at a fixed $C_2H_4$ partial pressure of 2.06 atm. Experimental data by [16].

while it underestimates the solubility at high pressure. Parallel to what observed in (*14*), also in this case deviations between the model calculation and experimental sorption level probably result from the dilation of the polymer matrix no longer being negligible at high $CO_2$ pressure. In the same figure the isotherm obtained from the dual mode model is also reported for the sake of comparison: it is useful to remember that in this case the dual mode isotherm is predicted on the basis of model parameters obtained from fitting the pure components solubility at the same temperature.

In Figure 3 the solubility of $C_2H_4$ in PMMA as a function of $CO_2$ partial pressure is reported. Again, the $C_2H_4$ partial pressure is fixed. In this case, the ethylene fugacity in the polymeric mixture is constant, but its content in the solid decreases due to the increasing sorption of $CO_2$. This effect is satisfactorily represented by the NELF model, as is clearly shown by the comparison in the plot. For the NELF model predictions, the range of the solubility variations associated with the uncertainty in the $C_2H_4$ partial pressure value (±0.08 atm) are also reported in the same figure. Interestingly, the width of the corresponding solubility range compares well with the uncertainty in the experimental values. It is worthwhile to note that in this case the dual mode prediction is significantly different from that of the NELF model and a better representation of the experimental data is obtained in the latter case.

## Conclusions

With a simple rheological model for the bulk of the polymeric material, the polymer density can be considered as a proper order parameter to characterise departures from equilibrium and a general relation is obtained between equilibrium and non-equilibrium Helmholtz free energy. The thermodynamic analysis for such systems also allows a generalisation of the relation for the chemical potential of penetrant components in non-equilibrium polymeric mixtures. By using the NELF model, the gas and vapour solubility in glassy polymers may be calculated when the pure components lattice fluid parameters and the polymer density in pseudo-equilibrium conditions are known. Even for the cases in which the dilation of the polymer matrix due to the sorption process is not available the model may be used, insofar as it predicts the solubility coefficient in the limit of low pressure, simply considering the pure polymer glassy density as the value of the order parameter.

It is interesting to note that a polymeric material may be prepared in the glassy state at the same temperature with different density values, as a consequence of different prehistories. According to the NELF model those states result in different pseudo-equilibrium solute contents.

In previous work (*1*, *13*, *14*) we considered the comparison between the predictions of the model and the experimental results for the solubility of pure gases and vapours in glassy polymers, and concluded that NELF is a definitely reliable model. In this work the comparison is extended to the case of solubility of mixed gases, by considering the solubility of $CO_2/C_2H_4$ mixtures in PMMA reported by Koros and co-workers (*4*, *15*, *16*). The model correctly represents the value and the trend of the penetrants solubility when the partial pressure of the penetrants changes. The results are obtained via a completely predictive procedure; the comparison is limited to the low pressure range since experimental data for the dilation of the polymer matrix during sorption are not available. The prediction of the solubility at higher pressures could be more reliably obtained if dilation data or a proper constitutive equation for the non-equilibrium swelling of the polymer sample were available.

**Acknowledgements**
This work has been partially supported by the Italian Ministry for University and Scientific and Technological Research (40% and 60% funds).

Table I Lattice fluid parameters for the S.L. model

| Component | ε*/R | M/r | v* | Source of PVT data |
|---|---|---|---|---|
| | *(K)* | *(kg/kmol)* | *(l/kmol)* | |
| PMMA | 703 | 14.7 | 11.6 | ref. 17 |
| $CO_2$ | 300 | 6.00 | 3.96 | ref. 18 |
| $C_2H_4$ | 295 | 4.60 | 6.72 | ref. 18 |

**Literature Cited**

1. Doghieri, F.; Sarti, G.C. *Macromolecules* **1996**, *29*, 7885.
2. Vimalchand, P.; Donohue, M.D. *Ind. Eng. Chem. Fundam.* **1985**, *24*, 246.
3. Chen, F.; Fredenslund, Aa.; Rasmussen, P. *Ind. Eng. Chem. Res.* **1990**, *29*, 875.
4. Sanders, E.S.; Koros W.J. *J. Polym. Sci., Polym. Phys. Ed.* **1986**, *24*, 175.
5. Larché, F.C.; Cahn, J.W. *Acta Metall.* **1982**, *30*, 1835.
6. Carbonell, R.G.; Sarti C.G. *Ind. Eng. Chem. Res.* **1990**, *29*, 1194.
7. Treolar, L.R.G., *The Physics of Rubber elasticity*, 3rd ed.; Clarendon: Oxford, UK, 1975.
8. Thomas, N.L.; Windle A.H. *Polymer* **1980**, *21*, 613.
9. Durning, C.J. *J. Polym. Sci.: Polym. Phys. Ed.* **1985**, *23*, 1831.
10. Flory, P.J.; Erman, B. *Macromolecules* **1982**, *15*, 800.
11. Billovits, G.F.; C.J.Durning, *Chem. Eng. Comm.* **1989**, *82*, 21.
12. Sanchez, I.C.; Lacombe, R.H., *Macromolecules* **1978**, *11*, 1145.
13. Sarti, G.C.; Doghieri, F.D. *Chem. Eng. Sci.* **1998**, in press
14. Doghieri, F.D. ; Sarti, G.C *J. Membrane Sci.* **1998**, in press.
15. Sanders, E.S.; Koros, W.J.; Hopfenberg, H.B.; Stannett, V.T. *J. Membrane Sci.* **1983**, *13*, 161.
16. Sanders, E.S.; Koros, W.J.; Hopfenberg, H.B.; Stannett, V.T. *J. Membrane Sci.* **1984**, *18*, 53.
17. Danner, R.P. and High, M.S., *Handbook of Polymer Solution Thermodynamics;* A.I.Ch.E.: New York, USA, 1992.

18. Vargaftik, N.B *Handbook of physical properties of liquids and gases*, 2nd ed.; Hemisphere Publ.: Washington, USA, 1975.
19. Sanchez, I.C.; Panayiotou, C.G. In *Models for Thermodynamics and Phase Equilibria Calculations*; Sandler, S.I., Ed., Dekker Pub.: New York, USA, 1994.
20. Olabisi, O.; Simha, R. *Macromolecules* **1975**, *8*, 76.
21. Brandrup J.; Immergut, E.H., *Polymer Handbook*; John Wiley and Sons: New York, USA, 1989.
22. Mollerup, J. *J. Chem. Soc., Faraday. I.* **1975**, *71*, 2351.

## Chapter 14

# Relation between Gas Permeabilities and Structure of Polyimides

**Yusei Hirayama, Toshimune Yoshinaga, Shunsuke Nakanishi, and Yoshihiro Kusuki**

**Polymer Laboratory (Chiba), UBE Industries, Ltd., 8-1 Goi-Minamikaigan, Ichihara, Chiba 290-0045, Japan**

In this study, several polyimides were prepared and the relation between their gas permeation properties and chemical structure was determined and compared with data from previous studies. Permeability coefficients and diffusion coefficients, D, of the polyimides were obtained by a time-lag method for He, $CO_2$, $O_2$, $N_2$, and $CH_4$. The correlation of log(D) vs. reciprocal fractional free volume was poor due to the polyimides having substituents, such as halogens, in the diamine unit. Better linear relations were observed for plots of log(D) vs. storage modulus and cohesive energy density for these polyimides. These results suggest that gas diffusivity is influenced by mobility of polymer chains due to intermolecular interaction. Gas molecules in the glassy polymers diffuse more easily as segmental mobility increases, and estimation of gas diffusivities may be possible by the use of factors such as cohesive energy density.

Many papers have been published regarding the gas permeation properties of polyimides(*1-7*). Polyimides are very useful polymers for the analysis of relations between gas permeation properties and polymer structure, because many different structures, based on diacids and diamines, can be prepared from existing compounds. We have previously reported relations between free volume or cohesive energy density and gas diffusivity(*8,9*). In this study, we prepared five kinds of polyimides and explored correlations between gas diffusivity and parameters such as storage modulus, cohesive energy density, and free volume. These results were compared with data from previous studies. We also discuss the merits of the use of each of these parameters to correlate gas permeability properties of different polymers.

Gas diffusivity and permeability are generally understood to depend sensitively on free volume, which corresponds to the amount of free space, in a polymer matrix(*10,11*). Generally, linear correlations between log(D) or log(P) and reciprocal free volume are observed(*1,3,11-21*). Gas diffusion coefficients, D, are often correlated with fractional free volume of polymers, $V_f$, as follows,

$$D=A_0\exp(-B_0/V_f) \quad (1)$$

 

where $A_0$ and $B_0$ are adjustable parameters(*20*). The fractional free volume was calculated as follows:

$$V_f=(V-V_0)/V \qquad (2)$$

where V is the specific molar volume at the temperature of the diffusivity measurements, and $V_0$ is the volume occupied by the repeat unit of the polymer. $V_0$ was estimated as $1.3V_w$, where $V_w$ is the van der Waals volume of the repeat unit estimated by Bondi's method(22).

The cohesive energy density, CED, of each polymer was calculated by the group contribution method of Fedors(*23*). The CED is expressed by Equation 3:

$$CED=\Delta E/V_c \qquad (3)$$

where $\Delta E$ is the hypothetical evaporative energy and $V_c$ is the molar volume of the polymer(*23*).

## Experimental

**Materials.** The polyimides were prepared from tetracarboxylic dianhydrides (such as BPDA and 6FDA) and various diamines as shown in Figure 1. Polymerization and film preparation protocol have been previously described(*8*). The polyimides were prepared from BPDA or 6FDA and the diamine by polycondensation in p-chlorophenol. This 10-20 wt % polyimide solution was cast onto a clean glass plate and the p-chlorophenol was evaporated at 100 °C in a clean oven. The film on the glass plate was annealed at 300 °C, and after being peeled off, dried at 80-100 °C for 5 more hours.

**Measurements.** The details of all measurements were described in previous papers(*8,9*). The permeation properties of the polyimide films were measured by a time-lag method(*21,24*) for He, $CO_2$, $O_2$, $N_2$, and $CH_4$ at a upstream pressure of about 2.5 $kg/cm^2$ and a downstream pressure of $10^{-5}$ torr in the temperature range from 35 to 100 °C.

An apparent permeability coefficient, P, was determined from the slope of downstream pressure vs. time curves at steady state conditions. An apparent diffusion coefficient, D, was calculated from the lag-time, $\theta$, using the following equation,

$$D=\ell^2/(6\theta) \qquad (4)$$

where $\ell$ is the film thickness. An apparent solubility coefficient, S, was calculated from Equation 5:

$$S=P/D \qquad (5)$$

Factoring the permeability into diffusivity and solubility terms, the permselectivity of components A and B may be expressed using Equation 5 as:

$$P_A/P_B=(D_A/D_B)(S_A/S_B) \qquad (6)$$

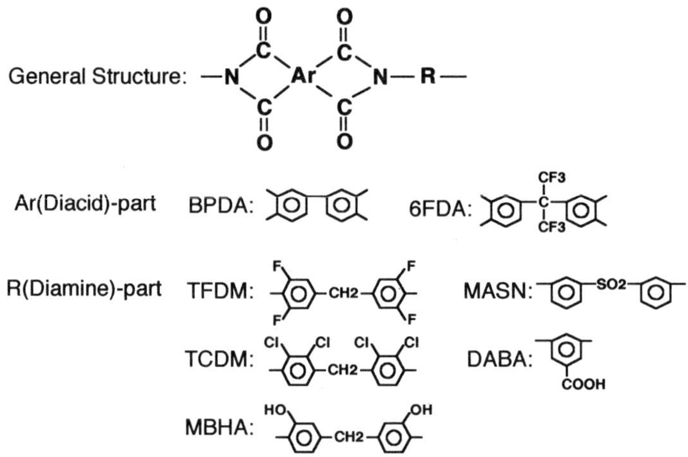


Figure 1. Chemical structures of polyimides

where $D_A/D_B$ is referred to as the mobility selectivity of components A and B, and $S_A/S_B$ is the solubility selectivity.

Dynamic thermal mechanical properties were analyzed using an autovibron dynamic mechanical visco-elastometer. The elastic modulus, E, is given by the following equation,

$$|E|^2=|E'|^2+|E''|^2 \tag{7}$$

where E' and E" are the storage modulus and loss modulus, respectively.

Density data were obtained at 35°C using a conventional density gradient column method. Wide angle X-ray diffraction (WAXD) patterns were measured using CuK$\alpha$ radiation having a wavelength of 1.54Å. d-Spacing values were obtained from the WAXD patterns.

## Results and Discussion

Physical properties of the polyimide films are listed in Tables I, II and III. Based on WAXD spectra and DSC results, the five polyimides were determined to be amorphous. Chemical structures and characteristics of the amorphous BPDA and 6FDA polyimide films from the previous study are shown in Figure 2, and Tables IV and V, respectively(*8,9*). In Table V, we present new permeability and diffusivity data at 80 °C of the polyimides reported in our previous study.

Table I. Characteristics of polyimide films

| No. | Polyimide | $T_g$ °C | d g/cm$^3$ | d-Spacing Å | $V_f$ - | E' GPa | CED J/cm$^3$ |
|---|---|---|---|---|---|---|---|
| 101 | BPDA-TFDM | 307 | 1.42 | 5.8 | 0.135 | 2.1 | 1090 |
| 102 | BPDA-TCDM | 304 | 1.45 | 5.8 | 0.135 | 2.5 | 1100 |
| 103 | BPDA-MBHA | - | 1.38 | 5.1 | 0.099 | 4.4 | 1390 |
| 104 | BPDA-MASN | 283 | 1.42 | 5.5 | 0.111 | 3.7 | 1050 |
| 105 | 6FDA-DABA | 309 | 1.51 | 5.3 | 0.153 | 2.3 | 850 |

Except for BPDA-MASN, the polyimides in this study had side chains containing polar substituents. The order of glass transition temperatures, $T_g$, was:

6FDA-DABA≒BPDA-TFDM>BPDA-TCDM>BPDA-MASN.

No peaks which could be ascribed to a glass transition were observed in the DSC thermogram of BPDA-MBHA. From WAXD, the order of d-spacing was:

BPDA-TCDM>BPDA-TFDM=BPDA-MASN>6FDA-DABA>BPDA-MBHA.

Based on the permeation studies, the order of the permeability and diffusion coefficients was:

Table II. Permeability coefficients and permselectivities of polyimide films

| No. | Polyimide | P at 35 °C, Barrer He | $CO_2$ | $O_2$ | Permselectivity at 35 °C $CO_2/CH_4$ | $O_2/N_2$ |
|---|---|---|---|---|---|---|
| 101 | BPDA-TFDM | 19.9 | 5.7 | 1.53 | 25 | 5.4 |
| 102 | BPDA-TCDM | 10.3 | 1.25 | 0.37 | 40 | 8.0 |
| 103 | BPDA-MBHA | 4.6 | 0.22 | 0.059 | - | 9.0 |
| 104 | BPDA-MASN | 4.4 | 0.173 | 0.51 | 60 | 11 |
| 105 | 6FDA-DABA | 31 | 3.4 | 1.01 | 63 | 8.0 |

1 Barrer is $10^{-10}$ $cm^3(STP) \cdot cm/(cm^2 \cdot s \cdot cmHg)$

| No. | Polyimide | P at 80 °C, Barrer He | $CO_2$ | $O_2$ | Permselectivity at 80 °C $CO_2/CH_4$ | $O_2/N_2$ |
|---|---|---|---|---|---|---|
| 101 | BPDA-TFDM | 35 | 9.8 | 3.0 | 16 | 4.2 |
| 102 | BPDA-TCDM | 18.2 | 1.99 | 0.74 | 22 | 6.0 |
| 103 | BPDA-MBHA | 9.2 | 0.56 | 0.187 | 29 | 6.5 |
| 104 | BPDA-MASN | 8.2 | 0.37 | 0.139 | 33 | 7.6 |
| 105 | 6FDA-DABA | 51 | 6.3 | 1.99 | 32 | 5.8 |

Table III. Diffusion coefficients and mobility selectivities of polyimide films

| No. | Polyimide | D at 35 °C $\times 10^{10}$, $cm^2/s$ $CO_2$ | $O_2$ | Mobility selectivity at 35 °C $CO_2/CH_4$ | $O_2/N_2$ |
|---|---|---|---|---|---|
| 101 | BPDA-TFDM | 75 | 210 | 7.0 | 4.7 |
| 102 | BPDA-TCDM | 19.8 | 70 | 8.3 | 5.7 |
| 103 | BPDA-MBHA | 2.4 | 10.4 | - | 6.9 |
| 104 | BPDA-MASN | 2.4 | 11.8 | 8.8 | 8.2 |
| 105 | 6FDA-DABA | 41 | 145 | 13 | 6.8 |

| No. | Polyimide | D at 80 °C $\times 10^{10}$, $cm^2/s$ $CO_2$ | $O_2$ | Mobility selectivity at 80 °C $CO_2/CH_4$ | $O_2/N_2$ |
|---|---|---|---|---|---|
| 101 | BPDA-TFDM | 290 | 930 | 3.9 | 3.8 |
| 102 | BPDA-TCDM | 89 | 290 | 5.9 | 4.3 |
| 103 | BPDA-MBHA | 15.1 | 60 | 5.8 | 6.0 |
| 104 | BPDA-MASN | 14.5 | 57 | 6.7 | 6.2 |
| 105 | 6FDA-DABA | 186 | 510 | 8.2 | 5.3 |

BPDA-TFDM>6FDA-DABA>BPDA-TCDM>BPDA-MBHA>BPDA-MASN.

A similar order was observed in the storage modulus and CED. Gas solubility coefficients of the five polyimide films were essentially independent of polymer structure and similar to those of the polyimide films in our previous study(*8*).

BPDA-MBHA had smaller values of $V_f$ and E' and larger diffusion coefficients than BPDA-HAB. This result suggests that introducing $-CH_2-$

Table IV. Characteristics of polyimides in previous studies(*8*,*9*)

| No. | | d-Spacing, Å | $V_f$ | E', GPa | CED, J/cm$^3$ |
|---|---|---|---|---|---|
| 1 | BPDA- DADM | 5.8 | 0.110 | 2.3 | 990 |
| 2 | DADS | 5.2 | 0.117 | 2.8 | 1050 |
| 3 | PASN | 5.5 | 0.117 | 3.5 | 1050 |
| 4 | HFIP | 6.2 | 0.163 | 2.9 | 740 |
| 5 | MFA | 5.8 | 0.163 | 3.1 | 1040 |
| 6 | MBA | 3.8 | 0.161 | 4.2 | 1030 |
| 7 | MCA | 6.1 | 0.164 | 3.0 | 1050 |
| 8 | BAPE | 4.9 | 0.151 | | 880 |
| 9 | BAPS | 5.2 | 0.135 | 3.2 | 900 |
| 10 | BAPP | 5.4 | 0.142 | 2.7 | 840 |
| 11 | HFBAPP | 6.0 | 0.162 | 2.2 | 850 |
| 12 | MDT | 5.5 | 0.173 | 3.5 | 910 |
| 13 | CDM | 5.9 | 0.176 | 2.6 | 1050 |
| 14 | MDX | 6.2 | 0.175 | 2.6 | 850 |
| 15 | HAB | 5.3 | 0.145 | 5.7 | 1480 |
| 16 | TSN | 5.8 | 0.118 | 4.6 | 1060 |
| 31 | 6FDA- DADE | 5.8 | 0.173 | 2.5 | 740 |
| 32 | DADM | 5.8 | 0.155 | 2.1 | 720 |
| 33 | MDT | 5.9 | 0.168 | 2.7 | 680 |
| 34 | MDX | 6.2 | 0.164 | 2.2 | 650 |
| 35 | CDM | 5.8 | 0.182 | 3.4 | 770 |
| 36 | TSN | 5.8 | 0.146 | 3.1 | 770 |
| 37 | BAPE | 5.5 | 0.170 | 2.2 | 700 |
| 38 | HAB | 5.9 | 0.182 | 4.0 | 980 |

E' is reportedat 35 °C.

between the two phenylene rings in the diamine component increased chain flexibility and, in turn, increased diffusion coefficients.

Gas diffusivities in BPDA-TCDM, which contains four bulky chloride substituents on two phenylene rings, were larger than those in the dichloro substituted analogs, BPDA-CDM and BPDA-MCA. Similarly, gas diffusivities in a tetrafluoro substituted polyimide, BPDA-TFDM, were larger than those of its disubstituted analog, BPDA-MFA. These tetrasubstituted polyimides had smaller values of $V_f$ and E' at 35 °C than their disubstituted analogs. The calculated CED values of the polyimides containing four bulky and polar substituents (i.e. BPDA-TFDM and BPDA-TCDM) increase little compared with those of the disubstituted analogs, BPDA-MFA and BPDA-MCA, respectively.

In tetramethyl and dimethyl substituted polymers, BPDA-MDX and BPDA-MDT, respectively, gas diffusion coefficients in the tetrasubstituted

Table V. P and D of the polyimides at 35 ℃ (from previous study(*8*)) and 80 ℃

| No. | | 35 ℃ P, Barrer He | $CO_2$ | $O_2$ | Permselectivity $CO_2/CH_4$ | $O_2/N_2$ | D×10$^{10}$, $cm^2/s$ $CO_2$ | $O_2$ | Mobility selectivity $CO_2/CH_4$ | $O_2/N_2$ |
|---|---|---|---|---|---|---|---|---|---|---|
| 1 | BPDA-DADM | 4.6 | 1.00 | 0.21 | 49 | 6.9 | 14.7 | 35 | 10 | 4.2 |
| 2 | DADS | 3.7 | 0.77 | | 47 | | 12.6 | | 6.9 | |
| 3 | PASN | 7.0 | 1.75 | 0.35 | 44 | 7.0 | 20 | 58 | 8.6 | 5.4 |
| 4 | HFIP | 34 | 16.8 | 3.8 | 37 | 5.0 | 130 | 550 | 7.7 | 5.8 |
| 5 | MFA | 4.3 | 0.54 | 0.143 | 42 | 7.0 | 9.0 | 27 | 8.8 | 4.7 |
| 6 | MBA | 3.6 | 0.32 | 0.092 | 58 | 9.0 | 4.8 | 29 | 8.7 | 7.0 |
| 7 | MCA | 3.7 | 0.34 | 0.096 | 61 | 8.9 | 5.7 | 27 | 8.8 | 8.6 |
| 8 | BAPE | 4.1 | 1.25 | 0.26 | 41 | 6.5 | 35 | 67 | 15 | 5.6 |
| 9 | BAPS | 6.0 | 1.85 | 0.37 | 47 | 6.8 | 27 | 92 | 8.4 | 5.6 |
| 10 | BAPP | 7.5 | 2.8 | 0.60 | 33 | 6.4 | 62 | 178 | 9.3 | 6.2 |
| 11 | HFBAPP | 18.3 | 7.3 | 1.67 | 34 | 5.5 | 112 | 370 | 8.2 | 4.8 |
| 12 | MDT | 8.9 | 1.41 | 0.37 | 51 | 7.6 | 17.1 | 55 | 11 | 6.2 |
| 13 | CDM | 6.9 | 0.98 | 0.26 | 60 | 8.5 | 12.3 | 49 | 10 | 5.7 |
| 14 | MDX | 32 | 22 | 4.7 | 27 | 5.2 | 155 | 490 | 5.8 | 4.5 |
| 15 | HAB | 1.09 | 0.031 | 0.0096 | 76 | 9.5 | 0.36 | 1.67 | 15 | 6.1 |
| 16 | TSN | 12 | 2.7 | 0.58 | 73 | 7.6 | 21 | 60 | 16 | 6.2 |
| 31 | 6FDA-DADE | 34 | 13.8 | 2.9 | 51 | 5.7 | 147 | 420 | 11 | 5.6 |
| 32 | DADM | 33 | 15.1 | 3.2 | 47 | 5.6 | 155 | 430 | 10 | 4.9 |
| 33 | MDT | 35 | 8.8 | 2.5 | 44 | 6.2 | 101 | 290 | 12 | 5.4 |
| 34 | MDX | 74 | 44 | 10.3 | 30 | 4.8 | 470 | 1270 | 8.3 | 4.3 |
| 35 | CDM | 30 | 6.8 | 1.81 | 56 | 6.7 | 84 | 230 | 14 | 5.3 |
| 36 | TSN | 104 | 56 | 14.2 | 38 | 5.0 | 430 | 1050 | 11 | 4.7 |
| 37 | BAPE | 18.3 | 7.9 | 1.73 | 37 | 5.5 | 98 | 250 | 8.7 | 4.5 |
| 38 | HAB | 30 | 4.9 | 1.22 | 86 | 7.3 | 39 | 136 | 21 | 7.1 |

| No. | | 80 ℃<br>P, Barrer<br>He | $CO_2$ | $O_2$ | Permselectivity<br>$CO_2/CH_4$ | $O_2/N_2$ | D×$10^{10}$, $cm^2/s$<br>$CO_2$ | $O_2$ | Mobility selectivity<br>$CO_2/CH_4$ | $O_2/N_2$ |
|---|---|---|---|---|---|---|---|---|---|---|
| 1 | BPDA-DADM | 10 | 1.99 | 0.54 | 24 | 5.3 | 100 | 220 | 6.5 | 3.3 |
| 2 | DADS | 8.6 | 1.59 | | 25 | | 84 | | 5.2 | |
| 3 | PASN | 13.6 | 3.2 | 0.78 | 25 | 5.2 | 95 | 290 | 5.1 | 4.4 |
| 4 | HFIP | 58 | 22 | 6.2 | 18 | 3.9 | 520 | 1650 | 4.9 | 3.5 |
| 5 | MFA | 9.7 | 1.29 | 0.42 | 22 | 5.2 | 54 | 126 | 6.6 | 3.8 |
| 6 | MBA | 8.2 | 0.73 | 0.27 | 27 | 6.7 | 38 | 300 | 6.7 | 10 |
| 7 | MCA | 8.6 | 0.80 | 0.29 | 24 | 6.2 | 42 | 290 | 5.9 | 10 |
| 8 | BAPE | 9.6 | 2.4 | 0.67 | 20 | 4.8 | | | | |
| 9 | BAPS | 13.3 | 3.6 | 0.91 | 22 | 5.0 | 158 | 460 | 6.3 | 4.4 |
| 10 | BAPP | 16.2 | 5.1 | 1.42 | 17 | 4.6 | 360 | 830 | 6.3 | 4.0 |
| 11 | HFBAPP | 34 | 11.8 | 3.3 | 18 | 4.1 | 750 | 1800 | 7.3 | 3.8 |
| 12 | MDT | 17.7 | 2.5 | 0.83 | 26 | 5.9 | 86 | 240 | 6.6 | 5.5 |
| 13 | CDM | 14.7 | 1.89 | 0.64 | 27 | 6.2 | 71 | 300 | 7.5 | 5.6 |
| 14 | MDX | 54 | 24 | 7.5 | 14 | 4.2 | 650 | 1750 | 4.5 | 3.4 |
| 15 | HAB | 3.0 | 0.105 | 0.037 | 45 | 7.5 | 2.5 | 9.2 | 9.8 | 4.6 |
| 16 | TSN | 22 | 4.0 | 1.13 | 34 | 5.7 | 83 | 250 | 8.1 | 5.1 |
| 31 | 6FDA-DADE | 59 | 19.8 | 5.5 | 24 | 4.3 | 710 | 1890 | 7.4 | 4.6 |
| 32 | DADM | 57 | 21 | 5.9 | 23 | 4.4 | 810 | 2040 | 7.9 | 4.3 |
| 33 | MDT | 59 | 12.7 | 4.3 | 23 | 4.7 | 340 | 1050 | 6.2 | 4.3 |
| 34 | MDX | 116 | 52 | 15.8 | 17 | 4.0 | 1510 | 4400 | 5.0 | 3.4 |
| 35 | CDM | 51 | 9.4 | 3.3 | 26 | 5.2 | 300 | 890 | 7.8 | 4.3 |
| 36 | TSN | 145 | 60 | 18.4 | 21 | 4.0 | 1110 | 3000 | 6.5 | 3.9 |
| 37 | BAPE | 34 | 12.2 | 3.4 | 19 | 4.3 | 430 | 1080 | 5.3 | 3.6 |
| 38 | HAB | 51 | 8.0 | 2.5 | 37 | 5.2 | 191 | 596 | 12 | 5.1 |

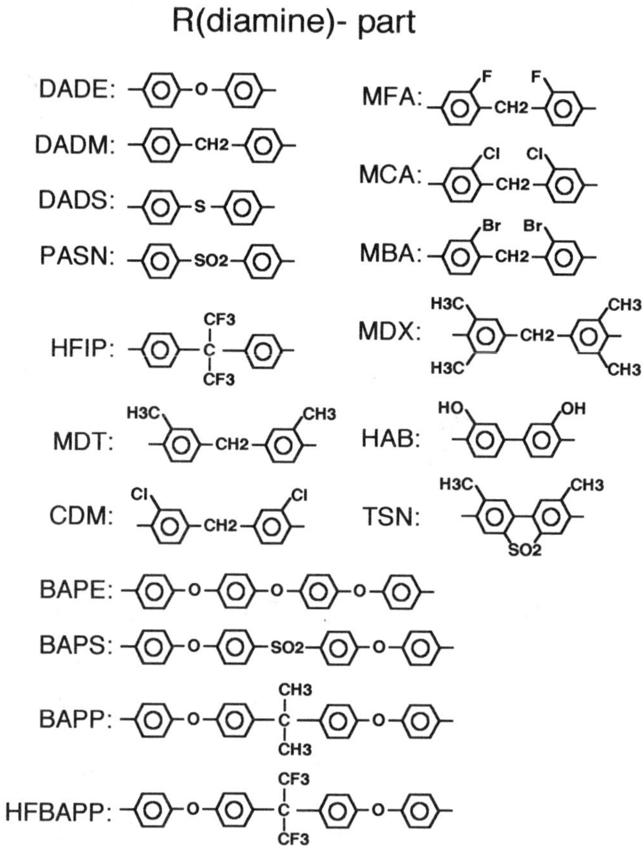


Figure 2. Diamines used in previous studies(*8,9*)

polymer, BPDA-MDX, were much larger than those of its disubstituted analog, BPDA-MDT. This result is interesting since these polymers have very similar fractional free volume values. The calculated CED and storage modulus values of BPDA-MDX were smaller than those of BPDA-MDT. These composite results suggest that, for gas diffusion in these substituted BPDA-DADM polyimides, effects related to increasing segmental mobility override those associated with decreasing free volume. CED values, which are sensitive to intermolecular interaction and values of storage modulus decrease with increasing segmental mobility despite the introduction of bulky and polar substituents such as halogens and fractional free volume values for substituted BPDA-DADM polyimides are almost the same.

**Relation between diffusivity and cohesive energy density.** Gas diffusion through polymers has been described by various models(*16*). For example, Meares(*25*) relates the energy required to open a gap in the polymer matrix large enough to permit a diffusion step to the energy of activation for diffusion, $E_d$, as follows:

$$E_d \fallingdotseq (\pi/4) \cdot \sigma^2 \cdot \lambda \cdot CED \tag{8}$$

where $\sigma$ is the mean penetrant size, and $\lambda$ is the length of the diffusional jump. This model presumes that as a gas molecule passes through a polymer matrix, it is hindered by polymer chains(*15,25,26*). These barriers to gas transport are the rate-limiting factors in penetrant diffusion. For the penetrant molecule to jump from one equilibrium position to another, a gap in the polymer matrix of at least the size of the gas molecule must be created. The bending of polymer chains to create the gap may be envisioned to scale roughly with the length of the rigid elements of the polymer backbone, (i.e. the distance between flexible bonds). The diffusion jump length, $\lambda$, will be determined by the size of these elements. Here, we assumed that $E_d$ and the apparent energy of activation for diffusion are equivalent. The Arrhenius behavior of D has been described as follows(*20,21,24,27*),

$$D=D_0\exp(-E_d/RT) \tag{9}$$

where R is the universal gas constant, and T is absolute temperature.
It is known that $D_0$ and $E_d$ are related as follows,

$$\ln(D_0)=aE_d-c \tag{10}$$

where a is approximately 0.001/R and c is a constant which is different for glassy and rubbery polymers(*21,27,28*). Based on Equations 8, 9 and 10, the CED, which is a measure of the cohesive forces inside the polymer matrix, should correlate with diffusivity.

Figures 3 and 4 present correlations between diffusion coefficients, log(D), and calculated CED. For all polyimides examined, including the polyimides in this study, good correlations between log(D) and CED were observed for each gas in spite of differences in measurement temperature.

Polyimides such as BPDA-CDM, -MFA, -MBA, -MCA, -HAB, -MBHA, 6FDA-CDM, -HAB, etc., exhibit relatively low diffusivities even though they have bulky substituents. The mobility of segments in these polyimides, which

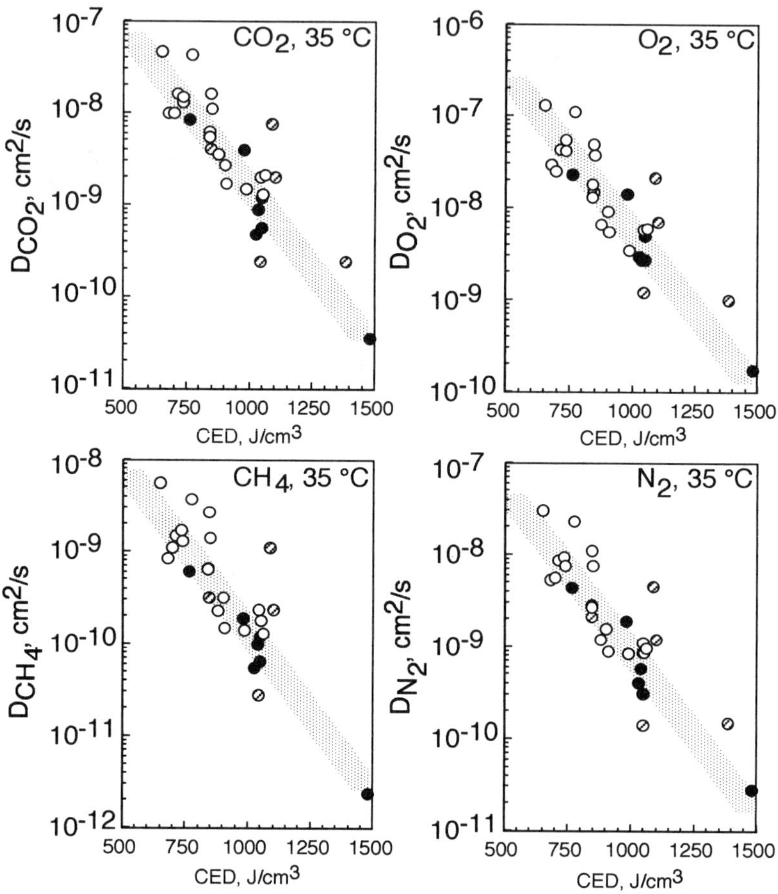


⊘: This study
○: Polyimides in previous studies(*8,9*)
●: Polyimides containing polar substituents(*8,9*)

Figure 3. Correlation between D and CED at 35℃

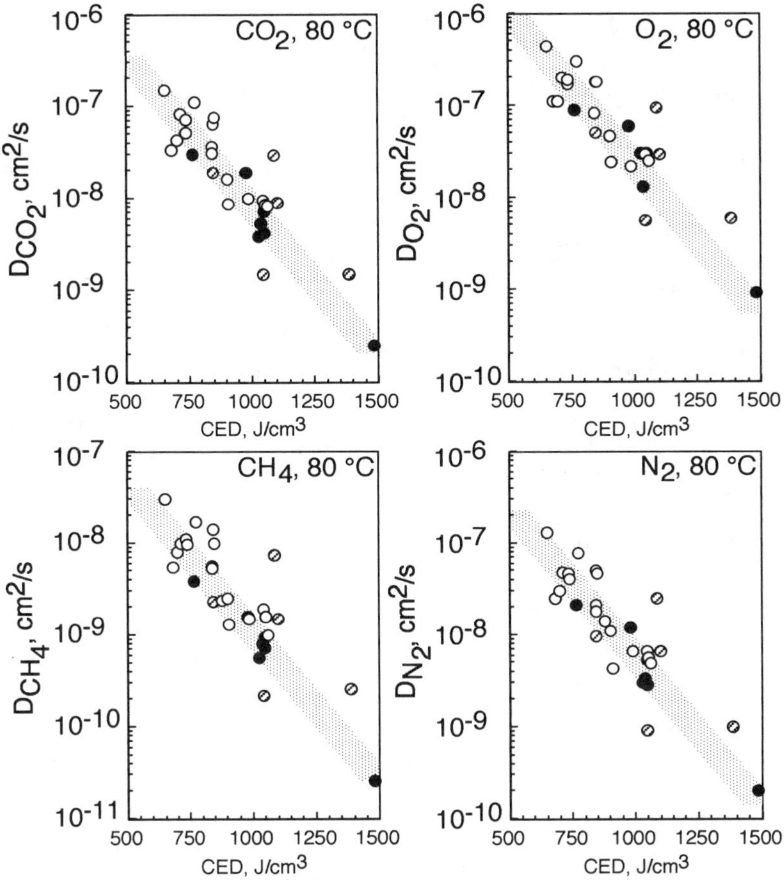


⊘ : This study
○ : Polyimides in previous studies(*8,9)*
● : Polyimides containing polar substituents(*8,9*)

Figure 4. Correlation between D and CED at 80℃

have polar substituents, is probably be lower than those of polymers without polar substituents because polar substituents enhance polymer segment interactions (i.e. increase CED). Based on these results, CED appears to be an effective and significant parameter for estimating gas diffusivity because CED is a measure of the energy required to open intersegmental gaps in the polymer matrix, which is the rate-limiting step in gas diffusion.

**Relation between diffusivity and elastic modulus.** For cases where E'>>E", which is the case for our polyimides, the elastic modulus, E, is nearly equal to E'. The elastic modulus of a polymer is a measure of flexibility and mobility of polymer chain segments(*29*). The elastic modulus is insensitive to molecular motions of both backbone and side chains. In a study of dynamic viscoelastic properties of films of random and block copolyimides containing semi-flexible polyimide groups and rigid rod polyimide groups, E' increases monotonically as the rigid rod polyimide content in the copolyimides increases(*30*). The elastic modulus was also expected to correlate with gas diffusivity because enhanced chain motion increases gas diffusion(*31*).

Figure 5 presents the correlation of log(D) and E' at 35 °C. The correlation between log(D) and E' was similar to the correlation between log(D) and CED. These results suggest that gas diffusivity is influenced by motion of not only the side chain but also that of the backbone chain in the polymer because E, which is nearly equal to E' in our study, includes effects of both primary dispersion based on microbrownian motion and secondary dispersion based on local relaxation modes or side chain motion. Gas diffusivity may increase as E' decreases due to an increase in segmental motion. Elasticity depends on the polymer structure, and E' is useful for the estimation and analysis of gas diffusion. Koros et al.(*17*), Yee et al.(*32*) and Muruganandam et al.(*33*) have studied the relation between diffusivities and results of viscoelastic measurements in polycarbonates. Their values of D and E' (E'>>E") agree well with our results in Figure 5. These results suggest that gas diffusivities may be strongly influenced by total motion of segments in many families of polymers, including polyimide.

**Relation between diffusivity and free volume.** Log(D) was not well-correlated with d-spacing, since d-spacing is not necessarily correlated with the amount of free space in the polymer. Correlations of log(D) or log(P) and $1/V_f$ in glassy polymers have been reported by many researchers. In these cases, $V_f$ includes a contribution from the excess volume in the glassy state. Figure 6 presents the correlation between log(D) and $1/V_f$ for the family of polyimides we have considered. 6FDA-polyimides had large values of D relative to BPDA-polyimides because the bulky $-CF_3$ substituents in 6FDA (the acid component in the polymer) hinder packing of polymer segments. Overall, these correlations seemed to be poor. Deviations were most significant for polyimides containing polar substituents. These polyimides had low diffusivities relative to their values of $V_f$. Similar behavior has been reported for polymers having side chains containing polar substituents(*17, 19, 33*). McHattie et al. reported that gas diffusion coefficients of polycarbonates containing polar substituents were low relative to their fractional free volume values(*17*). The relation between the diffusion coefficients of polar substituted polycarbonates by Muruganandam et al.(*33*) and fractional free volume values calculated from their densities also showed similar behavior. Puleo et al. reported that gas permeability coefficients of polystyrenes were low relative to

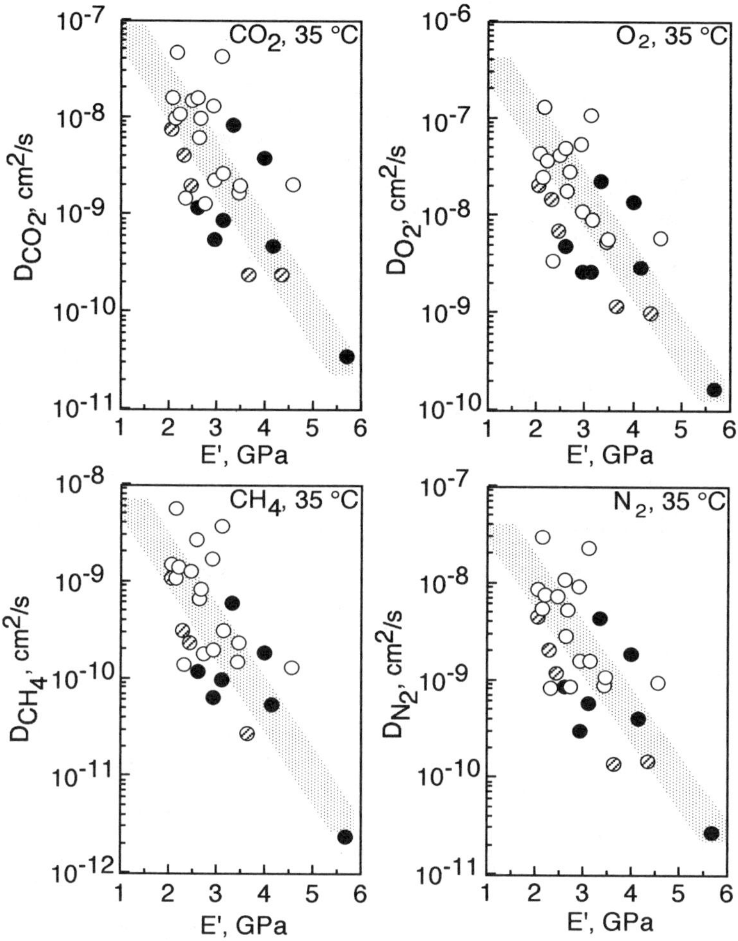


Ⓩ: This study
○: Polyimides in previous studies(*8,9*)
●: Polyimides containing polar substituents(*8,9*)

Figure 5. Relationship between D and E' at 35℃ for various gases

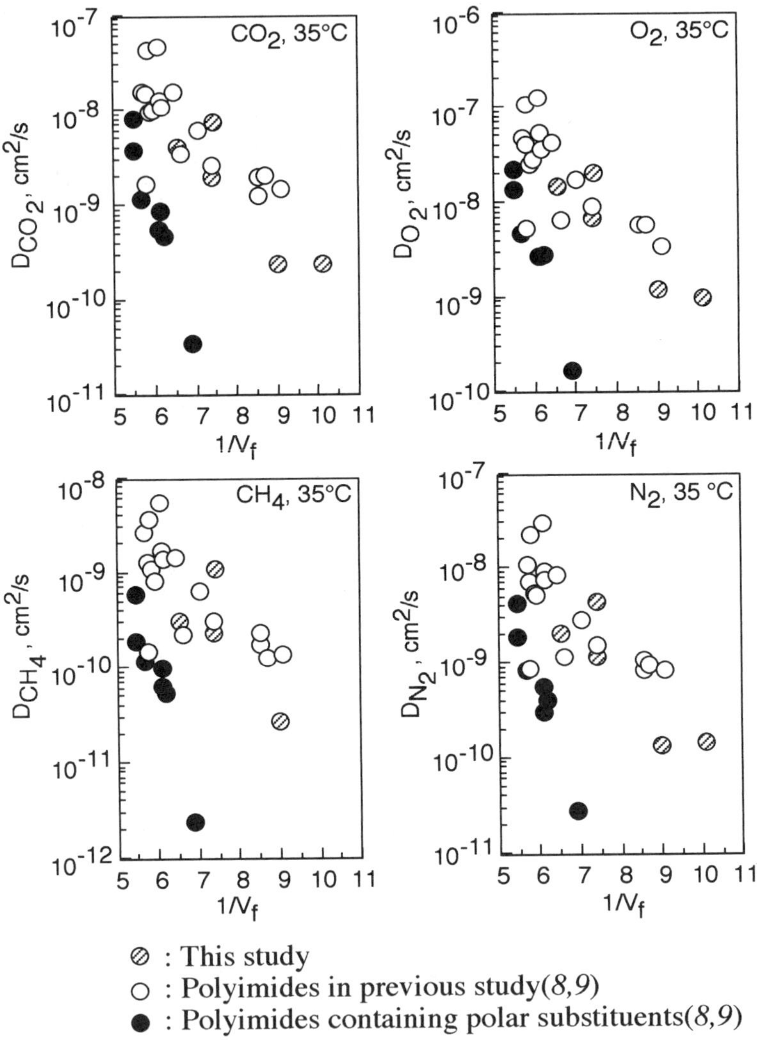


Figure 6. Relationship between D and $1/V_f$ at 35°C for various gases

the value anticipated based on the polymer's $V_f$ value as a result of the introduction of hydroxyl substituents(*19*). Within a subset of the polyimides (i.e. those which contained two polar substituents on two phenylene rings in the diamine component, such as BPDA-CDM, -HAB, 6FDA-CDM-MBHA and so on), the correlation between log(D) and $1/V_f$ was good. For a family of polyimides having ether bonds in the polymer backbone, such as BPDA-BAPS, -BAPE and -BAPP, the correlation of log(D) and $1/V_f$ was fairly good. This result suggests that increasing the number of ether bonds in the polymer increases $V_f$ and D due to increases in chain flexibility.

As above mentioned, the correlation between log(D) and $1/V_f$ for all of the polymers was not so good. However, the correlation between log (D) and $1/V_f$ was much better in families of polyimides of similar chemical structure. These results indicate that gas diffusivity values can not be rationalized solely on the basis of average values of static free volume.

**Prediction of gas diffusivity in glassy polymers.** CED is useful as it can be estimated conveniently without experimental measurements. In contrast, $V_f$ values require measurements of polymer density. Accordingly, we attempted to correlate CED and D for several families of glassy polymers. Figure 7 presents this correlation for polyimides(*1-5*), polycarbonates(*17,33*), polysulfones(*34*), polystyrenes(*19*), and the polyimides which we have studied. Overall, a good correlation between log(D) and CED was observed.

Correlations between log(D) and E', and between log(D) and CED were good. These results suggest that gas diffusivity is affected by the cohesive forces of polymers. Toi et al. and we have reported a correlation between log(D) and CED independently(*9,35-37*). Jia and Xu have reported correlations between gas permeability and a parameter which included contributions from both CED and $V_f$(*38*). CED depends on the square of the solubility parameter, which we treat as a total parameter with dispersive, polar, and hydrogen-bonding contributions. Van Krevelen(*39*) has indicated that a simple correlation between dielectric constant, $\varepsilon$, and CED is observed for polymers. This notion is expressed by Equation 11.

$$(CED)^{1/2} \fallingdotseq 7\ \varepsilon \tag{11}$$

Polymer dielectric constant has been correlated with log(P) by Matsumoto et al.(*40*) They indicate that the dielectric constant is a function of free volume and polarity.

As a factor related to chain motion, one may also consider polymer elastic modulus. As indicated in "Properties of Polymers"(*39*), at one time it was hoped that mechanical studies of polymers could be entirely replaced by electrical measurements. There are indeed close similarities between the general shapes and temperature-dependences of the mechanical and dielectric loss curves, but the quantitative connection between these phenomena is not as simple as was originally believed. Electrical measurements constitute a useful addition to, but not a substitute for, mechanical studies. However, the elastic modulus and the dielectric constant are related to similar physical behavior. These parameters are response functions obtained by stimulating a material and measuring the subsequent relaxation phenomena. Therefore, similar to the dielectric constant, the elastic modulus depends not only on chain motion but also on free space in the polymer matrix. In fact, relaxation phenomenon of

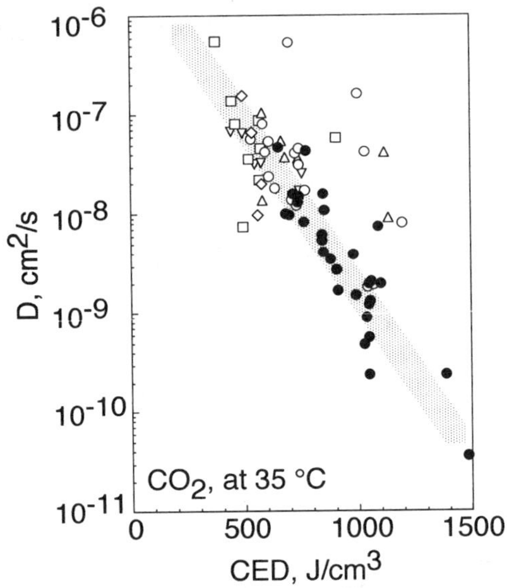


● : Polyimides in our study
○ △ : Polyimides by others(*1-5*)
□ : Polystyrenes(*19*)
▽ : Polycarbonates(*17,33*)
◇ : Polysulfones(*34*)

Figure 7. Correlation of D at 35℃ and CED of glassy polymers

mechanical dispersion is related to free volume in polymers(31). From Equation 11 and the previous discussion, CED also includes contributions related to both the mobility of polymer segments and the free space in the polymer matrix. Relative to free volume, CED and elastic modulus are more sensitive to polymer mobility. For gas diffusion, which proceeds by penetrant jumping from one hole to another, the free space (i.e. the holes) among segments is formed by chain motion in the glassy state. Both CED and elastic modulus are sensitive to chain motion to form free spaces. The elastic modulus has disadvantages for gas diffusion prediction. Disadvantages include the need to measure elastic modulus directly and the difficulty of measuring it accurately. In contrast, CED is useful for gas diffusion prediction in polymers because CED can be estimated without measurements.

On the other hand, there are more recent methods to predict the relation between permeability or diffusivity and polymer structure(*41,42*). Park and Paul reported a new group contribution method for estimating free volume(*41*). This approach provides a method for correlating data in a way that permits prediction of P or permselectivity behavior of polymer structures based on their chemical structure. This method assumes that the free volume is not constant for all gases for a specific polymer, and that the occupied volume depends upon both the gas and the structural unit. This group contribution method is useful because it does not require direct measurement of polymer density. We applied this group contribution method to our experimental data. The linear relation between ln(D) and $1/V_f$ is improved by using the new group contribution method. However, for polyimides such as BPDA-HAB and BPDA-MBHA, which have hydroxyl groups, the diffusion coefficients were low relative to their values based on the calculated fractional free volume(*41*) and deviated from the correlation line between D and $1/V_f$. These low diffusivity values reflect a rigid structure due to strong hydrogen bonding between hydroxyl groups(*19*). Additionally, we could not calculate $V_f$ for BPDA-MFA and -TFDM by their new group contribution method because the fluorine group is not included among the published group contribution units. On the other hand, as shown in Figure 7, diffusion coefficients in polycarbonates, polysulfones and polystyrenes are well-correlated with calculated CED.

Robeson et al. have also reported a group contribution approach to predict permeability and permselectivity for $O_2/N_2$ and $He/N_2$ gas pairs of aromatic polymers by determining the contribution of each structural unit of the polymer to the overall permeation behavior(*42*). Their ability to predict permeability/permselectivity values for polymers outside the database is also good. However, we could not apply their new group contribution method to our experimental data for polyimides because their group contribution units did not include all of the necessary structural units.

Figure 8 presents experimental permselectivity data plotted versus permselectivity estimated from the correlation between CED and permeability for $He/CH_4$ and $O_2/N_2$. Good agreement between experimental and predicted permselectivity for $O_2/N_2$ was demonstrated (similar to the results of Park et al. and Robeson et al.). Relative to $P_{O_2}/P_{N_2}$, the deviation between experimental and predicted results is slightly larger for $P_{He}/P_{CH_4}$ and $P_{CO_2}/P_{CH_4}$.

As shown by a comparison between CDM and MCA or between PASN and MASN, the calculation of CED using the group contribution method of Fedors(*23*) has some weaknesses. One of the weaknesses is that different

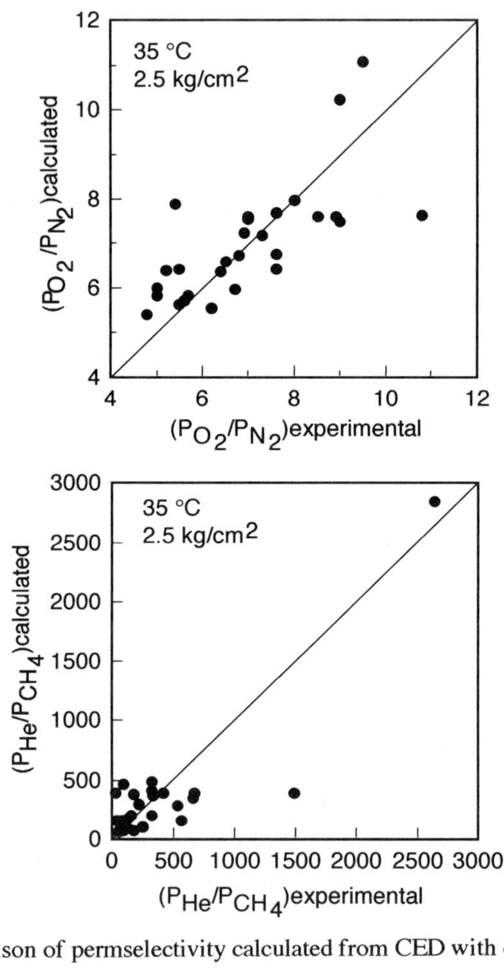


Figure 8. Comparison of permselectivity calculated from CED with experimental values

bond positions on the aromatic units (such as CDM and MCA) are not distinguishable. Another weakness is that interactions between polymer segments and penetrant molecules are not included in our analysis. Therefore, we recognize that the group contribution method to calculate CED should be further refined to provide more accurate estimates of CED. Although CED has some drawbacks, gas diffusivity prediction using CED is useful because CED can be calculated in various polymer structures without any measurements.

## Conclusions

We have investigated relations between apparent gas diffusion coefficients in glassy polymers and both calculated and measured physical parameters. Cohesive forces and polymer mobility influence gas diffusivities in our glassy polyimides. The correlation between D and common parameters (i.e. $V_f$, CED, E') of polyimides having side chain substituted polar groups differs fromthose of nonpolar polymers and weakly polar polymers. Gas diffusivity values were not well-correlated with free volume. Relative to free volume, CED and E' are more sensitive to polymer mobility. Estimation of gas diffusivities based on group contribution methods is possible by using factors, such as CED, which can be calculated without any measurements.

## Acknowledgments

The authors would like to express their sincere gratitude to Professor T. Sakakibara of Kagoshima University for synthesis and supply of the diamine monomers(i.e. MCA, MBA, MFA, TCDM and TFDM).

## Literature Cited

1. Tanaka, K.; Kita, H.; Okano, M.; Okamoto, K. *Polymer* **1992,** *33*, 585.
2. Coleman, M. R.; Koros, W. J. *J. Membrane Sci.* **1990,** *50*, 285.
3. Tanaka, K.; Kita, H.; Okamoto, K. *Sen-i Gakkaishi* **1990,** *46*, 541.
4. O'Brien, K. C.; Koros, W. J.; Husk, G. R. *J. Membrane Sci.* **1988,** *35*, 217.
5. Kita, H.; Okamoto, K. *Hyoumen* **1992,** *30*, 210.
6. Stern, S. A.; Mi, Y.; Yamamoto, H. *J. Polym. Sci., Polym. Phys. Ed.* **1989,** *27*, 1887.
7. Kim, T. H.; Koros, W. J.; Husk, G. R.; O'Brien, K. C. *J. Membrane Sci.* **1988,** *37*, 45.
8. Hirayama, Y.; Yoshinaga, T.; Kusuki, Y.; Ninomiya, K.; Sakakibara, T.; Tamari, T. *J. Membrane Sci.* **1996,** *111*, 169.
9. Hirayama, Y.; Yoshinaga, T.; Kusuki, Y.; Ninomiya, K.; Sakakibara, T.; Tamari, T. *J. Membrane Sci.* **1996,** *111*, 183.
10. Barbari, T. A.; Koros, W. J.; Paul, D. R. *J. Polym. Sci., Polym. Phys.Ed.* **1988,** *26*, 709.
11. Hensema, E. R.; Sena, M. E. R.; Mulder, M. H. V.; Smolders, C. A. *Gas Sep. Purif.* **1994,** *8*, 149.
12. Fujita, H. *Fortschr. Hocholym. Forsch.* **1961,** *3*, 1.
13. Gu, W.; Chern, R. T.; Chen, R. Y. S. *J. Polym. Sci., Polym. Phys. Ed.* **1991,** *29*, 1001.
14. Reimers, M. J.; Barbari, T. A. *J. Polym. Sci., Polym. Phys. Ed.* **1994,** *32*, 131.

15. Min, K. E.; Paul, D. R. *J. Polym. Sci., Polym. Phys. Ed.* **1988,** *26*, 1021.
16. Stern, S. A.; Trohalaki, S. *ACS Symposium Series* **1990,** *423*, 22.
17. McHattie, J. S.; Koros, W. J.; Paul, D. R. *J. Polym. Sci., Polym. Phys. Ed.* **1991,** *29*, 731.
18. Schult, K. A.; Paul, D. R. *J. Polym. Sci., Polym. Phys. Ed.* **1996,** *34*, 2805.
19. Puleo, A. C.; Muruganandam, N.; Paul, D. R. *J. Polym. Sci., Poltm. Phys. Ed.* **1989,** *27*, 2385.
20. Ghosal, K.; Freeman, B. D. *Polymers for Advanced Technologies*, **1994,** *5*, 673.
21. Naylor T. In *Comprehensive Polymer Science*, Allen, G., Eds.; Pergamon Press, New York, N.Y., **1989,** 643.
22. Bondi, A. *J. Phys. Chem.* **1964,** *68*, 441.
23. van Krevelen, D. W.; Hoftyzer, P. J. *Properties of Polymers*; 2nd Ed.; Elsevier: Amsterdam, The Netherlands, 1976; Chapter 7, pp130-140.
24. *Diffusion in polymers*; Crank, J.; Park, G. S. Eds.; Academic Press Inc.:London, England, 1968; Chapters 1, 6.
25. Meares, P. *J. Am. Chem. Soc.* **1954,** *76*, 3415.
26. Blume, I.; Smit, E.; Wessling, M.; Smolders, C. A. *Makromol. Chem., Macromol. Symp.* **1991,** *45*, 237.
27. Koros, W. J. *in Barrier Polymers and Structures*; Koros, W. J., Ed.; American Chemical Society: Washington, D.C., 1990; pp1.
28. van Krevelen, D. W.; Hoftyzer, P. J. *Properties of Polymers*; 2nd Ed.; Elsevier: Amsterdam, The Netherlands, 1976; Chapter 17, pp410-420.
29. Kuroda, N. Sen-i Gakkaishi **1979,** *35*, T413.
30. Oishi, Y.; Itoya, K.; Kakimoto, M.; Imai, Y. *Kobunnshi Ronbunshu* **1990,** *47*, 353.
31. Ferry J.D. *Viscoelastic properties of polymers*; 3rd Ed.; John Wiley & Sons, Inc.:NewYork, N.Y., 1980; Chapter 11, pp291-304.
32. Yee, A. F.; Smith, S. A. *Macromolecules* **1980,** *14*, 54.
33. Muruganandam, N.; Paul, D. R. *J. Membrane Sci.* **1987,** *34*, 185.
34. McHattie, J. S.; Koros, W. J.; Paul, D. R. *Polymer* **1991,** *32*, 2618.
35. Toi, K.; Suzuki, H.; Ito, T.; Ikemoto, I.; Kasai, T. *Polym. Preprints Jpn.* **1991,** *40*, 3401.
36. Hirayama, Y.; Yoshinaga, T.; Kusuki, Y.; Ninomiya, K.; Sakakibara, T.; Tamari, T. *Polym. Preprint Jpn.* **1992,** *41*, 642.
37. Toi, K.; Suzuki, H.; Ikemoto, I.; Ito, T.; Kasai, T. *J. Polym. Sci., Polym. Phys. Ed.* **1995,** *33*, 777.
38. Jia, L.; Xu, J. *Polym. J.* **1991,** *23*, 417.
39. van Krevelen, D. W.; Hoftyzer, P. J. *Properties of Polymers*; 2nd Ed.; Elsevier: Amsterdam, The Netherlands, 1976; Chapter 11, pp231-238.
40. Matsumoto, K.; Xu, P.; Nishikimi, T. *J. Membrane Sci.* **1993,** *81*, 15.
41. Park, J. Y.; Paul, D. R. *J. Membrane Sci.* **1997,** *125*, 23.
42. Robeson, L. M.; Smith, C. D.; Langsam, M. *J. Membrane Sci.* **1997,** *132*, 33.

Chapter 15

# Regiospecific Polyamide–Imides for Gas Separations

**Michael Langsam**

**Air Products and Chemicals, Inc., 7201 Hamilton Boulevard, Allentown, PA 18195–1501**

Regiospecific polyamide-imide (PAI) materials can be structurally represented as:

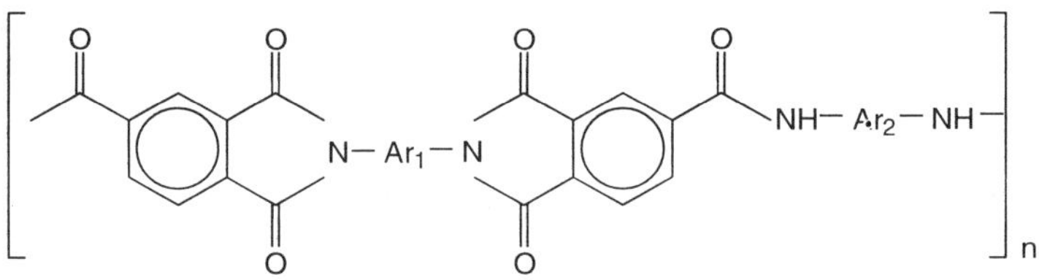


and are unusual in that the synthesis methodology will result in a polymer where the imide bridge ($Ar_1$) and the amide bridge ($Ar_2$) can be two different aromatic diamines. These polymers are prepared fully imidized under very mild solution techniques to form a high molecular weight material using trimellitic anhydride acid (TMA). By contrast the random polyamide-imides (i.e., Torlon; copolymer with ODA/*m*Pda) are prepared from trimellitic anhydride acid-chloride (TMAc) to the amide-amic acid stage and are then fabricated and imidized at high temperature. We have examined a group of alkylated 1,3 phenylene diamines in the regiospecific PAI structure to determine the effects of changes in the imide bridge ($Ar_1$) and in the amide bridge ($Ar_2$) on the properties of gas permeability (P) and gas selectivity ($\alpha$). As a general trend the imide bridge component controls selectivity while the amide bridge component controls permeability. The "optimum" structure has the smallest diamine in the imide component and the largest diamine in the amide component.

The separation of gas mixtures by polymeric membranes has become a commercially important methodology for a number of different systems (1). Several recent review articles have discussed the interaction between polymer structure and gas permeability properties (2,3). The quantification of the effect of polymer structure on gas transport properties recently has been reported (4,5) and it is now possible to optimize gas transport properties for well defined polymeric materials. For those materials which do not have a well defined data base it is necessary to prepare and measure the gas transport properties. The polyamide-imides (PAI) are a class of polymeric materials which do not have an extensive data base for gas transport properties (6,7). Work by Yamazaki and coworkers (8) demonstrated that PAI materials could be prepared easily and in a manner whereby the amide bond could be prepared from a phosphite activated carboxylic acid and an aromatic amine. Yang and coworkers (9-11) have shown that novel dicarboxylic acids could be prepared from trimellitic acid anhydride (TMA) and aromatic diamines and that these dicarboxylic acids could be coupled with a second diamine to form regiospecific PAI materials. Our focus was to examine the effects of a phenylene diamine and its alkylated analogs on the gas transport properties of regiospecific PAI materials and to identify those structures which maximized both permeability and selectivity.

## Experimental

**Reagents**. N,N Dimethylformamide (DMF); CAS #68-12-2; Aldrich anhydrous grade, was used as received. 1-Methyl-2-pyrrolidinone (NMP); CAS #872-50-4; Aldrich anhydrous grade, was used as received. *m*-Xylene; CAS #158-38-3; Aldrich anhydrous grade, was used as received. Calcium Chloride; CAS #10043-52-4; Aldrich Powder -30 to +80 Mesh, 97% pure was baked at 200°C under .1 micron pressure for 48 hours. Triphenylphosphite (TPP); CAS #101-02-2; Aldrich 97% active, was used as received. Pyridine; CAS #110-86-1; Aldrich anhydrous grade, was used as received. 1,2,4 Benzene tricarboxylic anhydride (TMA) CAS #552-30-7; Aldrich 97% active, was used as received. 1,3-phenylenedimine (*m*Pda); CAS #108452; Aldrich 99+%, was sublimed and stored under $N_2$. 2,4-Toluene diamine (2,4-TDA); CAS #95-80-7, a generous gift of R.C.V. Carr (>99.8% pure), was used as received. 2,6-Toluene diamine (2,6-TDA); CAS #823-40-5, a generous gift of R.C.V. Carr (>99.8% pure), was used as received. 2,4-Diaminomestylene (DAM); CAS #3102-70-3; EMS Dottikon, reagent grade (>99% pure), was used as received. Diethyl Toluene diamine [Isomer mix] (DETDA); CAS #68479-98-1, Albamarle Chemical, (>99% active), was used as received.

**$Ar_1$ DIDC Monomers: Preparation and Properties**. The general procedure is detailed for 2,6 TDA-DIDC. The following components were placed in a 500 ml 3-neck round bottom flask equipped with $N_2$ purge, mechanical stirrer and a dean-stark trap: TMA (25.0556 g 0.13 moles); DMF (75 ml). The mixture was heated to 60°C for 1 hour to dissolve the TMA. Then 2,6TDA (7.9411g; 0.065M), DMF (25 ml) and *m*Xylene (20 ml) were added to the reaction mixture and the dean-stark trap was filled with *m*Xylene. The temperature was raised to 190°C and water (2.5–3.0 ml) distilled off the reaction mixture with 70 ml of *m*Xylene/DMF solvent mixture over 3 hours. The 2,6TDA-DIDC monomer was soluble in the reaction mixture at 190°C but insoluble in DMF at room temperature. The solid was filtered, air dried for 24 hours then vacuum dried at 200°C for 48 hours. The dried monomer was analyzed using thermal gravimetric analysis (TGA) and differential thermal analysis (DTA) to determine residual solvent and melting point.

**Polymerization Techniques.** All of the experimental PAI materials used in this project were prepared by the Yamazaki technique (8). This procedure was employed by Yang and coworkers (9-11) in recent literature citations. The general procedure is detailed for 2,6 TDA-DIDC combined with DAM. The following components were placed in a 250 ml 3 neck round bottom flask equipped with a mechanical stirrer and a $N_2$ purge: 2,6 TDA-DIDC (9.4076 gm, 0.02 M); DAM (3.0044 gm, 0.02 M); $CaCl_2$ (10.1 gm); pyridine (20 ml, .25 M) and NMP (60 ml). The reaction mixture was heated to 100°C for 30 minutes. The reaction mixture was then cooled to 0°C with wet ice. Then TPP (13 ml, .05 M) was added with NMP (20 ml) and the reaction mixture was heated to 100°C for 3 hours with rapid stirring. After three hours the reaction mixture was cooled, diluted with 50 ml NMP and precipitated in $H_2O$ using a 1 gallon Waring blender. The solid was washed with water and then water/methanol mixtures at a ratio of 2/1; 1/2; and finally all methanol using the Waring blender for efficient extraction of inorganic salts and phosphate esters. The polymer was then air dried overnight and vacuum dried for 24 hrs at 150°C. The molecular weight was characterized by intrinsic viscosity measurements and the weight average molecular weight [Mw] was determined by gel permeation chromatography [GPC].

**Dense Film Formation and Gas Transport Measurements.** Polymer samples were converted into mechanically stable dense films using the following procedure: Polymer (~10 grams) was dissolved in ~70 grams of NMP to form an ~12.5% dope solution. The dope solution was passed through a Zitex® PTFE "Coarse" grade pressure filter. After degassing, the dope solution was cast into 150 mm casting rings and dried to a free standing film after 8 hours at 100°C and 200 mm pressure. This dense film was subsequently dried for 96 hours at .1 micron pressure and 295°C. All of these regiospecific PAI materials exhibited glass transition temperatures greater than 300°C. TGA analysis indicated less than 0.2% residual solvent. The dried dense films were stored under $N_2$ prior to testing. All gas transport measurements were performed in a microprocessor controlled manometric cell designed and constructed at Air Products. Four films were tested at the same time and the reported permeability values are an average of the four films. We used three gases: hydrogen, oxygen, and nitrogen. Each gas was given 24 hours for equilibrium to be established. Two consecutive tests were run at 24 hours and 48 hours for each gas. If the average of the two sets measurements were within 4% the gas feed to the test cells was changed. If the average of the two sets of measurements were greater than 4% the tests were continued in 24-hour increments until the average permeability values were within 4% in two consecutive sets of measurements.

**Polymer Synthesis for Regiospecific PAI Material.** As envisioned, the regiospecific PAI material would have the general structure of:

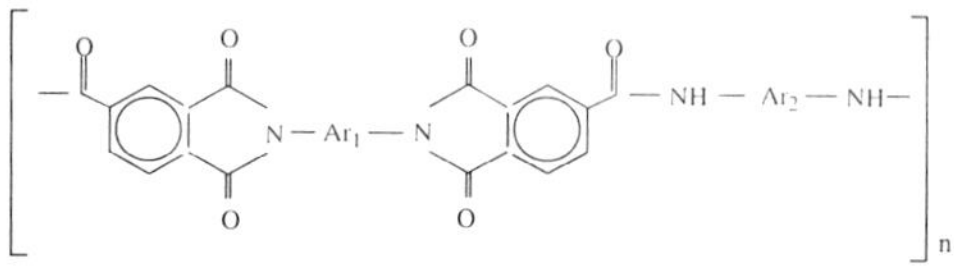

where $Ar_1$ and $Ar_2$ could be different aromatic centers. The overall synthesis would follow the following sequence:

**Preparation of $Ar_1$ DIDC.** The condensation of 2 moles of TMA with a mole of $H_2N\text{-}Ar_1\text{-}NH_2$ yields:

This $Ar_1$ DIDC monomer can be considered to be a modified isophthalic acid analog, where the distance and bond angle between the carboxylic acid groups are increased. Additionally the $Ar_1$ group should, depending upon structure, modify the packing and bond angles between the imide rings and thereby influence gas transport properties. The advantage of these $Ar_1$ DIDC monomers in the preparation of the regiospecific PAI materials will be emphasized later in this article. It is sufficient here to point out the solubility problems associated with Nomex and Kevlar, two commercial polyamides, to set the stage for later discussions.

**Polymerization of $Ar_1$ DIDC with $H_2NAr_2NH_2$.** The direct condensation between an aromatic amine and an aromatic carboxylic acid to form an amide bond is not a known reaction. The basicity of the amine is not sufficient to drive the reaction to completion. One of the components needs to be activated for the amide bond to form. Of the many ways to activate one of these components (either amine or carboxylic acid) we found that the Yamazaki coupling procedure offered the most straightforward route to high molecular weight aromatic polyamides for small scale synthetic procedures. In the Yamazaki Activation Technique (8) a triphenyl phosphite/pyridine activated complex interacts with the carboxylic acid group. The formation of the amide bond releases equimolar amounts of phenol and phosphorous oxy compounds into the reaction mixture. These side products subsequently are separated from the polymer by precipitation and multiple step washing of the polymer. The advantage of the Yamazaki reaction is that the precursor materials are all stable at room temperature and can be transferred with ease at atmospheric conditions.

By comparison, in the Higashi activation technique (12), a thionyl chloride/NMP complex is prepared at -10°C and used as an insitu complexing agent for the carboxylic group. The formation of amide bonds results in equimolar amounts of $SO_2$ and HCl being released. This procedure was discussed in great detail by Hong and coworkers (13). They highlighted the advantages of the insitu chlorination process both for activation of the DIDC monomer and for reduced costs.

Our first goal in these studies was to prepare a polyamide-imide of sufficient molecular weight so that mechanically stable dense films could be cast for permeability studies. The effects of experimental variables on molecular weight have not been extensively studied in this program. A recent publication by Russo and coworkers (14) describes the effects of experimental variables on the intrinsic viscosity of a *p*Pda/TPA polyamide using the Yamazaki coupling technique. The goal of these authors (14) was to maximize molecular weight. By comparison, we are seeking an intrinsic viscosity of ~0.5 dL/gm, which is sufficient to form strong films for permeability analysis.

**Diimidedicarboxylic Acid Monomers - Quality Control and Computer Modeling Studies.** The preparation of a number of candidate diimidedicarboxylic acid monomers ($Ar_1$ DIDC) followed the solution dehydration procedure previously noted in the literature (9-11). The details of the synthesis and purification of these $Ar_1$ DIDC monomers can be found in the Experimental Section. The monomers used in this study are noted in Table I. At present, the only quality control standards we have for these materials are TGA for residual solvent and DTA for melting point. There is a difference between the melting points reported in Table I and those cited in the literature (10) for *m*Pda based DIDC monomer. This may be due to an interpretation of the DTA derived values. At our laboratory we use the initial break in the DSC as the melting point of the monomer. In any event, the purity of our DIDC monomers is quite good, based on the high molecular weight of the subsequent PAI materials.

An illustration of a typical DTA trace for a DIDC monomer is illustrated in Figure 1 where 2,6 TDA-DIDC is used.

Please note that the TDA (IM) entry in Table I represents the DIDC monomer prepared from the 80/20 isomer mix of 2,4 TDA and 2,6 TDA. The melting point depression of the 2,4 isomer in the mixture and melting point elevation for the 2,6 isomer in the mixture is noted. At this point I have no explanation for this phenomenon. The diethyl TDA (IM) is an isomer mixture of the 2,6 placement of the methyl group (20%) and the 2,4 placement of the methyl group (80%).

All of the $Ar_1$ groups are in a 1, 3 (or meta) configuration between the TMA groups. The substituents on the $Ar_1$ unit should have an effect on the orientation of the $Ar_1$ ring between the two phenyl-imide fused rings. We estimate the equilibrium angle for the $Ar_1$ group using the Cerius2 molecular modeling program. These results are presented in Table II. The equilibrium bond angles increase as the bulk of the alkyl substitutents on the ring increases. This increase in bond angle can be illustrated by comparing two extreme cases: *m*Pda and the 2, 6 isomer of DETDA. Here the bond

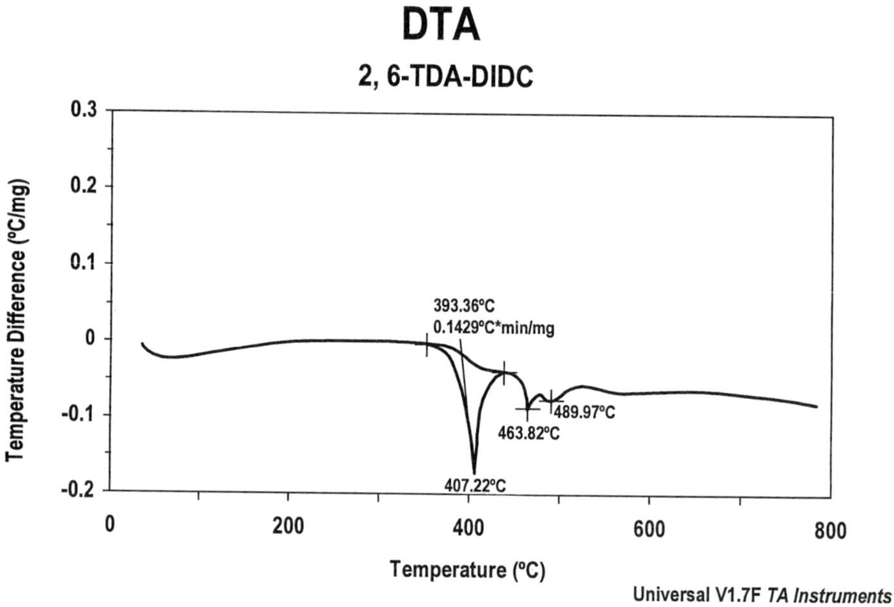


1: DTA for 2,6 TDA-DIDC

Table I
$Ar_1$ Diimide-Dicarboxylic Acid Candidate Monomers

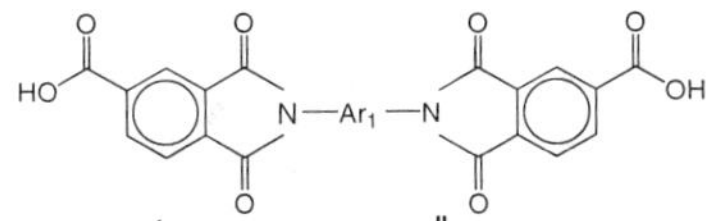


| $Ar_1$[1] | Melting Point[2] |
|---|---|
| *m*Pda | 315 (Lit [10] 321) |
| 2,6 TDA | 393 |
| 2,4 TDA | 334 |
| TDA (IM) | 329, 404 |
| DAM | 356 |
| DETDA (IM) | 352, 397 |

[1]structures associated with these monomer acronyms are in the Appendix
[2]in °C by DTA techniques

angles increase from 41° to almost 88°. These calculations were performed on the α,ω carboxylic acid terminated monomer. This rather large range in bond angles is reflected in a substantial change in the selectivity properties of the regiospecific PAI materials. This will be documented in Table VII where the imide ($Ar_1$) linkage is varied from *m*Pda to DETDA. We observe that as the bond angle between the $Ar_1$ and phthalimide ring increases, the selectivity in the bulk polymer decreases. While this is, at this point, only an observation, we are encouraged that these modeling results may offer rather direct insight into a physical measurement. A cartoon of the comparison of the *m*Pda-DIDC to that of the 2,6 diethyl TDA is illustrated in Figure 2.

**Computer Modeling of Amide Bond Structures**. The computer modeling of the imide bond demonstrated that the substituents on the $Ar_1$ structure had a large effect on equilibrium bond angles. The same analytical approach was taken for the corresponding amide linkage ($Ar_2$) to estimate the bond angles for these subunits. These results are reported in Table III. The overall results indicate that the $Ar_2$ group in the amide environment exhibits a uniformly higher bond angle than the $Ar_1$ in the imide environment. This observation of a uniformly higher bond angle between the $Ar_2$ ring and the phthalamide ring is reflected in the rather uniformly similar selectivity values that are observed for changes in the amide group which will be reported in Table V. It should be pointed out that these are only computer modeling results and, therefore, are only as good as the program modeling parameters. Additionally, we are using these computer generated bond angles for small segments of polymer chains to indicate behavior of penetrant molecules in three dimensional arrays of polymer chains in the solid state. While this represents a rather large extrapolation, we believe there is an interesting trend which will be seen in the gas transport values.

Table II
Bond Angles for Diimide Structures[1]

| $Ar_1$ | Bond Angle[2] |
|---|---|
| *m*Pda | 41.4 |
| 2,6 TDA | 60.2 |
| 2,4 TDA | 42.6[3] 59.7[4] |
| DAM | 80.6 |
| 2,6 DETDA | 83.5 |
| 2,4 DETDA | 85.3[5] 89.0[6] |

1 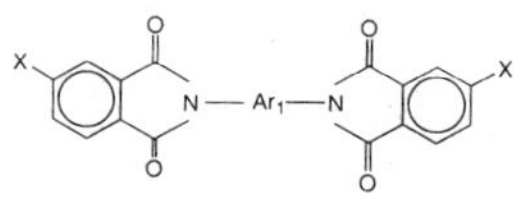 X ≡ carboxylic acid

2 bond angle phthalimide - $Ar_1$ rings (in degrees)
3 between H/H
4 between $H/CH_3$
5 between $CH_3/C_2H_5$
6 between $C_2H_5/C_2H_5$

Table III
Bond Angles for Diamide Structure[1]

| $Ar_2$ | Series A Bond Angle[2] |
|---|---|
| *m*Pda | 83.9 |
| 2,6 TDA | 84.7 |
| 2,4 TDA | 85.6[3] 85.5[4] |
| 2,6 DETDA | 83.9 |
| 2,4 DETDA | 81.8[5] 80.3[6] |

1 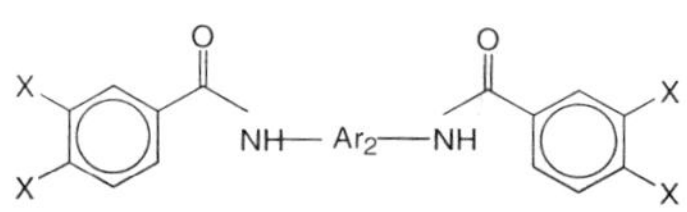

2 bond angle aromatic amide and $Ar_2$ rings (in degrees)
3 between H/H
4 between $H/CH_3$
5 between $CH_3/C_2H_5$
6 between $C_2H_5/C_2H_5$

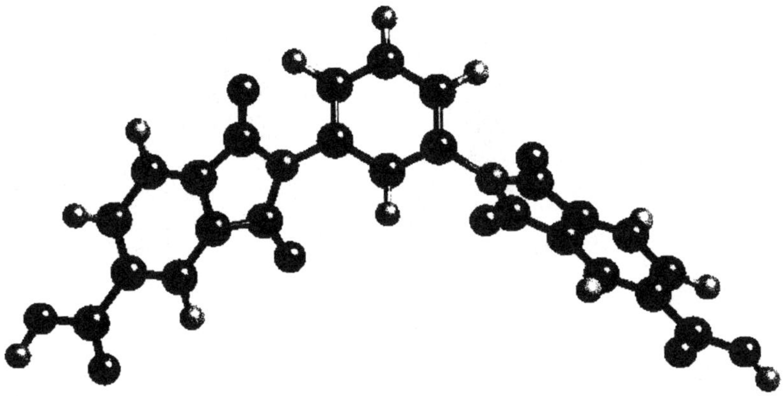

a

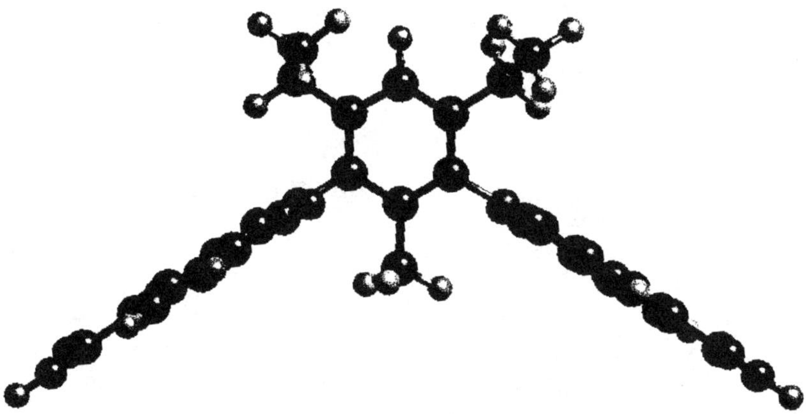

b

2: Cerius 2 molecular models of *m*Pda-DIDC and 2,6 DETDA-DIDC

**Gas Transport Properties for Regiospecific PAI Materials.** Before beginning a systematic synthesis study, we felt that we needed to understand solubility limitations for these regiospecific materials. Earlier it was pointed out that the commercial polyamide Nomex (*m*Pda/isophthalic acid) is soluble in hexamethyl phosphoramide, hot *m*-cresol and sulfuric acid. The all para form [Kevlar] (*p*Pda/terephthalic acid) is soluble only in concentrated sulfuric acid. With this understanding as to the limited solubility of unalkylated aromatic polyamides, we prepared the symmetric regiospecific PAI materials shown in Table IV. The initial material prepared was the all DETDA(IM) type. This had the bulkiest alkyl groups and had the highest likelihood for good solubility. Afterwards, we prepared the all DAM, all TDA (IM) and finally the all *m*Pa based material. The *m*PDA based material is the completely "bare" 1,3 placement material. As such, it can be considered a 1:1 combination of Nomex with *m*Pda/PMDA (polyimide). This material is completely soluble in NMP and forms a clear dense film on casting, thereby establishing that we should not experience solubility problems with any of the other 1,3/1,3 linked materials with ring alkyl groups.

Table IV
Intrinsic Viscosity of Symmetric Regiospecific PAI Materials

| $Ar_1 \equiv Ar_2$[1] | IV[2] |
|---|---|
| DETDA (IM) | .491 |
| DAM | .509 |
| TDA (IM) | .500 |
| *m*Pda | .562 |

[1] Organic structures associated with these monomer acronyms in Appendix
[2] Intrinsic viscosity - DMAc/30°C in dL/gm

**Variation in $Ar_2$ - Amide Bond.** A series of regiospecific PAI materials were prepared using DAM-DIDC and DETDA(IM)-DIDC with a number of different diamines in the amide ($Ar_2$) position. Permeability, selectivity, and intrinsic viscosity results for these polymers are summarized in Tables V and VI. The focus of this study was to examine the influence of size and volume of the substitutent in the amide ($Ar_2$) position for two different imide ($Ar_1$) linkages on permeability and selectivity properties. We anticipated, based on the volume effects of these imide ($Ar_1$) units, that there would be two blocks of experimental results with P increasing as we progressed from DAM to DETDA. The initial screening with DAM-DIDC (Table V) encompasses the widest range of amide ($Ar_2$) units. The DETDA-DIDC based polymers (Table VI) were only prepared with selected amide ($Ar_2$) units to permit a comparison of polymers with the two different imide ($Ar_1$) units.

The discussion begins with the DAM-DIDC based materials and demonstrates the effects of changes in the amide ($Ar_2$) structure on permeability (P)/selectivity ($\alpha$) properties. With the amide unit ($Ar_2$) varying from *m*Pda to DAM there is little change in $\alpha(O_2/N_2)$ and only a 2.4x increase in $P(O_2)$. This fits into our computer modeling observations regarding the amide ($Ar_2$) linkage reported in Table III. A working hypothesis to rationalize this observation is that increasing the ring methyl groups from 0 (*m*Pda) to 3 (DAM) increases the number of gas transport sites without increasing the average size [thereby reducing $\alpha(O_2/N_2)$] of the gas transport site. The effects for P(He) and $\alpha$(He/$N_2$) are not as sharply defined as for $P(O_2)$ and $\alpha(O_2/N_2)$, probably because of the larger size differential between the kinetic diameters of helium (2.60 $Å^2$) and nitrogen (3.64 $Å^2$) vs. oxygen (3.46 $Å^2$) and nitrogen (3.64 $Å^2$).

A comparison of 2,6 TDA, 2,4 TDA and TDA (IM) indicates that both $P(O_2)$ and $\alpha(O_2/N_2)$ follow the expected trends. The 2,6 isomer, being symmetric has lower $P(O_2)$ and higher $\alpha(O_2/N_2)$ when compared to the 2,4 isomer. The 80/20 Isomer Mix form of TDA has $P(O_2)$ and $\alpha(O_2/N_2)$ values which fall between the 2,4 and 2,6 with the expected bias towards the 2,4 isomer. We hypothesize that in the amide environment both of these diamines (i.e., *m*Pda and TDA(IM)) are similarly shielded by amide hydrogen bonding so that they behave in a similar fashion with regard to gas transport. It is only when more ring methyl groups (3 methyls: DAM) or bulkier substituents are used (DETDA) that there is a change in P and $\alpha$ properties. Increasing the number of ring alkyl groups (DETDA [IM]) of the $Ar_2$ unit leads to an increase in $P(O_2)$ and a loss in $\alpha(O_2/N_2)$. In these materials we hypothesize that there is an increase in the average size of gas transport sites.

Table V
Effects of $Ar_2$ Unit on Polymer Characteristics of Regiospecific PAI Based on DAM-DIDC

| $Ar_2$[(1)] | Permeability Coefficient[(3)] | | | Selectivity | | IV[(4)] |
|---|---|---|---|---|---|---|
| | $O_2$ | $N_2$ | He | $O_2/N_2$ | He/$N_2$ | |
| *m*Pda | 1.45 | .218 | 19.0 | 6.67 | 87.2 | 1.04 |
| 2,6 TDA | 1.52 | .230 | 18.2 | 6.61 | 79.1 | .630 |
| 2,4 TDA | 2.01 | .308 | 21.7 | 6.53 | 70.5 | .727 |
| TDA (IM)[(2)] | 1.85 | .281 | 20.3 | 6.58 | 72.3 | .672 |
| DAM | 3.47 | .536 | 36.3 | 6.47 | 67.7 | .509 |
| DETDA (IM)[(2)] | 4.75 | .808 | 37.4 | 5.88 | 46.3 | .412 |

(1) Organic structures associated with these monomer acronyms in Appendix
(2) IM = Isomer Mix; typically 80% 2,4 and 20% 2,6
(3) Permeability coefficients (P) in barrers $\left[\frac{scc}{sec}\cdot\frac{cm}{cm^2}\cdot\frac{1}{cm(Hg)}\times 10^{10}\right]$ measured at 25°C
(4) Intrinsic Viscosity - DMAc/30°C in dL/gm

When we examine the more limited case of experimental materials prepared with DETDA-DIDC (Table VI) and then compare the three similar amide ($Ar_2$) units we have a rather startling difference for $P(O_2)$ and $\alpha(O_2/N_2)$. Placing DETDA in the imide ($Ar_1$) segment increases $P(O_2)$ for each comparative member but decreases $\alpha(O_2/N_2)$. In both data sets, the polymers in which the $Ar_2$ units are *m*Pda or DAM have similar $\alpha(O_2/N_2)$ values, and polymers containing DAM have the higher $P(O_2)$ values. These results are similar to those for the DAM-DIDC materials (Table V). Increasing the degree of ring alkylation or size of the alkyl groups as with DETDA (IM) in the amide ($Ar_2$) unit decreases $\alpha(O_2/N_2)$ and increases $P(O_2)$. The rather uniform selectivity for the smaller amide units (*m*Pda and DAM) was observed for both sets of materials (Tables V and VI).

Table VI
Effects of $Ar_2$ Unit on Polymer Characteristics of Regiospecific PAI Based on DETDA-DIDC

| $Ar_2$[1] | Permeability Coefficient[3] | | | Selectivity | | IV[4] |
|---|---|---|---|---|---|---|
| | $O_2$ | $N_2$ | He | $O_2/N_2$ | $He/N_2$ | |
| *m*Pda | 2.61 | .454 | 21.4 | 5.75 | 47.1 | .637 |
| DAM | 4.94 | .881 | 35.9 | 5.61 | 40.7 | .480 |
| DETDA (IM)[2] | 6.49 | 1.28 | 38.4 | 5.07 | 30.0 | .491 |

(1) Organic structures associated with these monomer acronyms in Appendix
(2) IM = Isomer Mix; typically 80% 2,4 and 20% 2,6
(3) Permeability coefficients (P) in barrers $\left[\frac{scc}{sec}\cdot\frac{cm}{cm^2}\cdot\frac{1}{cm(Hg)}\times 10^{10}\right]$ measured at 25°C
(4) Intrinsic Viscosity - DMAc/30°C in dL/gm

These results demonstrate that $P(O_2)$ and $\alpha(O_2/N_2)$ vary depending on the amide ($Ar_2$) units and whether DAM or DETDA is in the imide ($Ar_1$) unit, the best combinations of P and $\alpha$ are observed with small units in the imide ($Ar_1$) unit and up to DAM in the amide ($Ar_2$) unit. Since there is little change in selectivity ($\alpha$) as the amide ($Ar_2$) portion is varied from *m*Pda to DAM the next set of experiments would investigate smaller imide ($Ar_1$) components with varying amide ($Ar_2$) components.

**Variations in $Ar_1$-Imide Bond**. In Table VII a second series of regiospecific PAI materials were prepared using DAM and other diamines in the amide ($Ar_2$) unit of the polymer while systematically varying the imide ($Ar_1$) portion of the polymer. Here we used DAM or bulkier diamines in the amide portion to increase the permeability characteristics of these materials. At the other extreme we used *m*Pda and TDA (IM) in the amide ($Ar_2$) portion to study the symmetrical materials. The all *m*Pda analog of these regiospecific PAI materials is completely soluble in the imide form in NMP casting solution. The gas transport properties for this all *m*Pda form are, as expected, very low. The $He/N_2$ selectivities are very attractive. This material should be an excellent candidate for separating hydrogen/methane mixtures. By comparison, the all TDA analog is more permeable and somewhat less selective than the all *m*Pda material. These results are provided in the first two entries in Table VII. In the remainder of Table VII we have varied the imide ($Ar_1$) portion while maintaining two different segments for the amide ($Ar_2$) portion. We will analyze the data for PAI homopolymers in two subgroups: $Ar_2$ = DAM; $Ar_2$ = DETDA.

Table VII
Effects of $Ar_1$ Unit on Polymer Characteristics of Regiospecific PAI

| $Ar_1$[(1)] | $Ar_2$[(1)] | Permeability Coefficient[(3)] | | | Selectivity | | IV[(4)] |
|---|---|---|---|---|---|---|---|
| | | $O_2$ | $N_2$ | He | $O_2/N_2$ | $He/N_2$ | |
| *m*Pda | *m*Pda | .0607 | .00715 | 2.77 | 8.49 | 380 | .562 |
| TDA(IM)[(2)] | TDA(IM)[(2)] | .239 | .0280 | 6.78 | 8.53 | 242 | .500 |
| *m*Pda[(5)] | DAM | .277 | .0305 | 7.61 | 9.09 | 249 | |
| 2,6 TDA | DAM | .747 | .0949 | 14.1 | 7.87 | 149 | |
| TDA(IM)[(2)] | DAM | .627 | .0830 | 11.4 | 7.55 | 137 | |
| DAM | DAM | 3.47 | .536 | 36.3 | 6.47 | 67.7 | .509 |
| DETDA[(2)] | DAM | 4.94 | .881 | 35.9 | 5.61 | 39.7 | .480 |
| *m*Pda | DETDA[(2)] | .528 | .0781 | 9.63 | 6.76 | 123 | |
| 2,6 TDA | DETDA[(2)] | 1.16 | .170 | 16.3 | 6.82 | 95.9 | .414 |
| TDA(IM)[(2)] | DETDA[(2)] | 1.03 | .156 | 13.6 | 6.60 | 87.2 | .412 |
| DETDA[(2)] | DETDA[(2)] | 6.49 | 1.28 | 38.4 | 5.07 | 30.0 | .491 |

(1) Organic structures associated with these monomer acronyms in Appendix 11.2
(2) IM = Isomer Mix; typically 80% 2,4 and 20% 2,6
(3) Permeability coefficients (P) in barrers $\left[\frac{scc}{sec}\cdot\frac{cm}{cm^2}\cdot\frac{1}{cm(Hg)}\times 10^{10}\right]$ measured at 25°C
(4) Intrinsic Viscosity - DMAC/30°C in dL/gm
(5) Polymer did not redissolve on casting, values based on extrapolation from copolymer data

**Using DAM as the amide ($Ar_2$) unit**. In this analysis of physical properties as a function of changes in the imide ($Ar_1$) bond let us focus first on entries three to seven in Table VII; *m*Pda to DETDA. Earlier we showed in Tables V and VI that for the range of amide ($Ar_2$) units, the $\alpha(O_2/N_2)$ was insensitive to the number of methyls from 0 (*m*Pda) to 3 (DAM); only $P(O_2)$ increased. Anticipating that we could maximize both $P(O_2)$ and $\alpha(O_2/N_2)$, we prepared the regiospecific PAI materials presented in Table VII, focusing on the effects of changes in the aromatic diamine in the imide ($Ar_1$) environment keeping the amide ($Ar_2$) as DAM. The *m*Pda-DIDC/DAM P/$\alpha$ properties are estimated from graphical extrapolation of copolymer data. This material was, after polymerization, drying and preparation of the casting dope, insoluble during the film formation step. This result was confirmed by preparing a new batch of this polymer, which was also insoluble. A more aggressive solvent (*m*-cresol) should be sufficient to achieve complete solubility but the estimated P/$\alpha$ properties for this material did not warrant such efforts. The 2,6 TDA form has very attractive gas transport properties for nitrogen. Using TDA (IM) as a starting material the DIDC form was reacted with DAM. The resulting polymer had a P($O_2$) of 0.627 barrers and an $O_2/N_2$ selectivity of 7.5. A comparison of the 2,6 TDA form with the TDA(IM) form indicates that both P($O_2$) and $\alpha(O_2/N_2)$ are lower in the isomer mix than in the symmetric 2,6 form. This effect is also observed for the DETDA analogs.

As we increase the ring alkylation we find that $O_2$ permeability increases and $O_2/N_2$ selectivity decreases. For the DAM/DAM form we now have 3 ring methyl substituents on $Ar_1$ and the oxygen permeability and $O_2/N_2$ selectivity decrease. Increasing the alkyl groups on $Ar_1$ to methyl/diethyl [DETDA] completes the series. With this material we have maximized permeability and minimized selectivity.

**Using DETDA as the amide ($Ar_2$) unit**. In the second set of four PAI materials with DETDA in the amide ($Ar_2$) position, we find that in all cases the permeability coefficient increases and the selectivity decreases compared to analogs containing DAM in the amide ($Ar_2$) position. The *m*Pda material is completely soluble in NMP. The 2,6 TDA and TDA (IM) polymers are also soluble in NMP. The polymer with TDA (IM) has P($O_2$) = 1.03 and $\alpha(O_2/N_2)$ = 6.60. The last entry for this subset is the DETDA-DIDC/DETDA. This material, which contains very bulky aromatic amines exhibits the highest permeability and lowest selectivity of all of the polymers in this table.

**Effects of Copolymerization on P/$\alpha$ Response**. Earlier we examined the effects of changing the structure of $Ar_1$ (imide) and $Ar_2$ (amide) and presented the working hypothesis that changes in the $Ar_1$ (imide) environment were more effective than changes in the $Ar_2$ (amide) environment on P/$\alpha$ properties. To corroborate this we prepared a variety of copolymers of $Ar_1$ type and of $Ar_2$ type and measured gas transport properties for those polymers. The results are presented in Table VIII.

The insolubility of *m*Pda-DIDC/DAM led us to consider the effects of changes in the average number of methyl group on the aromatic rings of both the imide portion ($Ar_1$) and amide portion ($Ar_2$) of these materials. The five materials focus on the effects of changing $Ar_1$ and $Ar_2$ with *m*Pda, and DAM (which are both forms of the 1,3 aromatic ring). This copolymerization procedure and the graphical display of the data also provides a qualitative indication of the precision of our P($O_2$) and P($N_2$) measurements. If the permeability values were subject to significant random errors during the measurement process, then we expect a degree of scatter of the points about the line of log P vs. composition. In actuality, the data array in Figures 3 and 4 [P($O_2$) and P($N_2$) vs. composition] were very well described by a straight line, indicating little scatter in the permeability coefficient measurements. As an added feature, this approach permits us to determine the P/$\alpha$ properties for *m*Pda DIDC with DAM. The copolymer with an $O_2$ permeability of .759 barrers and an $O_2/N_2$ selectivity of 8.25 has very attractive gas separation properties.

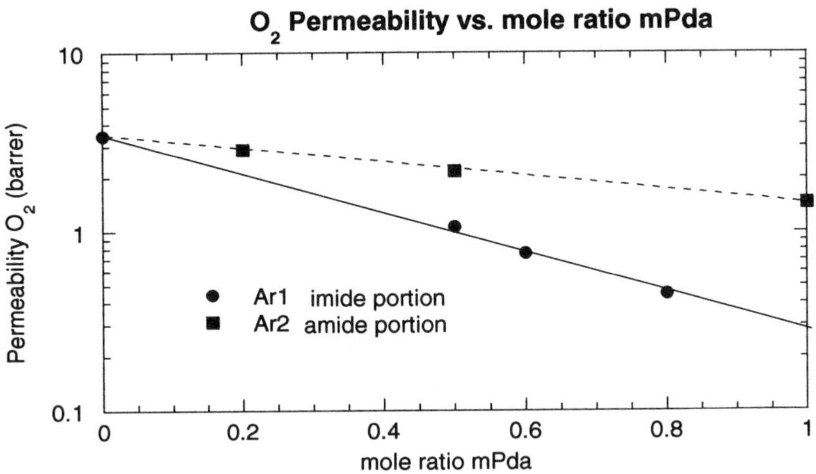


3: Oxygen Permeability vs Mole Fraction *m*Pda

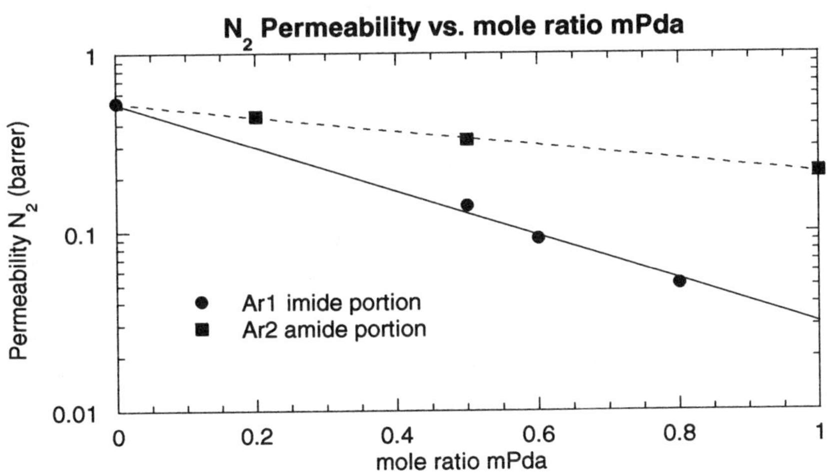


4: Nitrogen Permeability vs Mole Fraction of *m*Pda

Table VIII
Effects of Copolymerization in $Ar_1$ and $Ar_2$ on Polymer Characteristics of Regiospecific PAI

| $Ar_1$[1] | $Ar_2$[1] | Permeability Coefficient[3] | | | Selectivity | | IV[4] |
|---|---|---|---|---|---|---|---|
| | | $O_2$ | $N_2$ | He | $O_2/N_2$ | $He/N_2$ | |
| *m*Pda (.5)[a] DAM (.5)[a] | DAM | 1.07 | .141 | 17.2 | 7.59 | 122 | .453 |
| *m*Pda (.6)[a] DAM (.4)[a] | DAM | .759 | .0920 | 14.2 | 8.25 | 154 | .877 |
| *m*Pda (.8)[a] DAM (.2)[a] | DAM | .447 | .0512 | 10.3 | 8.73 | 201 | .843 |
| DAM | *m*Pda (.2)[a] DAM (.8)[a] | 2.91 | .449 | 31.8 | 6.48 | 70.8 | .535 |
| DAM | *m*Pda (.5)[a] DAM (.5)[a] | 2.19 | .331 | 25.2 | 6.52 | 74.6 | .580 |

(1) Organic structures associated with these monomer acronyms in Appendix
(2) IM = Isomer mix; typically 80% 2,4 and 20% 2,6
(3) Permeability coefficients (P) in barrers $\left[\frac{scc}{sec}\cdot\frac{cm}{cm^2}\cdot\frac{1}{cm(Hg)}\times 10^{10}\right]$ measured at 25°C
(4) Intrinsic Viscosity - DMAc/30°C in dL/gm
(a) The figures represent the mole fraction of aromatic diamines in these copolymer

The relative slopes of the log P vs. composition curves would be a good indication of the "sensitivity" of the number and distribution of methyl groups on the aromatic rings on gas transport properties. In this study we focus on the effects of substituting *m*Pda for DAM in both the imide ($Ar_1$) and amide ($Ar_2$) portions of the polymer. One of the end points will be the all DAM based material with $P(O_2) = 3.47$ barrers and $\alpha(O_2/N_2) = 6.47$. The extrapolation of the *m*Pda imide data to the all *m*Pda composition permitted us to estimate the P/$\alpha$ values reported in Table VII. The slope for the $Ar_1$ line is greater than for the $Ar_2$ line. All of these data indicate the following general trend:

A small $Ar_1$ unit (i.e,. *m*Pda) increases $\alpha$ ($O_2/N_2$) for the overall polymer.

A large $Ar_2$ unit (i.e., DAM) increases P ($O_2$) for the overall polymer.

## Conclusions

These regiospecific polyamide-imides allow us to control the number and type of alkyl substituents in both the imide ($Ar_1$) and amide ($Ar_2$) portions of this polymer and to determine the effects of these substituents on the properties of these materials.

Cerius 2 molecular models of both the diimide and diamide segments were estimated for different alkyl substituents. The diimide was more sensitive to variations in alkyl substituents, varying from 41° for no methyl groups (*m*Pda) to 88° for 1 methyl and 2 ethyl groups (2,4 DETDA). The permeability properties increased for all gases as the number and size of the alkyl groups increased in both the imide ($Ar_1$) and amide ($Ar_2$) portions of the polymer. In the amide portion of the polymer varying the substituents from none (*m*Pda) to three methyls (DAM) has no effect on selectivity of oxygen/nitrogen. Increasing the alkyl substituents to methyl/diethyl (DETDA) causes a drop in the selectivity of oxygen/nitrogen. In the imide ($Ar_1$) portion of the polymer increasing the alkyl substituent size and number increases permeability for all gases and decreases selectivity for all gas pairs. A combination of imide ($Ar_1$) segment with the minimum alkyl substituents and DAM in the amide ($Ar_2$) portion of the polymer should provide both high permeability with high selectivity.

## Appendix Acronyms/Chemical Structures

TMA

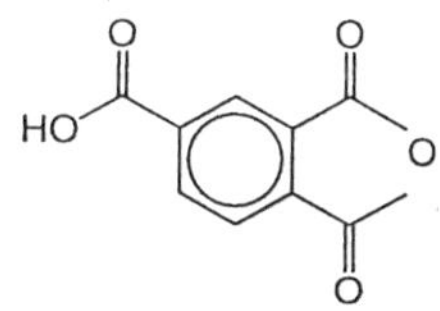


*m*Pda

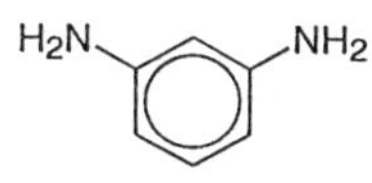


TDA (IM)

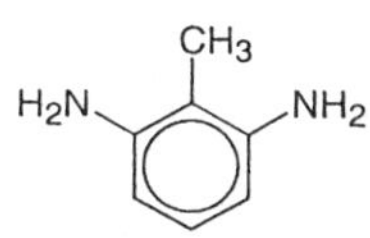


(2,6)

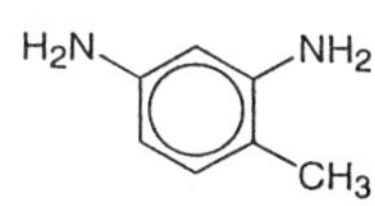


(2,6)

DETDA (IM)

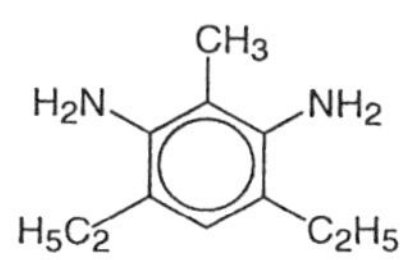


(2,6)

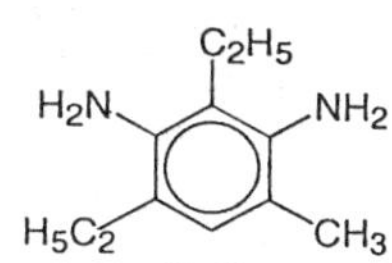


(2,6)

DAM

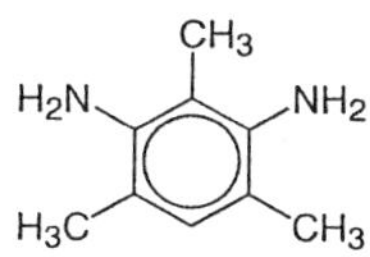

**References**

(1) Puri, P. S. *Membrane Journal*, **1996,** *6*, 117

(2) Stern, S. A. *J. Membrane Sci.*, **1994**, *94*, 1

(3) Langsam, M. In *Polyimides: Fundamentals and Applications*; Ghosh, M. K. and Mittal, K. L. Ed.; Marcel Dekker: New York, N.Y., 1996; pp 697-741.

(4) Park, J. Y.; Paul, D. R. *J. Memb. Sci.*, **1997**, *125,* 23

(5) Robeson, L. M.; Smith, C. D.; Langsam, M. *J. Memb. Sci.*, **1997**, *132*, 33

(6) Buys, H. C. W. M.; VanElven, A.; Jansen, A. E.; Tinnemans, A. H. A. *J. Appl. Polym. Sci*, **1990**, *41*, 1261

(7) Fritsch, D.; Peinemann, K-V. *J. Memb. Sci.*, **1995**, *99*, 29

(7a) Fritsch, D.; Avella, N. *Macromol Chem. and Phys.*, **1996**, *197*, 701

(8) Yamazaki, N.; Higashi, F.; Kawataba, J. *J. Polym. Sci., Polym. Chem.*, **1974**, *12*, 2149

(9) Yang, C-P.; Hsiao, S-H. *Makromol. Chem.*, **1989**, *190*, 2119

(10) Yang, C-P.; Hsiao, S-H.; Lin, J-H. *Makromol. Chem.*, **1992**, *193*, 1299

(11) Hsiao, S-H.; Yang, C.-P.; Wu, F-Y. *J. Polym. Sci., Polym. Chem.*, **1994,** *32*, 1481

(12) Higashi, F.; Nishi, T. *J. Polym. Sci., Polym. Chem.*, **1986**, *24*, 701

(13) Hong, Y-T.; Jin, M. Y.; Suh, D-H.; Lee, J. H.; Choi, K-Y., *Die Angewante Makromol. Chemie*, **1997**, *248*, 105

(14) Russo, S.; Mariani, A.; Mazzanti, L. E. *Macromol Symp.*, **1997**, *118*, 73

## Chapter 16

# Permselectivities of Poly(amide imide)s and Similar Poly(ester imide)s as Dense Films and Thin-Film Composite Membranes

**Detlev Fritsch**

**Institut für Chemie, GKSS Forschungszentrum GmbH, Max-Planck-Strasse, D-21502 Geesthacht, Germany**

Poly(amide imide)s (PAI's) are rarely tested as gas separation materials. We have recently discovered PAI's with and without 6F-containing groups which have reasonably high gas permeability and selectivity. To understand the influence on permselectivity of the amide and the ester bonds, PAI's and poly(ester imide)s (PEI's) differing only in this particular bond were synthesized, and their permeability and diffusivity to the gases He, $H_2$, $O_2$, $N_2$, $CO_2$ and $CH_4$ was determined. As an additional variation in structure, 4 methyl-groups were introduced in the ortho-position to the ester bonds. The para-linked PAI exhibits lower permeability than the analogous meta-linked PAI but offers higher selectivity (e.g. about 40% increase for $H_2/CH_4$). Gas permeability coefficients of PEI's are slightly higher than their PAI analogues. The introduction of 4 methyl-groups results in a marked increase in gas permeability (e.g. for $O_2$, permeability increases from 10.5 to 18.9 Barrer) and even a slight increase in selectivity. Due to the excellent film forming properties of the polymers, defect-free thin film composites with high fluxes and thicknesses as low as 200 nm were also prepared.

Aromatic polyimides have been identified as interesting materials for gas separation applications (*1-3*). They combine high permeability at considerable selectivity with superior mechanical strength and long term durability even at elevated temperature. Within this class of polymers, however, permeability covers several decades. For example, in the case of oxygen, permeabilities from < 0.001 to > 100 Barrer have been reported. The highest permeability was observed in polymers with rigid structures bearing bulky alkyl substituents ortho to the imide ring. Additional functional groups such as ether, ester, and amide may be introduced to vary the properties. Poly(amide imide)s (PAI's) are rarely tested for their gas permeation properties (*1, 3-7*) even though they can be tailored to form competitive membrane materials (*8*). Some recently synthesized PAI's offer the chance to prepare similar poly(ester imide)s (PEI's) which differ only in this particular bond. The permeability and selectivity of these polymers to He, $H_2$, $O_2$, $N_2$, $CO_2$ and $CH_4$ permits a determination of the influence of this structural variable on gas transport properties.

The basic gas permeation properties are not the only limiting factor in the use of polymers in gas separation. Film formation properties are as important as the gas permeability and selectivity of a polymer in determining its potential as a membrane material. Because the gas flux through a membrane depends most critically on thickness, the challenge is to make the membrane film - the active separation layer - as thin as possible while avoiding defects. The PAI and PEI polymers were, therefore, also tested for preparation and use as thin film composite membranes.

## Experimental

Monomer and polymer syntheses of PAI were performed as reported previously (*9*). Diacids with preformed imide groups were reacted with diamines to yield the PAI in one step (*9-11*). The diphenol 2,2-bis(4-hydroxyphenyl)hexafluoropropane for the PEI synthesis was obtained from Aldrich. The analogous diol with 4 methyl groups ortho to the hydroxy moiety (2,2-bis(3,5-dimethyl-4-hydroxyphenyl)hexafluoropropane) was synthesized by condensing hexafluoroacetone hydrate with 2 moles of 2,6-dimethylphenol following a patent (*12*). PEI's were obtained from diacids and diols in dry N,N-dimethylformamide (DMF)/pyridine (1/1) and an excess (2.6 mole/mole diacid) of tosyl chloride (*13*). Reaction time was 3 to 5 h at 120 °C under a dry argon flux. Reduced viscosities (0.5% at 30°C in N-methylpyrrolidone (NMP)) of > 0.35 dL/g were required for the preparation of free-standing, non-brittle, methanol stable films.

Solvent-free films about 30-90 µm thick were prepared (see ref. *8*). Their gas permeation properties were measured with pure gases using a self-built vacuum time-lag apparatus. Permeate pressure increase with time was recorded at 30°C by two MKS baratron pressure sensors (10 mbar maximum (permeate), 5 bar maximum (feed)) which were connected to a computer. Feed pressure was varied from 0.2 to 1 bar. Permeate pressure was $<10^{-3}$ mbar at the beginning of the experiment and was recorded up to 0.5-9 mbar, depending on the feed gas. Permeability was calculated from the slope of the permeate pressure versus time data in the steady state region. Apparent diffusion coefficients, $D_a$, were estimated from the time-lag $\theta$ by $D_a = l^2/6\theta$ ($l$ being the film thickness). Apparent solubility coefficients, $S_a$, were calculated from $S_a = P/D_a$.

Thin film composite membranes were prepared on in-house fabricated porous supporting membranes made of either polyetherimide (Ultem®) or polyacrylonitrile (PAN). Gas fluxes of composite membranes were measured at 1-5 bar feed pressure (pure gases) and room temperature using a soap-film bubble flow meter. In this case, the permeate pressure was the ambient atmospheric pressure.

## Results and Discussion

The PAI's were prepared as previously reported (*9*) following a direct condensation method (*10*) using diacids with built-in imide groups and sublimated diamines. The reaction was performed in NMP at 115 °C. 10 mmol of diacid in 30 ml of dry NMP were mixed with 16 ml of pyridine, 20 mmol of triphenyl phosphite (TPP) and 2 g of LiCl. 10 mmol of the diamine, dissolved in 30 ml of NMP, was rinsed in and the temperature was raised to 115 °C. It was stirred for 3 h at 115 °C under $N_2$ purge, cooled to room temperature and precipitated in water or methanol. The polymer was boiled twice with methanol to remove any LiCl and pyridine and dried overnight at 110 °C below 1 mbar. The yield was 75-95%. PEI's were synthesized from the same diacids but were reacted in DMF at 120 °C for 3-5 h with the corresponding diphenols and tosyl chloride as a condensation agent. The diphenols and the tosyl chloride were sublimated prior to use. Figure 1 reports the structure of the 6 polymers. The carboxyl group in 1 is para linked to the imide moiety. In 2, it is meta linked. Both polymers bear a methyl group ortho to the imide ring which hinders the rotation around this bond and therefore stiffens this part of the monomeric unit. The diacid in polymers 3 uses a similar strategy to stiffen the polymer backbone. It contains 4 methyl groups ortho to both of the imide rings. In 3c an additional 4 methyl groups shield the ester bonds and contribute to increasing bulkiness and overall chain stiffness.

The reduced viscosities are summarized in Table I. Assuming comparable degrees of polymerization within the PAI series (1 = p-PAI, 2a = m-PAI, 3a = PAI) the highest value was found for the PAI 3a (1.0-1.7 dL/g) reflecting the stiffness of the monomeric diacid building block. All of the polymers yield non-brittle, tough films from high boiling solvents like NMP, DMF or THF. Polymers 1, 2a and 2b are also soluble in methyl ethyl ketone (MEK) and acetone. No differences in solubility were detected between PAI and their related PEI structure.

The basic gas permeation data - permeability, apparent selectivity and diffusivity to various pure gases - are given in Tables II-V. The para-linked PAI 1 is less permeable to gases than the analogous meta-linked PAI 2a but offers higher selectivity. The only difference between the two structures is in the meta and para linkage of the imide benzene ring. This is visualized in the sketch in Figure 2. The ortho methyl group hinders the rotation around the imide bond. Therefore, two conformations are possible and exist in equal amounts in the polymers. Figure 2 illustrates the consequence of this structural feature to the free volume of each of the two conformations with the ortho methyl groups above and below the imide plane. In the case of para-linked PAI 1 (see Figure 2 above), the overall chain structure will not be changed by mixing the two conformers. With meta-linked PAI 2a, however, the two meta-linkages per monomer unit can induces a rather large change in chain orientation, thereby lowering the order in chain packing and increase the free volume. This result seems to be in contrast to previously reported behavior in pure polyimides (*3*) where polyimides containing para-linked imide groups exhibit gas permeability coefficients which are higher than those in meta-linked polyimides. In that case, however, no ortho methyl groups hinder the rotation at the imide bond and, consequently, two distinct conformations are not possible.

The PEI's (2b, 3b) are slightly more permeable than their PAI analogues (2a, 3a). The introduction of 4 methyl groups (3c = PEI-4-Me) results in a marked increase

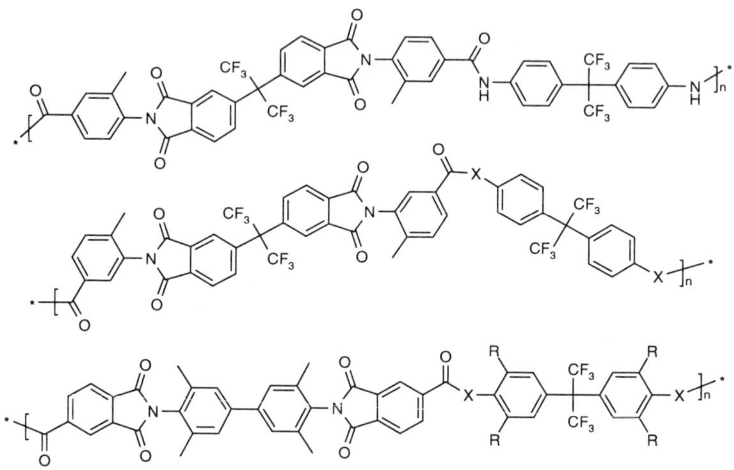

1: (p-PAI)

2a: X=NH (m-PAI)

2b: X=O (m-PEI)

3a: X=NH, R=H (PAI)

3b: X=O, R=H (PEI)

3c: X=O, R=$CH_3$ (PEI-4-Me)

Figure 1. Structures of the synthesized poly(amide imide)s and poly(ester imide)s.

**Table I. Reduced viscosities of the polymers.**

| *Polymer* | *Reduced Viscosity / (dL/g)* [a] |
|---|---|
| 1 | 0.64 - 0.86 |
| 2a | 0.80 - 1.2 |
| 2b | 0.35 - 0.47 |
| 3a | 1.0 - 1.7 |
| 3b | 0.35 - 0.41 |
| 3c | 0.63 |

[a] At 0.5%, 30 °C in N-methylpyrrolidone

**Table II. Ideal permeabilities of the polymers to various gases.**

| *Polymer* | *P* $O_2$ | *P* $N_2$ | *P He* | *P* $H_2$ | *P* $CO_2$ | *P* $CH_4$ |
|---|---|---|---|---|---|---|
| 1 | 6.1 | 1.1 | 72 | 64 | 34 | 0.51 |
| 2a | 8.3 | 1.6 | 81 | 76 | 44 | 0.98 |
| 2b | 8.8 | 1.8 | 80 | 72 | 47 | 1.1 |
| 3a | 9.4 | 2.1 | 67 | 78 | 56 | 1.6 |
| 3b | 10.5 | 2.5 | 64 | 72 | 60 | 2.3 |
| 3c | 18.9 | 4.3 | 107 | 137 | 110 | 3.6 |

Permeability (P) in Barrer ($cm^3$ (STP) cm $cm^{-2}$ $s^{-1}$ $cmHg^{-1}$ x $10^{-10}$).

**Table III. Ideal selectivities of the polymers to various gas pairs.**

| *Polymer* | $P(O_2/N_2)$ | $P(He/N_2)$ | $P(CO_2/CH_4)$ | $P(H_2/CH_4)$ |
|---|---|---|---|---|
| 1 | 5.5 | 65 | 67 | 125 |
| 2a | 5.2 | 50 | 45 | 78 |
| 2b | 5.0 | 45 | 43 | 65 |
| 3a | 4.5 | 32 | 36 | 50 |
| 3b | 4.2 | 26 | 27 | 31 |
| 3c | 4.4 | 25 | 31 | 38 |

**Table IV. Apparent diffusivities of the polymers to various gases.**

| *Polymer* | *D $O_2$* | *D $N_2$* | *D $CO_2$* | *D $CH_4$* |
|---|---|---|---|---|
| 1 | 3.8 | 0.84 | 0.99 | 0.11 |
| 2a | 4.7 | 1.2 | 1.1 | 0.18 |
| 2b | 5.7 | 1.4 | 1.6 | 0.25 |
| 3a | 5.1 | 1.4 | 1.4 | 0.26 |
| 3b | 7.0 | 2.1 | 2.0 | 0.48 |
| 3c | 11.6 | 3.1 | 3.2 | 0.66 |

D in: $cm^2\ s^{-1}$ x $10^{-8}$.

**Table V. Apparent solubilities of the polymers to various gases.**

| *Polymer* | *S $O_2$* | *S $N_2$* | *S $CO_2$* | *S $CH_4$* |
|---|---|---|---|---|
| 1 | 16 | 13 | 350 | 46 |
| 2a | 18 | 14 | 400 | 55 |
| 2b | 16 | 13 | 290 | 44 |
| 3a | 18 | 15 | 400 | 61 |
| 3b | 15 | 12 | 310 | 48 |
| 3c | 16 | 14 | 350 | 54 |

S in: $cm^3$ (STP) $cm^{-3}$ $cmHg^{-1}$ x $10^{-3}$.

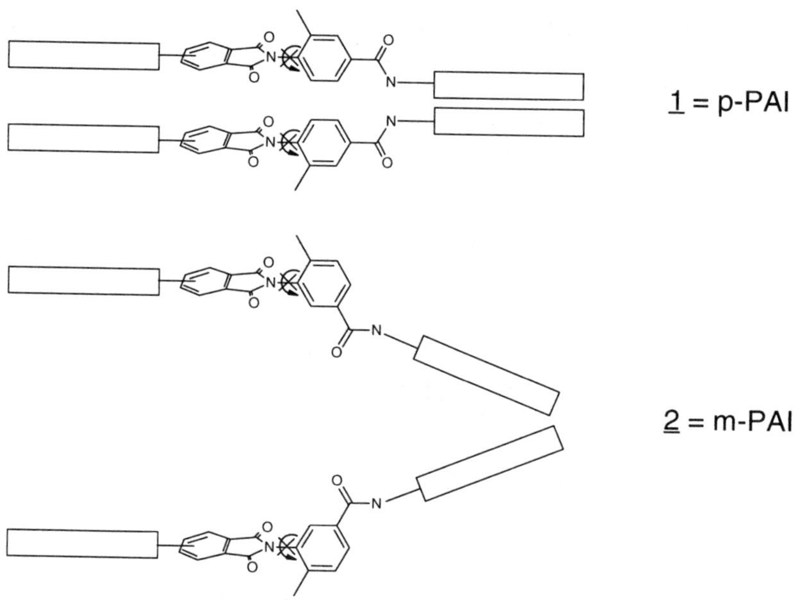


Figure 2. Possible conformations of para-linked PAI 1 (above) and meta-linked PAI 2a (below).

in gas permeability (e.g. $O_2$ permeability increases from 10.5 to 18.9 Barrer, as shown in Table II) while also slightly increasing the selectivity ($P(O_2/N_2)$ from 4.2 to 4.4 (see Table III).

Further insight into the structure/property relationships is obtained by comparing the changes in the apparent diffusivity ($D_a$) with structure. $D_a$ values of the gases in Table IV are always higher in the PEI's than in the PAI's. The apparent solubility ($S_a$) of the same gases, however, is higher in the PAI's than in the PEI's (e. g. $S_a$ values for $CO_2$ are 0.40 for the PAI (2a) and only 0.29 $cm^3$ (STP)/($cm^3$ cmHg) for the corresponding PEI (2b)). The product of $D_a$ and $S_a$, the permeability, is similar to the behavior of $D_a$ and is slightly higher in the PEI's than in the PAI's. In comparison to meta linked PAI 2a para linked PAI 1 is in all cases lower in $P$ and $D_a$ but is more selective (see Table III and the discussion above). So far it seems that PEI's may be more suitable than PAI's for gas separation membranes. In contrast to the previously mentioned gases ($O_2$, $N_2$, $CO_2$, $CH_4$), the fast permeating gases He and $H_2$ break the rule. Figure 3 reports the permselectivity of the polymers for $H_2/CH_4$. The permeability and selectivity are higher for the PAI's. The same trend is observed for $He/N_2$. Diffusivity data for He and $H_2$ could not be measured accurately because the time-lags are less than one second.

To test further the applicability of the polymers for gas separation in technical processes, thin film composite membranes were prepared on microporous polyetherimide supports. Table VI summarizes fluxes and selectivities to the pure gases $O_2$ and $N_2$. A final PDMS coating was applied to plug minor defects (*14*). Solvent/non-solvent mixtures (compare ref. *15*) of MEK/isopropanol (75/25) were found to be best for thin film composite preparation. Pure MEK (see Table VI, 1/2) is useful only at concentrations > 1% to yield defect-free coatings. After applying the PDMS coating on the composite membranes, at least the intrinsic selectivity of the polymer is realized. Formulations with non-solvent at 0.75 % polymer concentration or below yield membranes with higher flux (14-18 GPU, Table VI) and selectivity ($J(O_2/N_2)$ = 6.3, see Table VI) which is clearly above the intrinsic selectivity of the polymers (p-PAI 1 $P(O_2/N_2)$ = 5.5; m-PAI 2a $P(O_2/N_2)$ = 5.2). This result may be due to partial pore blocking. Some regions of the microporous polyetherimide support may be pore-plugged by the PAI polymer and result in a defect-free polyetherimide asymmetric membrane with $P(O_2/N_2)$ = 7 and a 'low' gas flux. (Compare ref. *14* for PDMS coating to plug defects of an asymmetric polysulfone membrane). The residual support may have a coating without pore plugging and yield a composite membrane with selectivity $P(O_2/N_2)$ = 5.2 (the intrinsic value of the m-PAI) and a 'high' gas flux. Because the flux of the entire membrane is composed of both regions the selectivity $P(O_2/N_2)$ must be between 5.2 and 7 depending on the fluxes in the two regions of the composite membrane. This argument is further supported by the actual thickness of the PAI coating. From the high resolution SEM micrograph the PAI coating is measured to be 200 nm. A calculation from the gas flux using the intrinsic polymer properties, however, yields an effective thickness of 300-350 nm.

Coatings from m-PEI 2b, mainly due to its lower viscosity (molecular weight), do not reach the polymer intrinsic selectivity but offers the highest flux (48 GPU; $J(O_2/N_2)$ = 4.4).

Figure 4 presents a high resolution SEM micrograph of composite 1/1 (data see Table VI); (A) before and (B) after the final PDMS coating. The PDMS layer is

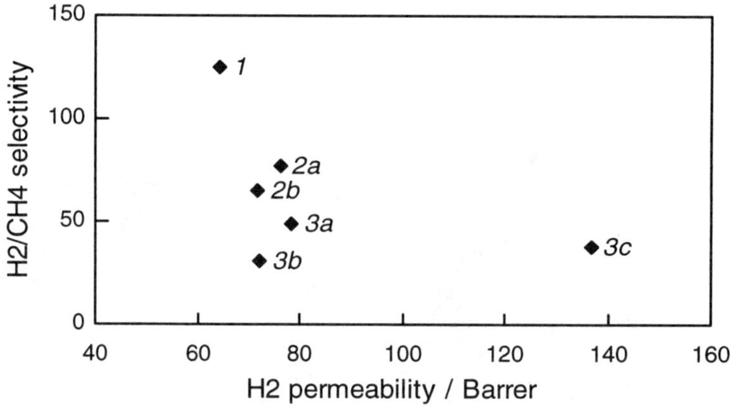


Figure 3. Permeability/Selectivity of $H_2$ and $H_2/CH_4$ of the PAI's 1, 2a, 3a and the PEI's 2b, 3b, 3c.

**Table VI. $O_2$-flux and $O_2/N_2$ selectivity of thin film composite membranes on a polyetherimide support.**

| *Sample* [a] | *Solvent* [b] | *Conc.* % | *$J(O_2/N_2)$* [c] | *$J\ O_2$* [d] PDMS | *$J(O_2/N_2)$* [e] PDMS |
|---|---|---|---|---|---|
| 1/1 | MEK/i-C3 (75/25) | 0.70 | 3.9 | 18 | 6.3 |
| 1/2 | MEK/ - (100/ -) | 1.0 | 4.1 | 5.9 | 5.5 |
| 2a/1 | MEK/i-C3 (75/25) | 0.75 | 5.4 | 14 | 6.3 |
| 2b/1 | MEK/i-C3 (75/25) | 1.0 | 3.0 | 48 | 4.4 |

[a]To be read as polymer No/sample No. [b]MEK= methyl ethyl ketone, i-C3= isopropanol. [c]Selectivity of the as-coated membranes. [d]Flux (J) in GPU= 1 x $10^{-6}$ $cm^3\ cm^{-2}\ s^{-1}\ cmHg^{-1}$ after PDMS coat. [e]Selectivity after PDMS coat.

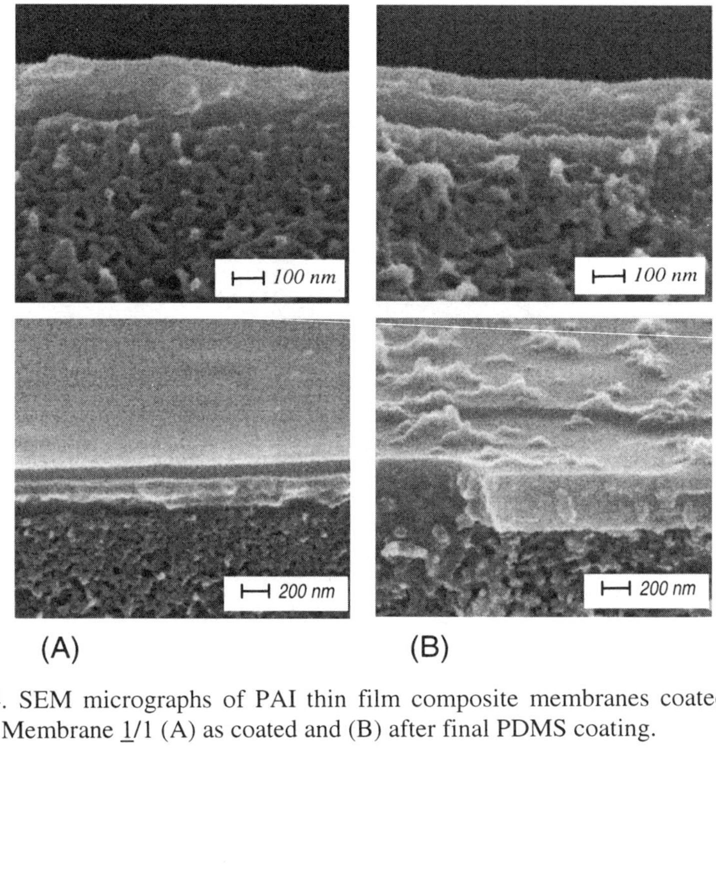


Figure 4. SEM micrographs of PAI thin film composite membranes coated on PEI support. Membrane 1/1 (A) as coated and (B) after final PDMS coating.

detected as a curled film of about 50 nm in thickness on the top surface. The curling is most probably due to preparation artifacts necessary for SEM.

## Conclusion

PAI and PEI polymers were identified as polymers suitable for membrane gas separation. Both classes of polymers possess excellent film forming properties at reduced viscosities > 0.35 dL/g (0.5%, NMP, 30 °C). Preparation of defect-free composite membranes with thicknesses of about 200 nm have been demonstrated. Further work is necessary to investigate the interaction of the support and active separation layer in long term use of the composite membranes.

## Acknowledgment

Mrs. R. Just is thanked for skillful technical assistance; Mrs. M. Aderhold and Mr. M. Schossig-Tiedemann are thanked for taking the SEM micrographs.

## Literature Cited

1. Langsam, M. In *Polyimides, Fundamentals and Applications;* Plastics engineering 36; Ghosh, M. K.; Mittal, K.L. Eds.; Marcel Dekker: New York , USA, 1996; Ch. 22, pp 697-741.
2. Robeson, L. M.; Burgoyne, W. F.; Langsam, M.; Savoca, A. C.; Tien, C. F. *Polymer* **1994,** *35*, 4970.
3. Stern, S. A. *J. Membrane Sci.* **1994**, *94*, 1.
4. Goa, X.; Lu, F. *J. Appl. Polym. Sci.* **1994**, *54*, 1965.
5. Steinhauser, N.; Mülhaupt, M.; Hohmann, M.; Springer, *J. Polym. Adv. Technol.* **1994**, *5*, 438.
6. Mohite, S.; Yamada, Y.; Furukawa, M. *Polym. Mat. Sci. Eng.* **1993**, *68*, 185.
7. Langsam, M. *Polym. Mat. Sci. Eng.* **1997**, *77*, 289.
8. Fritsch, D.; Peinemann, K.-V. *J. Membrane Sci.* **1995**, *99*, 29.
9. Fritsch, D.; Avella, N. *Macromol. Chem. Phys.* **1996**, *197*, 701.
10. Yamazaki, N.; Matsumoto, M.; Higashi, F. *J. Polym. Sci., Polym. Chem. Ed.* **1981**, *16*, 1.
11. Yang, C.-P.; Lin, J.-H. *J. Appl. Polym. Sci.* **1994**, *32*, 2653.
12. US 4,358,624 (Nov. 9, 1982) to General Electric.
13. Higashi, F.; Takahashi, I.; Akiyama, N. *J. Polym. Sci., Polym. Chem. Ed.* **1984**, *22*, 3607.
14. Henis, J. M. S.; Tripodi, M. K. *J. Membrane Sci.* **1981**, *8*, 233.
15. Peinemann, K.-V.; Li, S.-G. The 1996 International Congress on Membranes and Membrane Processes - ICOM '96 - August 18-23, **1996**, 287, Yokohama, Japan.

Chapter 17

# Crosslinked Diacetylene-Functionalized Polyimides for Gas Separation Membranes: Polymer Reactivity and Resultant Properties

**Birgit Bayer[1,3], Birgit Schöberl[1,4], Karthik Nagapudi[2], Mary E. Rezac[1,5], and Haskell W. Beckham[2]**

**[1]Schools of Chemical Engineering and [2]Textile and Fiber Engineering, Polymer Education and Research Center, Georgia Institute of Technology, Atlanta, GA 30332**

A diacetylene-functionalized polyimide homopolymer and blends of a diacetylene-functionalized oligomer with a nonreactive polyimide base were crosslinked by thermal annealing. Gas transport properties at 35 °C are reported for these materials. Permselectivity data on the blend containing 25 wt% reactive oligomer are provided for temperatures up to 300 °C. Crosslinking does not significantly affect gas transport. For the reactive homopolymer, the influence of annealing time and temperature were examined by differential scanning calorimetry. A characteristic diacetylene reaction exotherm can be used to monitor residual diacetylene content following arbitrary annealing treatments. Preliminary solid-state NMR data suggest the crosslinking involves a 1,2-addition reaction of the diacetylene groups.

Expansion of membrane applications into the separation of chemically aggressive materials and/or at temperatures outside the current operating window offers exciting possibilities (*1*). To be applied successfully in these situations, a membrane must be resistant to chemical attack, thermally stable, mechanically strong, easily processed, and able to selectively separate gases while providing for a high throughput of one component. Few commercial polymers possess this combination of properties (*2*).

Crosslinking has been employed to impart increased physical strength, dimensional stability, and solvent resistance to a polymer by restricting the mobility of polymer chains (*3,4*). Material flexibility typically decreases as the degree of crosslinking in the matrix increases. Since diffusion of gas molecules through a polymer involves the cooperative motion of chain segments, crosslinking may retard the permeation process (*5*). Indeed, the literature on crosslinked polyimides indicates a significant decrease in permeability results from this

[3]Current address: Rheno-Westphalian University of Technology, Aachen, Turmstrasse D-52056, Germany.
[4]Current address: Royal Dutch Shell, Hamburg, Germany.
[5]Corresponding author.

process (*6-9*). One exception is our recent work on crosslinkable polyimide blends. A 6FDA-IPDA polyimide was solution-blended with a diacetylene-containing 6FDA-based imide oligomer. Solid-state crosslinking was activated by thermal annealing. The resultant material became insoluble for blends containing 10 wt% and greater of the reactive diacetylene-containing oligomer. Most importantly, no measurable decrease in permeability was observed upon crosslinking (*10*).

Following crosslinking, the bulk densities of the blends increased by less than 1%. It is presumed that the low shrinkage is related to the characteristics and reaction chemistry of the carbon-carbon triple bond of the diacetylene groups. Diacetylenes incorporated along linear oligomers and polymers are capable of a topochemically controlled solid-state reaction with diacetylenes on adjacent molecules to give polydiacetylene crosslinks (*11-13*). This cross-polymerization occurs via a 1,4-addition of appropriately arranged diacetylene groups. Diacetylenes can also "hypercrosslink" via 1,2-additions when the packing arrangement is not suitable for cross-polymerization (*14, 15*).

The reaction chemistry of the diacetylene groups in the crosslinkable polyimide blends needs to be examined in order to understand how these materials can be crosslinked without reducing gas permeability. This report presents the initial results toward this goal. Differential scanning calorimetry and solid-state NMR data are presented for the reactive diacetylene-containing homopolymer. Before these data are described, permeability data on the crosslinkable polyimide blends are presented that expand upon our previously published work on these materials.

## Experimental

**Polymers.** This study involved the investigation of a series of polyimide blends composed of a nonreactive component with attractive processing and transport properties into which was blended a crosslinkable component. The polyimide from 4,4'-isopropylidene dianiline (IPDA) and 4,4'-(hexafluoroisopropylidene)diphthalic anhydride (6FDA) was employed as the nonreactive component. This polyimide, 6FDA-IPDA (Figure 1), was evaluated because of its outstanding physical properties, processability, high permeability and high permselectivity (*16*). The polymer is soluble in methylene chloride, can be prepared in high molecular weight, and is thermally stable to over 300 °C (*16*). The 6FDA-IPDA employed here was synthesized following the procedure of Husk, et al. (*17*).

The reactive polyimide developed and evaluated contains both a 6FDA moiety and an aliphatic diacetylene segment (1,1-6FDA-DIA, Figure 1). Structural similarity between the reactive and inert components was designed to increase the probability of forming miscible polymer blends. The 1,1-6FDA-DIA was prepared by the oxidative coupling of the *N,N'*-dipropargyl diimide monomer synthesized from two equivalents of propargyl amine and one equivalent 6FDA. Details of the synthesis have been provided (*18*). By controlling the reaction conditions, the molecular weight of this component was kept low to enhance miscibility with the high-molecular-weight 6FDA-IPDA. End-group analysis by $^1$H NMR spectroscopy revealed a number-average-degree of polymerization of 5. The pentamer was dried to a constant weight and stored in a desiccator prior to use. We refer to these diacetylene-containing materials as 1,1-6FDA-DIA to signify the anhydride from which they were produced (6FDA), and the number of methylene spacers (1 and 1) separating the diacetylene groups (DIA) from the imide nitrogens.

Films of the pentamer-containing blends were continuous and self-supporting at compositions of 25 wt% 1,1-6FDA-DIA and less. The properties of the 1,1-6FDA-DIA homopolymer were evaluated from a high molecular weight sample. End-group analysis as measured by NMR suggests that the average degree of polymerization for the homopolymer was 75. This material was dried to a constant weight and stored in a desiccator prior to use. Details of the polymer synthesis are provided in a companion paper (*19*).

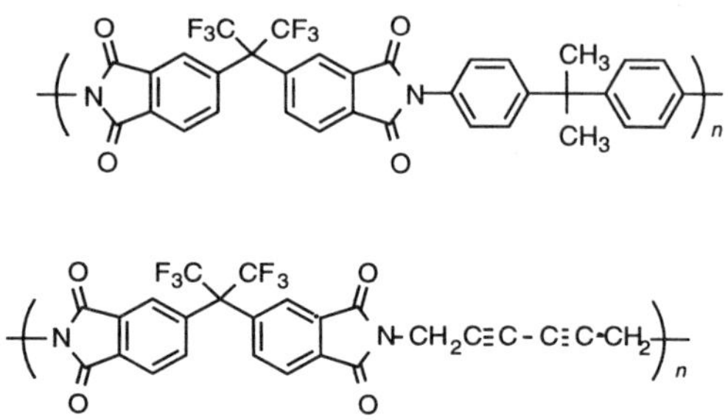

Figure 1: Repeat-unit structures of the 6FDA-IPDA (above) and 1,1-6FDA-DIA (below).

**Chemicals.** The 1,1-6FDA-DIA homopolymer was cast from acetone (Aldrich, 99.5%). Blend samples were cast from methylene chloride (Aldrich, 99%). Solvents and pressurized gases (Air Products) were used as received. Gas purities were at least 99.0%.

**Sample Preparation.** Film samples were prepared by dissolving the weighed quantities of the polymer in solvent. The solutions were filtered through a 0.45-µm filter (Whatman) into a casting ring on a leveled glass plate. Films were dried for a minimum of 24 hours before removal from the casting plate. Films were further dried under atmospheric conditions to facilitate the removal of residual solvent. Because the 1,1-6FDA-DIA can be thermally reacted, care was taken to dry the films at room temperature. Samples of the thickest film were analyzed for the presence of solvent. Following 10 days drying, differential scanning calorimetry showed an endotherm at approximately 50 °C (near the boiling point of the solvent) and an exotherm at approximately 250 °C, corresponding to the crosslinking reaction. The magnitude of the endotherm decreased with drying time until no trace of solvent was detected following 42 days of drying. All samples were dried for a minimum of 50 days prior to analysis.

**Crosslinking.** Crosslinking was achieved by thermal annealing. Samples were placed in an oven preheated to a determined temperature. The oven chamber was evacuated through the use of an attached vacuum pump (with a cold trap to prevent back-diffusion of oil vapors). Samples were annealed for a set time and subsequently removed to the atmosphere. Unless stated otherwise, blends and homopolymers were crosslinked by thermal annealing at 205 °C for 8 hours.

## Material Evaluation

**Gas Transport Measurement** Permeation measurements were made using a constant volume-variable pressure apparatus (*20*) at 25 – 300 °C and 10 atmosphere feed pressure with a permeate pressure of less than 10 torr. The permeability of a gas through a polymer can be described by equation 1:

$$P_A = \frac{N_A \ell}{\Delta p_A} \quad (1)$$

where $P_A$ is the permeability of gas A through the polymer of interest, $N_A$ is the flux of A, $\ell$ is the layer thickness, and $\Delta p_A$ is the partial pressure driving force of A. The ideal selectivity of the polymer for a given gas pair can be defined as:

$$\alpha^*_{A/B} = \frac{P_A}{P_B} \quad (2)$$

**Differential scanning calorimetry (DSC).** Powder samples were evaluated on a Seiko 220 DSC using the following heating profile: 25 – 100 °C at 15 °C/minute, 15-minute hold, 100 – 500 °C at 15 °C/minute. All runs were made under nitrogen purge. The intensities of the DSC signals are off-set for clarity.

**Solid-State Nuclear Magnetic Resonance (NMR).** Solid-state $^{13}C$ NMR spectra were measured on a Bruker DSX 300 operating at 75 MHz. CP/MAS spectra were collected using cross-polarization times of 2 ms, recycle delays of 3 s, $^{1}H$ 90° pulses of 4.5 μs, and a sample spinning speed of 6 kHz. 10k transients were signal-averaged for each spectrum. The total suppression of spinning side bands (TOSS) pulse sequence (*21*) provided spectra with only isotropic chemical shifts. Following analysis, the 1,1-6FDA-DIA hompolymer was annealed in the ceramic sample rotor at 205 °C for 8 hours and a second spectrum was collected.

## Results & Discussion

**Thermal Stability.** The thermal stability of the crosslinked 1,1-6FDA-DIA homopolymer and a blend containing 25 wt% 1,1-6FDA-DIA/75 wt% 6FDA-IPDA are presented in Figure 2. Thermogravimetric analysis (TGA) was completed on a Seiko 320 TGA at a heating rate of 15 °C/minute. The blend has

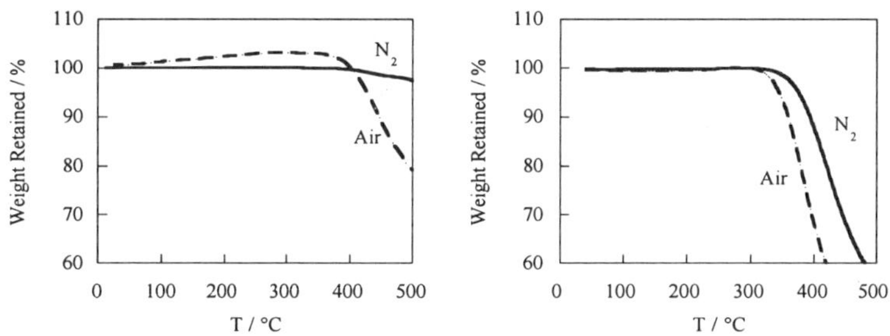


Figure 2: Thermal stability of crosslinked 25 wt% 1,1-6FDA-DIA/75 wt% 6FDA-IPDA blend (left) and 1,1-6FDA-DIA homopolymer (right).

somewhat higher degradation temperatures as a result of the inherently higher thermal stability of the 6FDA-IPDA. Nevertheless, the stability of these materials is comparable with that of other commercially available polyimides and polyetherimide (*21*). Negligible mass loss is observed at temperatures below 300 °C.

**Permeability.** We have previously reported on the permeation properties 1,1-6FDA-DIA/6FDA-IPDA blends with less than 25 wt% of the reactive component at 35 °C (*10*). The major findings from this study were as follows:

- Permeabilities and selectivities were composition independent for samples with 1, 2, 5 or 10 wt% 1,1-6FDA-DIA.
- For DIA contents of 25 wt%, a reduction in permeabilities (as compared to the blends with up to 10 wt% DIA) and an increase in oxygen/nitrogen and hydrogen/methane selectivities were observed.
- For a given blend composition, the permeability of permanent gases before and after crosslinking were approximately equivalent.
- For a given blend composition, the selectivities before and after crosslinking were approximately equivalent.

These results are summarized in Table I. The permeation properties of the crosslinked 1,1-6FDA-DIA homopolymer are reported for comparison.

Table I: Gas Transport Properties of 1,1-6FDA-DIA/6FDA-IPDA Blends (10 atm, 35 °C)

| Blend | Permeability Before Crosslinking (Barrer) | | | Permeability After Crosslinking (Barrer) | | | Selectivity Before Crosslinking | | Selectivity After Crosslinking | |
|---|---|---|---|---|---|---|---|---|---|---|
| wt % DIA* | $H_2$ | $O_2$ | $CO_2$ | $H_2$ | $O_2$ | $CO_2$ | $H_2/CH_4$ | $O_2/N_2$ | $H_2/CH_4$ | $O_2/N_2$ |
| 1 | 41.2 | 4.3 | 18.5 | 40.3 | 4.1 | 15.0 | 70 | 4.3 | 86 | 4.7 |
| 2 | 45.6 | 4.5 | 19.2 | 38.2 | 4.0 | 15.0 | 58 | 4.4 | 85 | 5.0 |
| 5 | 46.0 | 4.2 | 20.4 | 46.9 | 4.7 | 18.1 | 94 | 5.1 | 100 | 5.4 |
| 10 | 36.8 | 3.3 | 14.5 | 40.2 | 4.2 | 14.3 | 99 | 5.5 | 62 | 4.7 |
| 25 | - | - | - | 37.0 | 3.4 | 12.2 | - | - | 120 | 5.5 |
| 100 | - | - | - | 31.4 | 1.3 | 3.9 | - | - | 157 | 6.9 |

Blend data from reference (*10*) All were crosslinked at 240 °C for 4 h.
*1,1-6FDA-DIA, balance 6FDA-IPDA
1 Barrer = $10^{-10}$ $cm^3(STP)cm/cm^2$ s cm Hg

Unfortunately, the lack of integrity of the 1,1-6FDA-DIA homopolymer film prohibited transport evaluation in the uncrosslinked state. Crosslinking improved mechanical strength and allowed for transport analysis. As shown in Table I, the crosslinked 1,1-6FDA-DIA homopolymer shows decreased permeabilities and increased selectivities when compared to the blends with lower 1,1-6FDA-DIA contents.

Each blend has properties which fall near Robeson's upper bound (see Figure 3) (*23*). Variation in the content of DIA allows one to move nearly parallel to the upper bound curve and control the permeation/separation properties.

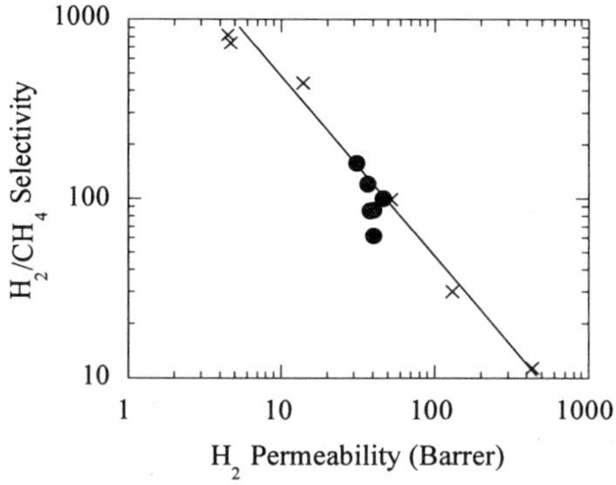


Figure 3. Comparison of the properties of crosslinked polyimide blends (•) with Robeson's "Upper Bound" materials (x) from reference (*23*) for hydrogen/methane separation. All data at 10 atm and 35 °C.

As these materials are potentially interesting for gas separations at elevated temperatures, the influence of temperature on permeation properties has been measured. Evaluation of the crosslinked 25 wt% DIA blend was made because this material possesses an attractive combination of properties. Results are presented in Figure 4 for temperatures up to 300 °C. The slope of the plot represents the activation energy for permeation divided by the gas constant, R (*24*). The calculated activation energies for permeation are presented in Table II. The activation energies for permeation through these crosslinked polyimides are on the same order as those reported for other polymers, including polyimides (*25,26*).

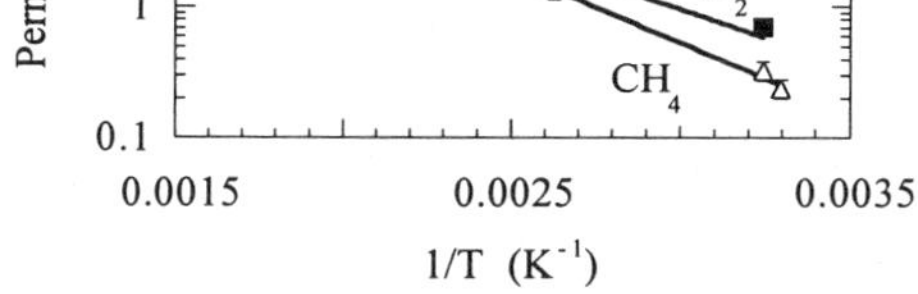


Figure 4: Influence of temperature on gas transport in crosslinked 25 wt% 1,1-6FDA-DIA/75 wt% 6FDA-IPDA blends (10 atm feed pressure).

Table II: Activation energy for permeation through crosslinked blends containing 25 wt% 1,1-6FDA-DIA (10 atm feed pressure).

| Gas | ΔEp (kcal/mol) |
|---|---|
| Nitrogen | 3.9 |
| Methane | 5.0 |
| Carbon Dioxide | 1.6 |
| Helium | 2.1 |

Note that the thickness of the polymer sample is required for the calculation of permeabilities (see equation 1). As the temperature is increased, the polymer volume, and therefore the thickness, increases. We have not measured volume expansions over this temperature range. Therefore, the data in Figure 4 have not been corrected for thickness variations. Structurally similar polyimides have been shown to increase in thickness by approximately 5% over this temperature range (*25*).

Analysis of Figure 4 indicates that the permeabilities increase by a factor of approximately 10 for the slow gases of nitrogen and methane, and about 5 for the fast gases of carbon dioxide and helium as the temperature is increased. As a result of the differences in temperature dependence, the separation selectivities are temperature dependent. The influence of temperature on separation factor is presented in Table III.

Table III: Influence of temperature on the selectivities of crosslinked 25 wt% 1,1-6FDA-DIA/75 wt% 6FDA-IPDA blends (10 atm feed pressure).

| Temperature | Selectivity | | | |
|---|---|---|---|---|
| (°C) | $He/CH_4$ | $CO_2/CH_4$ | $He/CO_2$ | $N_2/CH_4$ |
| 35 | 113 | 33 | 3.4 | 2.1 |
| 100 | 34 | 10 | 3.3 | 1.3 |
| 200 | 26 | 5 | 5.6 | 1.1 |
| 300 | 11 | 3 | 4.6 | 0.9 |

## Polymer Reactivity

**DSC.** Upon thermal annealing of the 1,1-6FDA-DIA homopolymer, the diacetylene groups react to give a crosslinked material. The diacetylene groups in the most reactive environments will be activated first. At longer annealing times, or at higher annealing temperatures, successively less reactive groups can be activated. This crosslinking reaction can be monitored via a characteristic exotherm in a DSC thermogram; the area under the exotherm is related to the content of reactive diacetylene groups. Following an arbitrary annealing treatment, the residual exotherm gives a measure of the unreacted diacetylene groups and thereby the overall conversion *(27)*. To illustrate this point, the DSC traces of the 1,1-6FDA-DIA homopolymer are shown in Figure 5 for increasing annealing temperatures, and in Figure 6 for increasing annealing times.

Careful analysis of Figure 5 indicates that the unannealed sample (dried at 100 °C for 8 hours to remove solvent) has a broad and complex exotherm. A minor shoulder is observed in the 120 – 200 °C range. The most prominent peak is centered about 250 °C and overlaps a broad exothermic event extending to 450 °C (centered at 350 °C). Annealing at 175 °C resulted in an elimination of the exothermic shoulder below 200 °C. The prominent peak at 250 °C was reduced in magnitude and shifted to a higher peak temperature. Annealing at 205 °C for 8 hours apparently results in the complete elimination of the initial peaks, leaving a single broad residual exotherm extending from 250 °C to 450 °C.

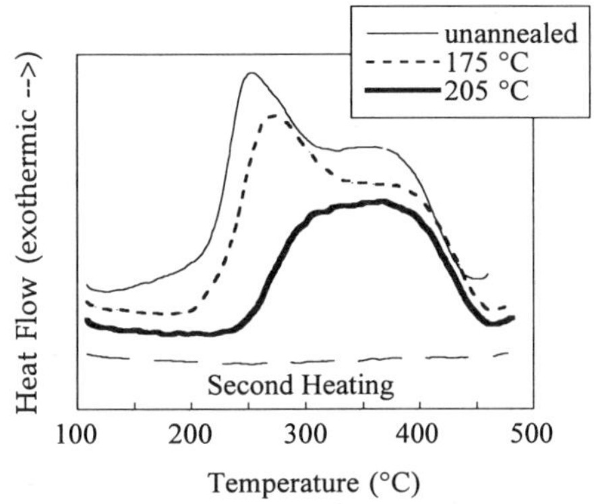


Figure 5: Influence of annealing temperature on reaction exotherm for 1,1-6FDA-DIA homopolymer. Legend presents the temperature at which annealing was completed for 8 hours prior to thermal analysis. DSC traces have been shifted on the y-axis for clarity. One DSC trace of a second heating is also shown.

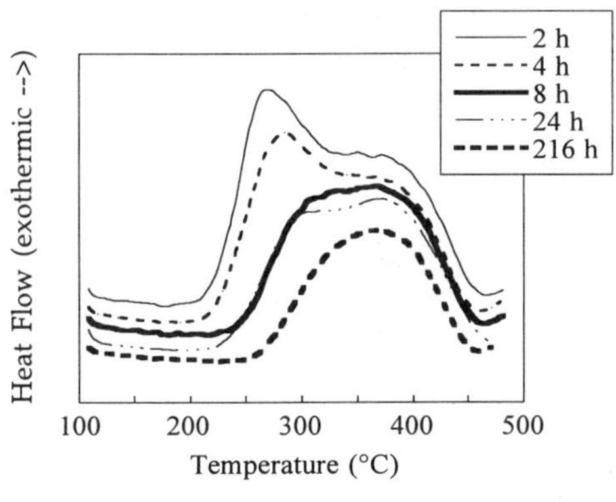


Figure 6: Influence of annealing times on residual exotherm for 1,1-6FDA-DIA homopolymer. Legend presents the times for which the samples were annealed at 205 °C prior to thermal analysis. DSC traces have been shifted along the y-axis for clarity.

The influence of reaction time at a constant annealing temperature of 205 °C is presented in Figure 6. At short annealing times (2 and 4 hours), the residual exotherms contain at least two overlapping components (centered at 250 and

350 °C). As annealing time is increased, the peak centered at 250 °C decreases in magnitude and shifts to higher temperature; after 8 hours annealing, this peak is nearly eliminated. Yet, even after 216 hours annealing, the material still shows a significant exotherm centered near 350 °C.

The DSC data indicate the presence of at least two primary reactions, one giving rise to the prominent exotherm centered at about 250 °C and another reaction giving rise to the broad exotherm extending to 450 °C. For increasing annealing times and for increasing annealing temperatures, the 250 °C exotherm progressively disappears (Figures 5 and 6). The broad higher-temperature exotherm appears to remain during this process.

Reactions of terminal ethynyl groups (*28,29*) are probably not significant contributors to the exotherm since this material is high molecular weight polymer. Indeed, DSC on the ethynyl-terminated monomer (data not shown) shows a single isolated exothermic peak centered at about 300 °C. Rather, the 250-°C exotherm most probably corresponds to the crosslinking reactions and the exotherm centered at 350 °C results, at least in part, from a high-temperature degradation reaction. The TGA curves of Figure 2 indicate the onset of weight loss around 330 °C.

Variations in the annealing temperature and time control the extent of crosslinking achieved. The chemical resistance of the resultant polymer is directly related to the presence of these crosslinks. Thus, the properties of the resultant material can be tuned by proper manipulation of the polymer blend composition, the crosslinking temperature, and the annealing time. Higher concentrations of the diacetylene-containing material, higher reaction temperatures, and longer annealing times should all result in a material with higher crosslink density. Although the influence of crosslink density on transport properties appears to be a fairly weak one for this particular polymer system, it has been shown that increasing crosslink density will eventually lead to a reduction in permeabilities and an increase in permanent gas selectivities (*30*).

**NMR.** The chemical nature of the crosslinking reaction is being investigated using a variety of solid-state spectroscopies. $^{13}$C solid-state NMR spectra are shown in Figure 7 for the 1,1-6FDA-DIA polymer before and after annealing at 205 °C for 8 hours. The spectra shown in Figure 7 were measured with $^{1}$H decoupling but no $^{19}$F decoupling, so the carbons of the hexafluoroisopropylidene group (see Figure 1) are not observed (the NMR solution-state spectra show that even the quaternary carbon of this group is weakly coupled to the fluorines). Peaks are clearly seen for the methylene carbons at 27 ppm, the diacetylene carbons at 65 – 78 ppm, the aromatic carbons at 124 – 138 ppm and the carbonyl carbons at 166 ppm. Upon annealing, the relative intensities of the methylene peak and the diacetylene peaks decrease. The diacetylene peaks decrease because they are involved in the crosslinking reaction. The methylene peak at 27 ppm decreases because it represents the methylenes $\alpha$ to the diacetylene groups; upon annealing, these methylenes are no longer $\alpha$ to diacetylene groups and therefore no longer resonate at 27 ppm.

While these spectra clearly indicate the crosslinking reaction involves the diacetylene carbons, they do not reveal the identity of the products. If the reaction were a 1,4-addition polymerization of diacetylenes to polydiacetylenes (PDA) (*31*), new peaks should appear around 107 ppm for the *sp*-carbons and around 130 ppm for the $sp^2$-carbons (*13,32*). Since no peak appears in the region around 107 ppm for the annealed sample, the crosslinking probably does not occur via this route. This conclusion is supported by the lack of order in these materials as determined by wide-angle x-ray diffraction (molecular order is needed for the solid-state 1,4-addition polymerization of diacetylenes). The more likely reaction products are those containing carbon-carbon double bonds. These would result from a 1,2-addition reaction or perhaps even a Diels-Alder-type reaction of the diacetylene groups (*33, 34*). Peaks for these $sp^2$-carbons would appear between 120 and 140 ppm, and overlap with the existing peaks of the aromatic groups. In fact, the relative intensity of the peaks in this region are greater than all other peaks in the spectrum of the crosslinked sample. Continuing NMR studies will include spectral editing for more complete structure determination of crosslink reaction products.

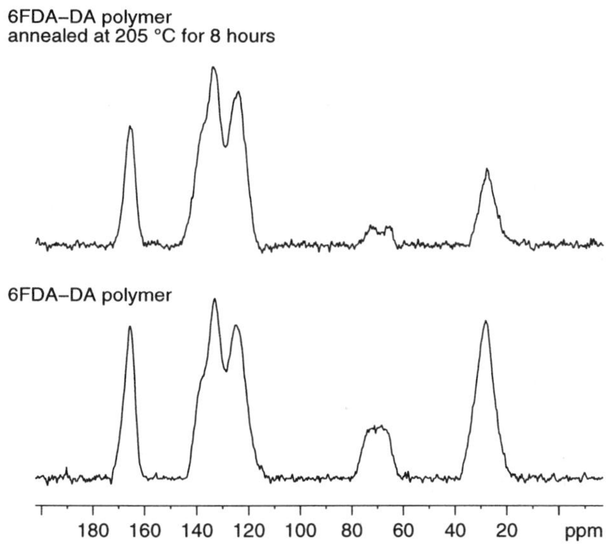


Figure 7: $^{13}C$ solid-state NMR spectra of 1,1-6FDA-DIA polymer recorded before and after annealing.

## Conclusions

Blends of a diacetylene-functionalized imide oligomer and a 6FDA-based nonreactive polyimide can be crosslinked by thermal annealing. The crosslinked blend containing 25 wt% of the crosslinkable component exhibits gas transport properties near Robeson's upper bound. The activation energies for gas permeation through this blend (measured from 35 - 300 °C) were similar to those of other polyimides. Both the blend and diacetylene-functionalized homopolymer show negligible mass loss below 300 °C as revealed by thermogravimetic analysis. Differential scanning calorimetry can be used to measure the residual content of reactive diacetylene groups following an arbitrary annealing treatment; for increasing annealing temperature and time, the diacetylene reaction exotherm becomes progressively smaller. This information could be used to control the extent of diacetylene reaction and thereby the ultimate crosslink density in diacetylene-functionalized materials used in gas separation membrane applications.

**Acknowledgments**

Although the research described in this article has been funded in part by the United States Environmental Protection Agency under assistance agreement number R824727 to MER and HWB, it has not been subjected to the Agency's peer and administrative review and therefore may not necessarily reflect the views of the Agency and no official endorsement should be inferred. B. Bayer and B. Schöberl acknowledge financial support from the Ernest Solvay Foundation. K. Nagapudi acknowledges financial support from the Office of Naval Research's Molecular Design Institute and NSF (DMR-9502246). Special thanks to N. Karangu for synthesizing the 1,1-6FDA-DIA and Dr. Tim Wang for assistance with synthesis of the 6FDA-IPDA.

**References**

1 Henis, J. M. S. In *Polymeric Gas Separation Membranes*, Paul, D. R. and Yampol'skii, Y. P., Eds., CRC Press, Boca Raton, FL, 1994.

2 Haubs, M.; Herold, F.; Krieg, C.; Meyer-Blumenroth, U.; Schneider, J.; Wagener, R.; Wildhardt, J. *Makromol. Chem., Makromol. Symp.* **1991,** *50,* 67.

3 Young R. J.; Lovell, P. A. *Introduction to Polymers*, Chapman & Hall Publishers, New York, NY, 1991.

4 Lin, A. A.; Sastri, V. R.; Tesoro, G.; Reiser, A.; Eachus, R. *Macromolecules* **1988**, *21*, 1165.

5 Sefcik, M. D.; Schaefer, J.; May, F. L.; Raucher, D.; Dub, S. M. *J. Polym. Sci. Polym. Phys. Ed.* **1983**, *21*, 1041.

6 Burgoyne, W. F., Jr.; Langsam, M.; Ford, M. E.; Casey, J. P. US Patent 4,931,182, **1990**.

7 Meier, I. K.; Langsam, M.; Klotz, H.C. *J. Membrane Sci.* **1994**, *94*, 195.

8 Meier, I. K.; Langsam, M. *J. Polym. Sci., Polym. Chem.* **1993**, *31*, 83.

9 Hayes, R. A. US Patent 4,717,393, **1988.**

10 Rezac, M. E.; Sorensen, E. T.; Beckham, H. W. *J. Membrane Sci.* **1997**, *136*, 249.

11 Wegner, G. *Die Makromolekulare Chemie* **1970**, *134*, 219.

12 Beckham, H. W.; Rubner, M. F. *Macromolecules* **1993**, *26,* 5192.

13 Beckham, H. W.; Spiess, H. W. *Macromol. Chem. Phys.* **1994**, *195*, 1471.

14 Dawson, D. J.; Fleming, W. W.; Lyerla, J. R.; Economy, J. *Reactive Oligomers*; ACS Monograph Series 282; 63 (1982).

15 Yamazaki, S.; Nakamura, K.; Kato, J.; Tokushige, K. Jpn. Patent 63-96144, **1988.**

16 Kim, T. H.; Koros, W. J.; Husk, G. R.; O'Brien, K. C. *J. Membrane Sci.* **1988**, *37*, 45.

17 Husk, G. R.; Cassidy, P.E.; Gebert, K. L. *Macromolecules* **1988**, *21*, 1234.

18 Karangu, N.; Rezac, M. E.; Beckham, H. W. *Polym. Prepr., Am. Chem. Soc. Div. Polym. Mater.* **1997**, *76,* 316.

19 Karangu, N.; Rezac, M. E.; Beckham, H.W. *Chemistry of Materials* **1997**, *10,* 567.

20 O'Brien, K. C.; Koros, W. J.; Barbari, T. A.; Sanders, E. S. *J. Membrane Sci.* **1986**, *29*, 229.

21 Dixon, W. T. *J. Chem. Phys.* **1982**, *77*, 1800.

22 Bayer, B. *DiplomarbeitThesis*, The University of Karlsruhe, **1997**.

23 Robeson, L. *J. Membrane Sci.* **1991**, *62*, 165.

24 Van Amerongen, G. J. *Rubber Chem. Technol.* **1964**, *37*, 1065.

25 Costello, L. M.; Koros, W. J. *I&EC Research* **1991**, *31*, 2708.

26 Buys, H.C.W.M.; Van Elven, A.; Jansen, A.E.; Tinnemans, A.H.A. *J. Appl. Poly. Sci.* **1990**, *41*, 1261.

27 Beckham, H. W.; Rubner, M. F. *Macromolecules* **1989**, *22*, 2130.

28 Grenier-Loustalot, M. F.; Sanglar, C. *High Perform. Polym.* **1996**, *8*, 315.

29 Swanson, S. A.; Fleming, W. W.; Hofer, D. C. *Macromolecules* **1992**, *25*, 582.

30 Andrady, A.L.; Sefcik, M. D. *J. Appl. Pol. Sci.* **1984**, *29*, 3561.

31 Enkelmann, V. *Adv. Polym. Sci.* **1984**, *63*, 91.

32 Beckham, H.W.; Rubner, M.F. *Polymer* **1991**, *32*, 1821.

33 Dawson, D. J.; Fleming, W. W.; Lyerla, J. R.; Economy, J. *Thermally Stable Polymers for Electronic Applications*; 282 ed.; Dawson, D. J.; Fleming, W. W.; Lyerla, J. R.; Economy, J., Eds.; American Chemical Society: St. Louis, MO, 1984, pp 63–79.

34 D'Alelio, G. F.; Waitkus, P. A. *Monomeric Diimides Having Terminal Conjugated Diacetylene Groups*; D'Alelio, G. F.; Waitkus, P. A., Eds.; Plastics Engineering Company: U. S., 1982.

## Chapter 18

# Indan-Containing Polymers for Gas Separation Membranes


**Gerhard Maier[1], Martin Wolf[1], Miroslav Bleha[2], and Zbynek Pientka[2]**

**[1]Technische Universität München, Lichtenbergstraße 4, D-85747 Garching, Germany**
**[2]Institute of Macromolecular Chemistry, Heyrovsky Sq. 2, Prague 6 CZ-16206, Czech Republic**



The permselectivity properties of two series of poly(ether ketone)s and polyimides containing various bulky indan groups in the main chain are described. Hydrogen, carbon dioxide, oxygen, and nitrogen permeabilities were determined. Despite their bulkiness, the indan groups did not lead to increased permeability coefficients or selectivities. This result is explained by the presence of relatively flexible phenylether linkages between the rigid indan groups.


Amorphous polymers with relatively high glass transition temperatures such as polyimides, poly(aryl ether)s, or polycarbonates are generally promising candidates for gas separation membranes (*1-4*). This view is based on a widely accepted model for the transport of gas molecules through a polymer matrix (*1-4*). According to this model, gas transport occurs by diffusional "jumps" of the penetrant molecules as a result of thermally activated motions of segments of the polymer backbone. These segmental motions open "jump channels", through which the gas molecules can pass. For a binary gas pair, selection based on penetrant size occurs by the opening of channels which are wide enough for passage of only the smaller penetrant, but are too narrow for the larger one. The channel width depends on the polymer backbone flexibility. More flexible chains allow wider openings, resulting in decreased selectivity. From this model, a general principle for simultaneous improvement of permeability coefficients and selectivity was deduced in order to overcome the undesired common "trade-off" behavior (*1-4*) between permeability and selectivity. Increased rigidity of the polymer main chain is expected to increase selectivity. At the same time, the introduction of bulky, rigid groups leads to increased fractional free volume and hence increased permeability coefficients.

Many attempts have been made to optimize polymer structures using this strategy. The results of the structural modifications could be qualitatively understood in most cases on the basis of main chain flexibility, bulkiness of certain structural elements, packing density, and fractional free volume. However, quantitative predictions of how a certain polymer structure modification will affect its permeability behavior are still very difficult. In a recent paper (*5*), increments for the permeability coefficients of a number of structural elements were deduced. The application of these increments for the calculation of permeability coefficients of a number of polymers for oxygen, nitrogen and helium was quite successful. However, the authors state that increments for new groups can not be deduced simply based on the chemical structure. Thus, predictions for new polymer structures are not possible with this approach.

In our opinion, a more detailed understanding of the relationship between the chemical structure of a polymer and its permselectivity characteristics can be achieved by identification of the segments which control diffusional jumps. For this purpose, we are studying various series of polymers containing rigid, bulky, substituted indan groups in the main chain, linked by phenylether segments of controlled flexibility. Ideally, such polymer architectures should lead to a situation where the diffusional jumps are controlled predominantly by the flexible phenylether segments. If this hypothesis is confirmed, modifications of the chemical structure of these segments should allow optimization of the flexibility and hence selectivity for any gas pair, without reducing the permeability coefficients. This paper describes the results of permeability measurements of two series of polyimides and poly(ether ketone)s with indan groups in the main chain.

## Polymer Structures

The synthesis and characterization of the polymers have been described earlier (*6-8*) and will only be outlined briefly. First, the indan group **3** is prepared by cyclodimerization of an $\alpha$-substituted styrene derivative **2** or the corresponding tertiary alcohol **1** under acidic conditions, as shown in Scheme 1 below.

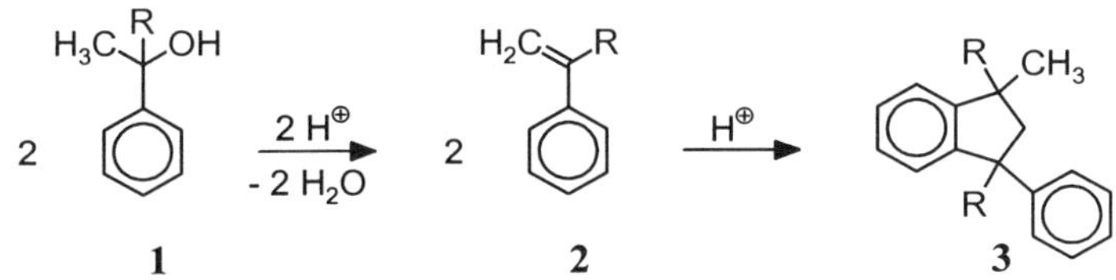


**1** **2** **3**

R = methyl, cyclohexyl

Scheme 1: Synthesis of indan groups

The 4-fluorobenzoyl groups are introduced by Friedel-Crafts acylation of the 3-phenylindan derivative with 4-fluorobenzoic acid (Scheme 2).

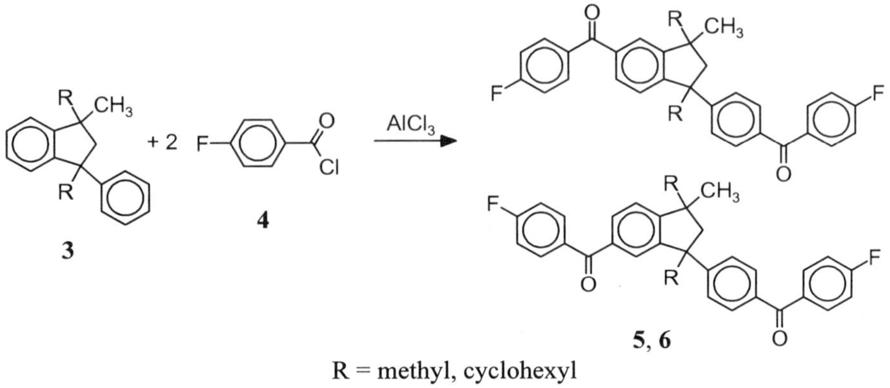


Scheme 2: Synthesis of monomers **5** (R = methyl) and **6** (R = cyclohexyl)

Nucleophilic replacement of the fluorine atoms in both monomers by 4-hydroxy-aniline results in the diamine monomers **7** and **8** (Scheme 3).

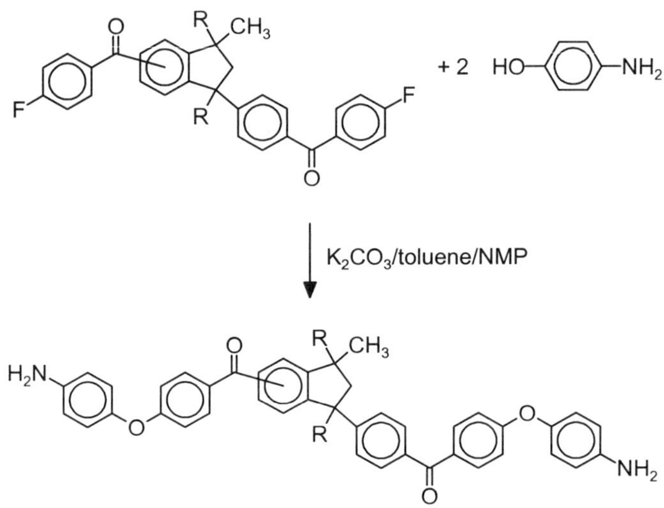


Scheme 3: Synthesis of monomers **7** (R = methyl) and **8** (R = cyclohexyl)

The poly(ether ketone)s **9** were prepared by condensation of monomer **6** with various bisphenols under the usual reaction conditions (*6*). Polyimides **10** (R = methyl) and **11** (R = cyclohexyl) were synthesized via the poly(amic acid) and thermal imidization of the films by condensation of monomers **7** and **8** with various dianhydrides (*7,8*). The structures of the polymers are shown in Scheme 4.

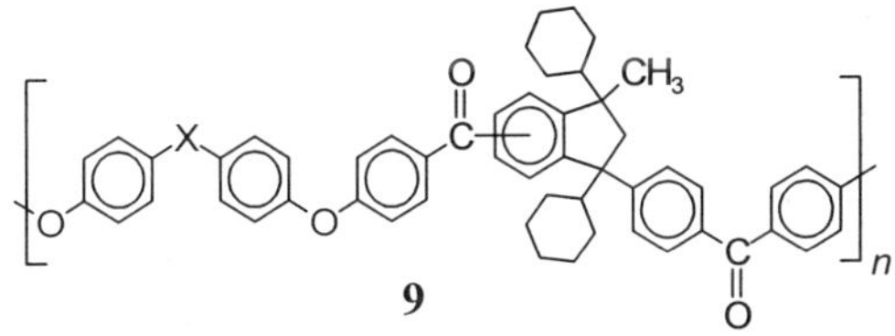

**9**

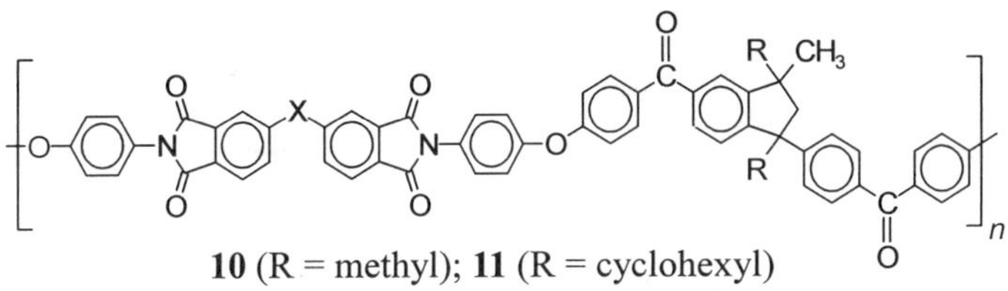

**10** (R = methyl); **11** (R = cyclohexyl)

Scheme 4: Structure of poly(ether ketone)s **9** and polyimides **10** and **11**

As expected, the presence of the isomeric structures has no effect on properties such as glass transition temperature or solubility (*6*). Both isomers introduce bends into the polymer chain due to a dihedral angle between the phenyl substituent in the 3-position of the indan group and the aromatic ring within the indan system. Also, the indan group can not be viewed as a classic cardo group as introduced by fluorene or phthalein groups.

Table I. Properties of poly(ether ketone)s **9**

| No. | -X- | $T_g$ [a)] [°C] | $\bar{M}_n$ [b)] [g · mol$^{-1}$] | $V_f$ [c] ($O_2$) | $V_f$ [c] ($N_2$) |
|---|---|---|---|---|---|
| 9a | -O- | 232 | 16 600 | 0.124 | 0.123 |
| 9b | $-C(CF_3)_2-$ | 235 | 17 900 | 0.175 | 0.174 |
| 9c | — | 255 | 13 200 | 0.139 | 0.137 |
| 9d | $-C(CH_3)_2-$ | 234 | 12 700 | 0.140 | 0.137 |
| 9e |  | 236 | 8 400 | 0.136 | 0.132 |
| 9f | -S- | 218 | 15 600 | 0.134 | 0.131 |
| 9g | 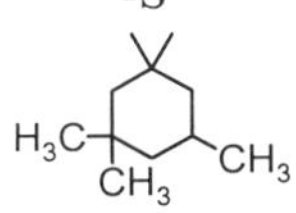 | 255 | 14 200 | 0.133 | 0.130 |
| 9h | -CO- | 229 | 12 700 | 0.130 | 0.128 |
| 9i | $-SO_2-$ | 251 | 17 200 | 0.149 | 0.147 |

a) DSC, 20 K/min;

b) GPC in THF, polystyrene calibration;

c) fractional free volume ($V_f = \frac{V_s - V_0}{V_s}$; $V_S$: specific volume (from density, determined by floating film in aq. $CaCl_2$-solution), $V_0$: specific volume at 0 K (from increments by Park and Paul (*9*))

**Properties of the Polymers**

Molar masses, glass transition temperatures, and fractional free volume (calculated from densities and Paul and Park's series of increments) of the polymers are summarized in Tables I and II (*6-8*). Fractional free volume ($V_f$) calculated by the method of Park and Paul (*9*) differs for different gases, depending on the diameters of the gas molecules. Here, the values for $O_2$ and $N_2$ are given as examples. As expected, the values for $V_f$ increase with decreasing kinetic diameter of the gas molecules.

Table II. Properties of polyimides **10** and **11**

| Polymer No. | -R | -X- | $T_g$ [a] [°C] | $\bar{M}_n$ [b] [$g \cdot mol^{-1}$] | $V_f$ [c] ($O_2$) | $V_f$ [c] ($N_2$) |
|---|---|---|---|---|---|---|
| **10a** | -methyl | $-C(CF_3)_2-$ | 241 | 28 000 | 0.161 | 0.157 |
| **10b** | -methyl | $-SO_2-$ | 251 | 20 800 | 0.139 | 0.133 |
| **10c** | -methyl | -CO- | 232 | 21 300 | 0.124 | 0.119 |
| **10d** | -methyl | -O- | 233 | d) | 0.119 | 0.114 |
| **10e** | -methyl | — | 246 | 31 500 | 0.132 | 0.126 |
| **11a** | -cyclohexyl | $-C(CF_3)_2-$ | 269 | 30 700 | 0.169 | 0.165 |
| **11b** | -cyclohexyl | $-SO_2-$ | 272 | 8 100 | 0.144 | 0.138 |
| **11c** | -cyclohexyl | -CO- | 261 | d) | 0.127 | 0.121 |
| **11d** | -cyclohexyl | -O- | 264 | d) | 0.122 | 0.116 |
| **11e** | -cyclohexyl | — | 274 | 25 800 | 0.135 | 0.129 |

a) DSC, 20 K/min
b) GPC in $CHCl_3$, polystyrene calibration;
c) fractional free volume ($V_f = \frac{V_s - V_0}{V_s}$; $V_s$: specific volume (from density, determined by floating film in aq. $CaCl_2$-solution), $V_0$: specific volume at 0 K (from increments by Park and Paul (*9*))
d) insoluble in $CHCl_3$

Although the molar masses were determined by GPC, the values can be compared within a series of polymers of similar chemical structure. The molar masses are quite similar among the series of poly(ether ketone)s **9**, with the exception of **9e**. However, since all polymers **9** form flexible, free standing films, the assumption that the molar masses in all cases are in the range where physical properties become independent of chain length appears justified. The situation within the series of polyimides is comparable, with **11b** being the exception.

The glass transition temperatures follow the well known trends. Rigid or strongly polar groups increaser $T_g$, while flexible groups decrease it. The polyimides show higher $T_g$'s than the poly(ether ketone)s due to the polar imide groups. Introduction of two cyclohexyl groups at the indan system increases the glass transition temperatures by approximately 30 °C. $T_g$ can be taken as a measure of global chain stiffness (*10*), which leads to the expectation of relatively high selectivities.

## Results of the Permeability Measurements

Gas permeabilities of all polymers **9**, **10**, and **11** were measured for $H_2$, $CO_2$, $O_2$, and $N_2$. The results are shown in Tables III - VI.

Table III: Gas permeability coefficients of poly(ether ketone)s **9** [Barrers]

| Polymer No. | X | P ($H_2$) | P ($CO_2$) | P ($O_2$) | P ($N_2$) |
|---|---|---|---|---|---|
| **9a** | -O- | 11.6 | 4.18 | 1.08 | 0.21 |
| **9b** | $-C(CF_3)_2-$ | 17.3 | 7.7 | 1.64 | 0.32 |
| **9c** | — | 16.0 | 6.34 | 1.35 | 0.32 |
| **9d** | $-C(CH_3)_2-$ | 17.7 | 9.36 | 1.83 | 0.33 |
| **9e** |  | 16.7 | 7.39 | 1.57 | 0.29 |
| **9f** | -S- | 13.5 | 5.7 | 1.12 | 0.2 |
| **9g** | 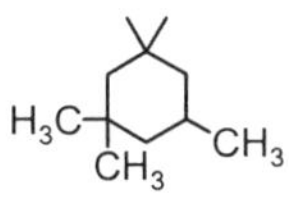 | 36 | 18.9 | 3.58 | 0.64 |
| **9h** | -CO- | 12.5 | 5 | 1.03 | 0.16 |
| **9i** | $-SO_2-$ | 14.5 | 7.2 | 1.32 | 0.23 |

Table IV: Selectivities of poly(ether ketone)s **9**

| Polymer No. | X | $\alpha(O_2/N_2)$ | $\alpha(H_2/N_2)$ | $\alpha(CO_2/N_2)$ | $\alpha(H_2/CO_2)$ | $\alpha(CO_2/O_2)$ | $\alpha(H_2/O_2)$ |
|---|---|---|---|---|---|---|---|
| **9a** | -O- | 5.1 | 55.2 | 19.9 | 2.8 | 3.9 | 10.7 |
| **9b** | $-C(CF_3)_2-$ | 5.1 | 54.1 | 24.1 | 2.2 | 4.7 | 10.5 |
| **9c** | — | 4.2 | 50.0 | 19.8 | 2.3 | 4.4 | 10.1 |
| **9d** | $-C-(CH_3)_2-$ | 5.5 | 53.6 | 28.4 | 1.9 | 5.1 | 9.7 |
| **9e** |  | 5.4 | 57.6 | 25.5 | 2.3 | 4.7 | 10.6 |
| **9f** | -S- | 5.6 | 67.5 | 28.5 | 2.4 | 5.1 | 12.1 |
| **9g** | 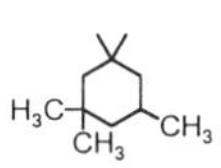 | 5.6 | 56.2 | 29.5 | 1.9 | 5.3 | 10.1 |
| **9h** | -CO- | 6.4 | 78.1 | 31.3 | 2.5 | 4.9 | 12.1 |
| **9i** | $-SO_2-$ | 5.7 | 63.0 | 31.3 | 2.0 | 5.5 | 11.0 |

As shown in Tables III and V, the permeability coefficients of the poly(ether ketone)s and the polyimides are comparable. Only polymer **9g**, which contains the bulky, relatively rigid trimethylcyclohexyl linker in the bisphenol unit, exhibits significantly increased permeability coefficients relative to the other polymers in this study. Also, it is clear that the substantial increase in permeability coefficients that was initially expected from the introduction of the indan group, especially with the two cyclohexyl substituents, is not observed. Figures 1 and 2 show the common plots of log$\alpha$ vs. log P for the poly(ether ketone)s and the polyimides.

Table V: Gas permeability coefficients of polyimides **10** and **11** [Barrers]

| Polymer No. | -R | -X- | P ($H_2$) | P ($CO_2$) | P ($O_2$) | P ($N_2$) |
|---|---|---|---|---|---|---|
| **10a** | -methyl | -C($CF_3$)$_2$- | 21.3 | 5.58 | 2.05 | 0.32 |
| **10b** | -methyl | -$SO_2$- | 12.7 | 3.6 | 0.91 | 0.16 |
| **10c** | -methyl | -CO- | 8.3 | 1.96 | 0.47 | 0.08 |
| **10d** | -methyl | -O- | 10.5 | 3.05 | 0.68 | 0.09 |
| **10e** | -methyl | — | 8.8 | 2.51 | 0.57 | 0.11 |
| **11a** | -cyclohexyl | -C($CF_3$)$_2$- | 23.1 | 9.64 | 1.92 | 0.33 |
| **11b** | -cyclohexyl | -$SO_2$- | 10.2 | 3.2 | 0.75 | 0.16 |
| **11c** | -cyclohexyl | -CO- | 9.2 | 1.68 | 0.91 | 0.22 |
| **11d** | -cyclohexyl | -O- | 13.23 | 3.76 | 0.9 | 0.08 |
| **11e** | -cyclohexyl | — | 13.3 | 5.68 | 1.11 | 0.28 |

## Discussion of Permselectivity Characteristics

For a discussion of the permselectivities, we assume that the overall selectivities $\alpha$ represent trends in diffusivity selectivities $D_X/D_Y$ rather than solubility selectivities $S_X/S_Y$. Data in the literature (*11-14*) show that within a series of polymers, such as polyimides, polycarbonates or poly(ether sulfone)s, differences in solubility selectivities do not exceed 10 % in most cases, as long as only one linking group in the repeating unit is changed. Therefore, the above mentioned assumption appears justified.

The permeabilty coefficients of the polymers **9** depend roughly on the bulkiness of the linking group X in the bisphenol unit. The most bulky group in this series, the trimethylcyclohexyl group in **9g**, results in the highest permeability coefficients, while the smallest linkers, the ether bridge in **9a**, the sulfide group in **9f**, and the carbonyl group in **9h** result in the lowest values (Table III).

From Figure 1, one can see that within the series of poly(ether ketone)s **9**, selectivities for $H_2/N_2$, $CO_2/N_2$ and $O_2/N_2$ vary very little with the nature of the connecting

Table VI: Selectivities of polyimides **10** and **11**

| Polymer No. | -R | -X- | α ($O_2/N_2$) | α ($H_2/N_2$) | α ($CO_2/N_2$) | α ($H_2/CO_2$) | α ($CO_2/O_2$) | α ($H_2/O_2$) |
|---|---|---|---|---|---|---|---|---|
| **10a** | -methyl | -$C(CF_3)_2$- | 6.4 | 66.6 | 17.4 | 3.8 | 2.7 | 10.4 |
| **10b** | -methyl | -$SO_2$- | 5.7 | 79.4 | 22.5 | 3.5 | 4.0 | 14.0 |
| **10c** | -methyl | -CO- | 5.9 | 103.7 | 24.5 | 4.2 | 4.2 | 17.7 |
| **10d** | -methyl | -O- | 7.6 | 116.7 | 33.9 | 3.4 | 4.5 | 15.4 |
| **10e** | -methyl | — | 5.2 | 80 | 22.8 | 3.5 | 4.4 | 15.4 |
| **11a** | -cyclohexyl | -$C(CF_3)_2$- | 5.8 | 70 | 29.2 | 2.4 | 5.0 | 12.0 |
| **11b** | -cyclohexyl | -$SO_2$- | 4.7 | 63.7 | 20.6 | 3.1 | 4.4 | 13.6 |
| **11c** | -cyclohexyl | -CO- | 4.1 | 41.8 | 7.6 | 5.5 | 1.8 | 10.1 |
| **11d** | -cyclohexyl | -O- | 11.2 | 165.4 | 47 | 3.5 | 4.2 | 14.7 |
| **11e** | -cyclohexyl | — | 4.0 | 47.5 | 20.3 | 2.3 | 5.1 | 12.0 |

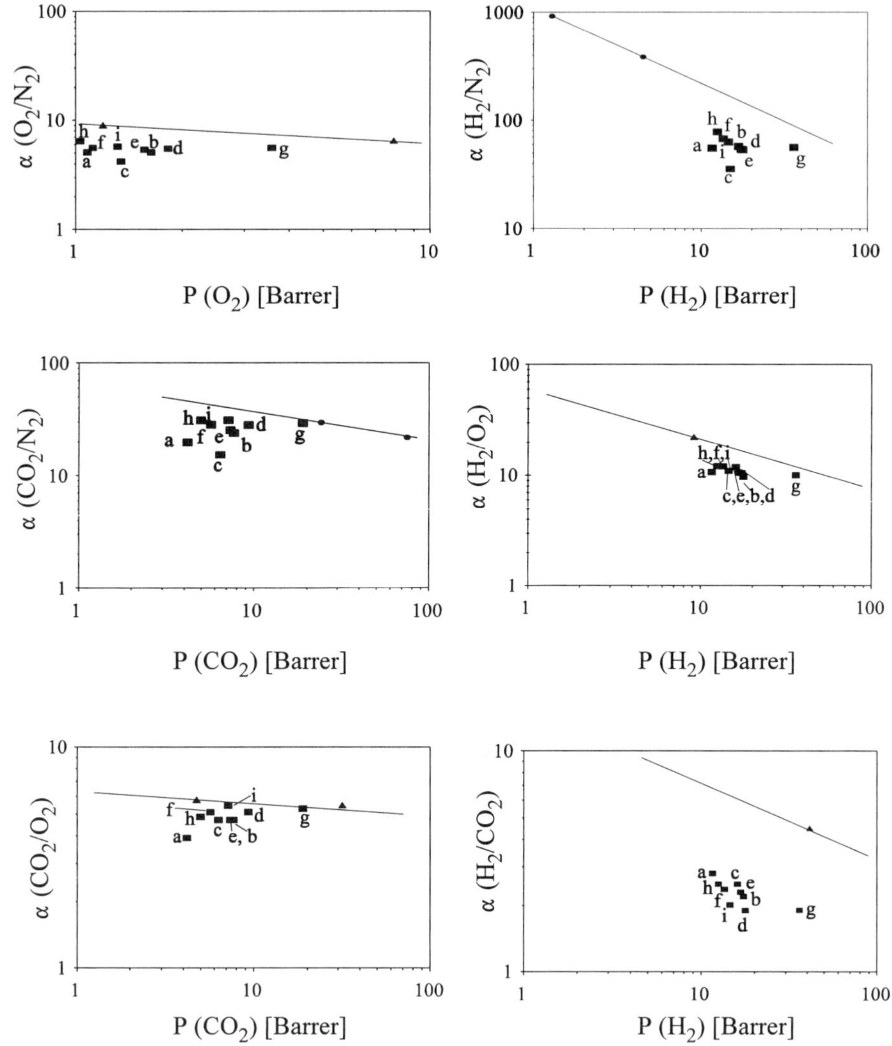


Figure 1. Selectivity-permeability-plots of poly(ether ketone)s **9** (upper-bound lines and some literature data ▲,● added for orientation)

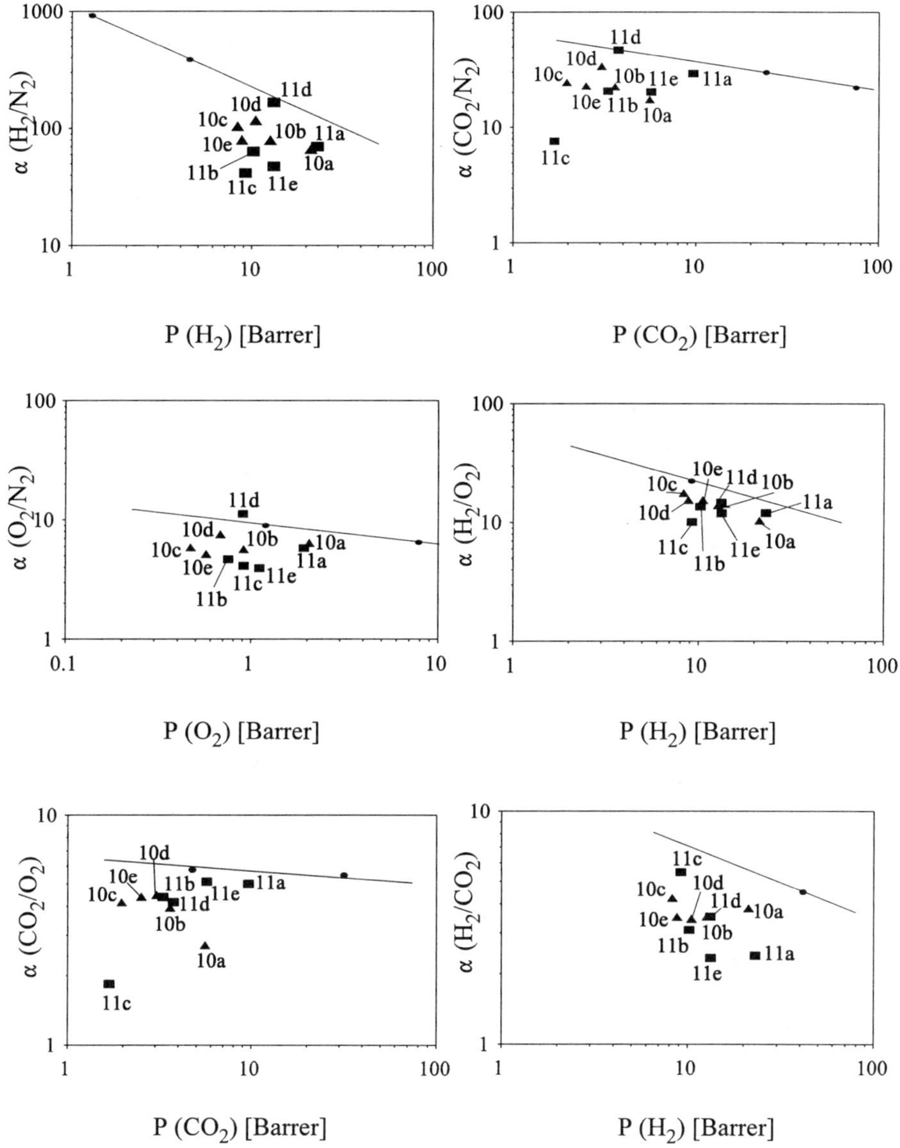


Figure 2. Selectivity-permeability-plots of polyimides **10** and **11** (upper-bound lines and some literature data ▲,● added for orientation)

group X in the bisphenol unit. Also, polymers with the highest permeability coefficients (generally **9d** and **9g**) are not those with the highest fractional free volume. This is in contrast to the common notion that permeability coefficients correlate strongly with $V_f$. In fact, only a relatively small portion of the free volume is actually accessible to gas molecules (*15*). Thus, not all structural modifications of a polymer chain which increase $V_f$ necessarily result in increased permeability coefficients. If many small voids are created, which are inaccessible to the penetrant molecules, a large fractional free volume can result without actually increasing the portion that is accessible to the gas molecules and contributes to *P*.

Despite the rather high global chain stiffness and resulting high glass transition temperatures, the selectivities of the poly(ether ketone)s are only average. This result agrees with our initial idea of short, rather flexible phenylether segments within the repeat unit. If the transport of penetrant molecules is indeed predominantly controlled by these segments, the overall rigidity of the polymer chain, that is, the indan part, should not have any positive effect on selectivity since the gas molecules would mostly pass through openings created by thermally activated motions of the more flexible segments rather than passing through the rigid, bulky indan groups.

This model can explain the unexpectedly low selectivity of polymer **9c** for $H_2/N_2$, $CO_2/N_2$, and $O_2/N_2$. Scheme 5 shows a schematic representation of the flexible segment which we believe is responsible for most of the transport jumps.

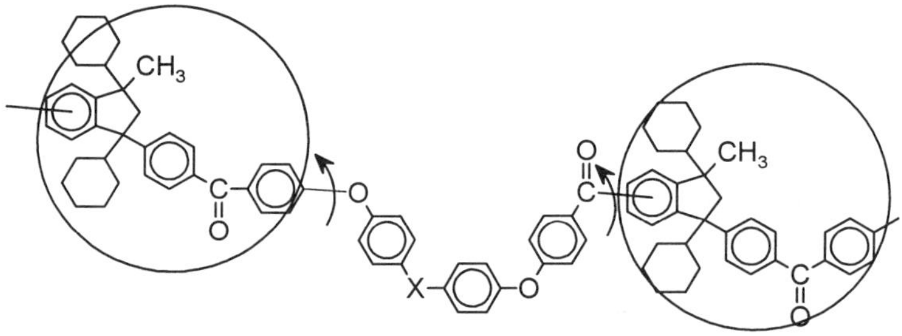


Scheme 5: Proposed flexible segment controlling diffusional jumps

All polymers **9** contain a connecting group X which allows a conformation similar to the one shown in Scheme 5. The only exception is polymer **9c**, which contains a biphenyl unit, i.e. a single bond in place of X. Therefore, in polymer **9c** the phenylether segment has to assume a different conformation. Because the biphenyl unit is all-*para* linked, this conformation is likely to be more extended than in the other polymers. Consequently, motions of this extended segment will open larger passages, resulting in decreased selectivity. CPK- or computer models of the polymers **9** show that the "cavity" defined by the phenylether segment indicated in Scheme 5 is too small for passage of $O_2$ or $N_2$ when all phenyl rings are coplanar. The opening is large enough for both kinds of molecules when all phenyl rings are rotated $90^o$ out of the plane. In reality, any conformation between these two extremes can be assumed, allowing a distinction between gas molecules based on their size.

Tentative results of dynamic mechanical testing and solid state NMR indicate considerable mobility of the aromatic rings at room temperature. Dielectric spectroscopy (10 kHz) using polymer **9g** also shows considerable mobility of the carbonyl groups (the strongest dipole in **9g**) at room temperature (a maximum in dielectric loss occurs around room temperature at a frequency of 10 kHz). These preliminary results corroborate our assumption of flexible phenylether segments. However, the exact structural elements which are moving have not been identified yet. This will be the subject of further work, where deuteration of certain phenyl rings and solid state NMR spectroscopy will be used for closer investigation of this situation.

Within the series of polyimides **10** and **11**, the polymers with the hexafluoroisopropyl linkage in the dianhydride unit (**a**) tend to exhibit the highest permeability coefficients. Similar observations have been made many times before and are explained by the bulkiness of this group. The polyimides with carbonyl linkage (**c**) tend to have the lowest permeability coefficients. Regarding the selectivities, there are no clear trends. Also, the transition from polymers **10** with methyl substituents at the indan group to the polymers **11** with two cyclohexyl groups has no uniform effect on permeability coefficients and selectivities. We expected cyclohexyl groups to add to the bulkiness and chain stiffness and hence to improve both parameters simultaneously. However, it appears that the inherent conformational flexibility of the cyclohexyl rings causes less decrease in chain packing density than expected. In any way, the overall chain stiffness is strongly increased by the cyclohexyl substituents, as is evidenced by the considerable increase in $T_g$ of polymers **11** relative to polymers **10**. Apparently, however, this increase in rigidity does not take place in the segments which dominate the formation of channels for diffusional jumps.

## Conclusions

Indan groups of differing bulkiness were introduced into several series of poly(ether ketone)s and polyimides. The indan groups were linked by relatively flexible ether containing segments. Contrary to expectations, the polymers showed no unusually high permeability coefficients or selectivities. This result can be attributed to the presence of the flexible phenylether segments, especially in poly(ether ketone)s **9**. Based on this observation, the transport of penetrant molecules is mostly controlled by thermally activated motions of the flexible segment between the indan groups. Thus, changes in bulkiness or rigidity of the indan groups has less effect than expected. Preliminary results from solid state NMR, dynamic mechanical measurements and dielectric spectroscopy indicate considerable mobility of the phenyl rings in films at room temperature. For further support of our model, more detailed investigations to identify the flexible parts of the repeating unit, additional structural modifications, especially in the segments between the indan groups are necessary.

## Acknowledgement

For solid state NMR measurements and dielectric spectroscopy as well as valuable discussions, we wish to thank Prof. Spieß and his group, especially Volker Schädler and Manfred Wilhelm, from the Max-Planck-Institute for Polymer Research in Mainz.

We gratefully acknowledge financial support of this work by Deutsche Forschungsgemeinschaft DFG (Ma 1256/10-1)

## Literature Cited

1. Koros, W.J.; Fleming, G.K. *J. Membrane Sci.* **1993**, *83,* 1.
2. Stern, S.A. *J. Membrane Sci.* **1994,** *94,* 1.
3. Paul, D.R; Yampol'skii, Yu.P. *Polymeric Gas Separation Membranes*, CRC Press, Boca Raton, 1994.
4. Kesting, R.E.; Fritzsche, A.K. *Polymeric Gas Separation Membranes,* John Wiley & Sons, New York 1993.
5. Robeson, L.M.; Smith, C.D.; Langsam, M. *J. Membrane Sci.* **1997,** *132,* 33.
6. Maier, G.; Wolf, M. *Macromol. Chem. Phys.* **1997,** *198,* 2421.
7. Maier, G.; Yang, D.; Wolf, M.; Nuyken, O. *High Performance Polym.* **1994,** *6,* 335.
8. Maier, G.; Wolf, M. *Macromol. Chem. Phys.* **1996,** *197,* 781.
9. Park, J.Y.; Paul, D.R. *J. Membrane Sci.* **1997,** *125,* 23.
10. Elias, H.-G. *Makromoleküle,* Hüthig & Wepf, Basel, 1990, Vol. 1, p. 847.
11. McHattie, J.S.; Koros, W.J.; Paul, D.R. *Polymer* **1991,** *32,* 2618.
12. McHattie, J.S.; Koros, W.J.; Paul, D.R. *Polymer* **1991,** *32,* 840.
13. Tanaka, K.; Kita, H.; Okano, M.; Okamoto, K. *Polymer* **1992,** *33,* 585.
14. Hellums, M.W.; Koros, W.J.; Husk, G.R.; Paul, D.R. *J. Membrane Sci.* **1989,** *46,* 93.
15. Shah, V.M.; Stern, S.A.; Ludovice, P.J. *Macromolecules* **1989,** *22,* 4660.

## Chapter 19

# Gas Separation Properties of Modified Poly(ether ketone[sulfone]) with Phthalic Side Group

## Role of Pendent Groups and Physical Cross-Linking on Gas Permeation Properties of Polymers

**Jiping Xu, Zhonggang Wang, and Tianlu Chen**

**Changchun Institute of Applied Chemistry, CAS, 159 People's Street, Changchun, 130022 Peoples Repubic of China**

Phenolphthalein based polyetherketone and polyethersulfone (PEK-C & PES-C) were modified to improve their gas permeation properties. The influence of side groups on gas permeation properties was studied by using dimethylphenolphthalein (DMPPH), tetramethylphenolphthalein (TMPPH) and diisopropyldimethyl-phenolphthalein (DIDMPPH) to synthesize a systematic series of polymers. The dimethyl substituted polymers show lower gas permeabilities than unsubstituted ones, while the TM- and DIDM- polymers give enhanced permeabilities. In some cases, simultaneous increases of both permeability and permselectivity were observed. The reasons were discussed in terms of free volume. The lactone ring of the pendent phthalic group of phenolphthalein and its derivatives was opened by reduction to the corresponding phenolphthalin(PPL)s and their polyetherketone(sulfone) analogs were synthesized. Introduction of pendent carboxylic groups greatly enhanced the permeability and/or permselectivity. Ionomers obtained from salt formation of the PPL polymers show very high gas permselectivity.

The Chinese patented polyetherketone and polyethersulfone (PEK-C & PES-C) are amorphous polymers with high glass transition temperatures ($T_g$=228℃ and 260℃ respectively)*(1-3)*. They were synthesized by nucleophilic polycondensation of phenolphthalein (PPH) and dichlorobenzophenone or dichlorodiphenylsulfone in sulfolane in the presence of potassium carbonate. They were found to be good membrane materials for ultrafiltration and microfiltration*(4)* and could also be used as charged membranes (after sulfonation or chloromethylation/quaternization) for nanofiltration and electrodialysis processes.

However, the gas separation properties of these materials are moderate, similar to those of polysulfone. In order to further improve their gas permeation properties, modification of the pendent phthalic group was studied: the side lactone ring was transformed into a lactam ring in order to introduce hydrogen bonding between polymer chains. Membranes from PEK-H and PES-H (having chemical structures similar to PEK-C and PES-C with the -O- in lactone ring of the phthalic group changed to -NH- or -NR-) and their copolymers show enhanced permselectivities*(5)*. However, ammoniation of PPH under pressure has limited their practical application.

From numerous studies of structure/gas permeation property relationships in polymers, two general conclusions can be reached: introduction of side alkyl groups to the aromatic ring usually increases the free volume of the polymer and hence enhances gas permeability while selectivity is lowered; and cross-linking of the polymer material by covalent bonds and also by hydrogen bonding or ionic bonds can increase permselectivity greatly. These principles were exploited in this work. First, dimethylphenolphthalein (DMPPH), tetramethylphenolphthalein (TMPPH) and diisopropyl-dimethylphenolphthalein (DIDMPPH) were used to synthesize the corresponding polymers: DMPEK-C, TMPEK-C, DIDMPEK-C, DMPES-C, TMPES-C and DIDMPES-C, and their gas permeation properties were studied. Interestingly, the DM-substituted polymers show lower gas permeabilities than unsubstituted ones, while the TM- and DIDM- polymers give enhanced permeabilities. This can be explained based on the free volume concept. In some cases, simultaneous increases of both permeability and permselectivity were observed. Second, the lactone rings of the pendent phthalic groups of phenolphthalein and its derivatives were reduced via ring opening to the corresponding phenolphthalin(PPL)s, and their polyether-ketone(sulfone)s were synthesized. Introduction of pendent carboxylic groups greatly enhanced permeability and/or permselectivity. Ionomers obtained from salt formation of the PPL polymers show very high gas permselectivity: around 300 for hydrogen/nitrogen, >11 for oxygen/nitrogen and 276,000 for water vapor/nitrogen.

## PEK-C and PES-C Derivatives with Side Subsituent Groups

Based on preliminary experiments, the polymer from dimethylphenolphthalein and dichlorodiphenylsulfone (DMPES-C) has lower permeability to gases than PES-C, (which is in contradiction with common rules). Similar results have been observed for polysulfone and dimethylpolysulfone*(6)*. Hence, a series of substituted polymers were prepared for a systematic study of the role of substituent groups.

**Monomers.** DMPPH was obtained from Beijing Reagents Co., TMPPH and DIDMPPH were obtained from Fluka, 4,4'-difluoro-diphenylsulfone(DFDPS) was obtained from Aldrich, 4,4'-dichlorodiphenylsulfone (DCDPS) was obtained from Shanghai Reagents Corp., and 4,4'-dinitrobenzophenone (DNBP) was synthesized in this laboratory*(7,8,9)*.

**Polymers.** The following polymers were prepared by solution polycondensation of bisphenol, potassium carbonate and DFDPS, DNBP or DCDPS in DMSO or sulfolane. Structures were characterized by NMR (Unity-400) and FT-IR (Digilab FTS-20E).

**Membranes.** Dense membranes of the polymers were cast onto glass plates from 7% solution in $CHCl_3/CH_2ClCHCl_2$ (4:6 vol.), dried 24 h at room temp., then dried for 6 h at 60℃, lifted from the glass plate in water, dried at 100℃ for 4 h and finally dried for one week in a vacuum oven at 10 torr and 120℃. For polymers insoluble in the mixed solvent, membranes were cast from 7% DMF solution.

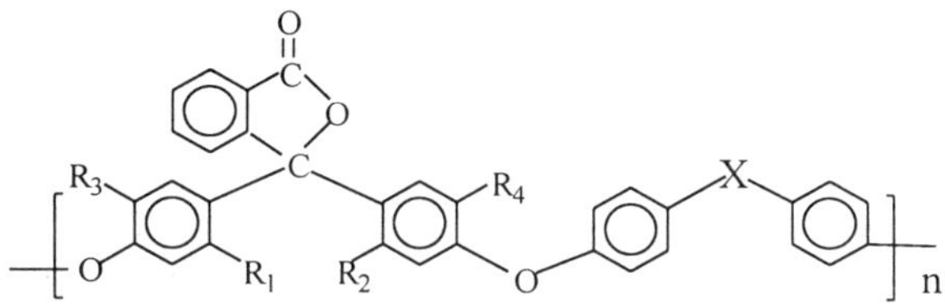

| Polymer | R1 | R2 | R3 | R4 | X |
|---|---|---|---|---|---|
| PEK-C | H | H | H | H | C=O |
| DMPEK-C | H | H | $CH_3$ | $CH_3$ | C=O |
| TMPEK-C | $CH_3$ | $CH_3$ | $CH_3$ | $CH_3$ | C=O |
| DIDMPEK-C | $CH_3$ | $CH_3$ | $CH(CH_3)_2$ | $CH(CH_3)_2$ | C=O |
| PES-C | H | H | H | H | $SO_2$ |
| DMPES-C | H | H | $CH_3$ | $CH_3$ | $SO_2$ |
| TMPES-C | $CH_3$ | $CH_3$ | $CH_3$ | $CH_3$ | $SO_2$ |
| DIDMPES-C | $CH_3$ | $CH_3$ | $CH(CH_3)_2$ | $CH(CH_3)_2$ | $SO_2$ |

Gas permeability coefficients for $CO_2$, $H_2$, $O_2$, $N_2$ and $CH_4$ were measured by a vacuum manometric method with an upstream pressure of 5 atm. in the temperature range from 30 to 100℃.

The gas permeation data of the PEK series of polymers are summarized in Table I, and those for the PES series are recorded in Table II.

Table I. Gas permeation properties of polyetherketones at 30℃

| Polymer | $P_{H_2}$ | $\alpha\ H_2/N_2$ | $P_{O_2}$ | $\alpha\ O_2/N_2$ | $P_{CO_2}$ | $\alpha\ CO_2/CH_4$ |
|---|---|---|---|---|---|---|
| PEK-C | 11.7 | 75.6 | 0.95 | 6.2 | 2.73 | 33.2 |
| DMPEK-C | 10.3 | 95.2 | 0.87 | 8.9 | 2.60 | 37.1 |
| TMPEK-C | 21.5 | 114.4 | 1.55 | 8.2 | 5.44 | 32.0 |
| DIDMPEK-C | 42.5 | 45.3 | 4.85 | 5.2 | 19.3 | 17.7 |

P in barrers.

Table II. Gas permeation properties of polyethersulfones at 30℃

| Polymer | $P_{H_2}$ | α $H_2/N_2$ | $P_{O_2}$ | α $O_2/N_2$ | $P_{CO_2}$ | α $CO_2/CH_4$ |
|---|---|---|---|---|---|---|
| PES-C | 12.1 | 72.6 | 0.95 | 6.2 | 5.74 | 40.1 |
| DMPES-C | 10.8 | 98.1 | 0.87 | 7.6 | 3.12 | 42.9 |
| TMPES-C | 16.9 | 84.1 | 1.55 | 7.5 | 7.69 | 34.2 |
| DIDMPES-C | 30.6 | 43.8 | 4.85 | 5.5 | 19.4 | 20.9 |

P in barrers.

Based on the data in Tables I and II, the gas permeability coefficients of the DM-based polymers is the lowest, similar to the results observed by Moe et al,*(6)*. The introduction of two methyl groups to the repeat unit of PEK(S)-C decreased permeability and increased permselectivity. Introduction of four methyl groups or two methyl and two isopropyl groups enhances the permeability. These results can be explained using the free volume concept. We suppose the bulky phthalic pendent group of PEK(S)-C provides some free volume to the polymer. The two methyl groups introduced in DMPEK(S)-C fill up some fraction of the free volume produced by the phthalic group, thus decreasing the total free volume of the polymer. A further increase in the number of methyl or isopropyl substituent groups enlarges the interchain distance and the free volume greatly. This explanation is supported by density measurement experiments to characterize free volume (*cf.* Table III).

Table III. Density and Free Volume of PEK-C's

| Polymer | ρ (g/cm$^3$) | V(cm$^3$/g) | $V_O$(cm$^3$/g)* | $V_F$(cm$^3$/g) | CED(kJ/cm$^3$)** |
|---|---|---|---|---|---|
| PEK-C | 1.249 | 0.801 | 0.677 | 0.124 | 458.4 |
| DMPEK-C | 1.247 | 0.805 | 0.693 | 0.112 | 474.8 |
| TMPEK-C | 1.195 | 0.833 | 0.706 | 0.131 | 429.1 |
| DIDMPEK-C | 1.140 | 0.877 | 0.718 | 0.159 | 418.6 |

* $V_O$ is calculated by the method of Sudgen*(10)*.

** CED is estimated by a group contribution method*(11)*.

In the case of TMPEK(S)-C, both permeability and permselectivity were increased simultaneously relative to PEK(S)-C. More detailed studies are being pursued to clarify the origin of this effect.

## PEK(S)-C Polymers with Side Carboxylic Group

Cross-linking polymers by radiation, chemical or photochemical means can increase permselectivity by decreasing the permeability of larger gases more than that of smaller gases. The permselectivity can be increased as much as thirty times (α $H_2/N_2$ increased from 47 to 1400) by photochemical crosslinking of a polyether-imide made from benzophenone dianhydride and diphenyletherdiamine*(12)*. After crosslinking, polymers lose their solubility in solvents and cannot be processed anymore. On the other hand, polymers with carboxylic or other pendent groups, which are capable of forming hydrogen or ionic bonds between the polymer chains may be regarded as polymers with physical crosslinks. Such polymers are still soluble in solvents and hydrogen bonds formed at room temperature can usually be broken by increasing the temperature to processing temperature. The ability of hydrogen bonding in PEK(S)-C to enhance permselectivity has already been shown for the PEK(S)-H family of polymers*(5)*.

In this study, phenolphthalein and diisopropyldimethylphenolphthalein were reduced in alkaline solution by Zn dust with lactone ring opening to phenolphthalin (PPL) and diisopropyldimethylphenolphthalin (DIDMPPL) *(7,8)* with good yield. In the process of polycondensation, the pendent carboxylic groups were protected by formation of the K salt. It was shown that no pendent carboxylic group took part in the nucleophilic polycondensation with dinitrobenzophenone or dichlorodiphenylsulfone in DMSO to give the following polymers:

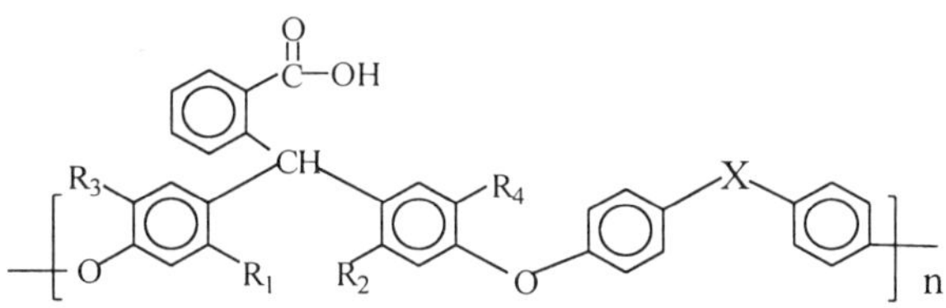


| Polymer | $R_1$ | $R_2$ | $R_3$ | $R_4$ | X |
|---|---|---|---|---|---|
| PEK-L | H | H | H | H | C=O |
| DIDMPEK-L | $CH_3$ | $CH_3$ | $CH(CH_3)_2$ | $CH(CH_3)_2$ | C=O |
| PES-L | H | H | H | H | $SO_2$ |
| DIDMPES-L | $CH_3$ | $CH_3$ | $CH(CH_3)_2$ | $CH(CH_3)_2$ | $SO_2$ |

Table IV. Gas permeation properties of polyetherketone(sulfone)s with side carboxylic group at 30℃

| Polymer | $P_{H_2}$ | $\alpha\ H_2/N_2$ | $P_{O_2}$ | $\alpha\ O_2/N_2$ | $P_{CO_2}$ | $\alpha\ CO_2/CH_4$ |
|---|---|---|---|---|---|---|
| PEK-C | 11.7 | 75.5 | 0.95 | 6.2 | 2.73 | 33.2 |
| PES-C | 12.1 | 72.6 | 0.95 | 6.2 | 5.74 | 40.1 |
| PEK-L | 7.26 | 150 | 0.49 | 10.1 | 1.81 | 43.6 |
| PES-L | 7.22 | 165 | 0.44 | 9.24 | 2.25 | 57.0 |
| DIDMPEK-L | 49.7 | 42.1 | 7.20 | 6.10 | 22.8 | 16.9 |
| DIDMPES-L | 49.6 | 65.5 | 4.53 | 5.98 | 25.1 | 22.3 |

P in barrers.

The procedures for membrane casting and gas permeability measurement were the same as above. The permeability of water vapor was measured separately according to GB 1037-70 (China). The gas permeation properties of PEK(S)-L and DIDMPEK(S)-C are shown in Table IV.

In comparison with the PEK(S)-C's, the PEK(S)-L's have much higher gas selectivity (as expected) at the expense of a slight decrease in permeability. The DIDM-substituted PEK(S)-L's, on the other hand, are 5~10 times more permeable than their unsubstituted analogs and exhibit permselectivity vallues similar to those of PEK(S)-L. These materials are promising candidates for the next generation of gas separation membrane materials.

**Ionomers from PEK(S)-L.** When PEK(S)-L membranes were soaked in KOH or NaOH solution overnight and then washed with water and dried, ionomer type membranes were obtained. Analyses show complete salt formation (i.e. no free acid groups), and the monovalent alkali metal ion, $K^+$ or $Na^+$, was coordinated with several carboxylic groups to form clusters.

The gas permeation properties of these ionomer membranes are recorded in Table V. As shown in Table V, the ionomers achieve the highest permselectivity of all of the modifications of PEK(S)-C studied: around 300 for the hydrogen/nitrogen gas pair, >11 for oxygen/nitrogen and 276,000 for water vapor/nitrogen.

Table V. Gas permeation properties of Ionomers from PEK(S)-L at 30℃

| Polymer | $P_{H_2}$ | $P_{N_2}$ | $\alpha\ H_2/N_2$ | $P_{O_2}$ | $\alpha\ O_2/N_2$ | $P_{H_2O}$ | $\alpha\ H_2O/N_2$ |
|---|---|---|---|---|---|---|---|
| PEK-L | 7.26 | 0.049 | 150 | 0.49 | 10.0 | 1349 | 27800 |
| PEK-Na | 4.67 | 0.023 | 202 | 0.24 | 10.4 | 5860 | 254000 |
| PEK-K | 3.95 | 0.013 | 302 | 0.155 | 11.8 | | |
| DIDMPEK-L | 49.7 | 1.18 | 42.1 | 7.20 | 6.10 | | |
| PES-L | 7.22 | 0.044 | 165 | 0.404 | 9.24 | 2840 | 65000 |
| PES-Na | 5.05 | 0.026 | 242 | 0.281 | 10.7 | 7260 | 276000 |
| PES-K | 4.89 | 0.018 | 269 | 0.208 | 11.4 | | |
| DIDMPES-L | 49.6 | 0.76 | 65.5 | 4.53 | 5.98 | 5050 | 6670 |

P in barrers.

The $P_{O_2}$ vs. $\alpha\ O_2/N_2$ and $P_{H_2}$ vs. $\alpha\ H_2/N_2$ plot for the modified PEK(S)-C polymers together with common polymers are presented in Figures 1 and 2. As can be seen from the plot, most of the polymers synthesized in this work are located above the "upper bound line" *(13)*.

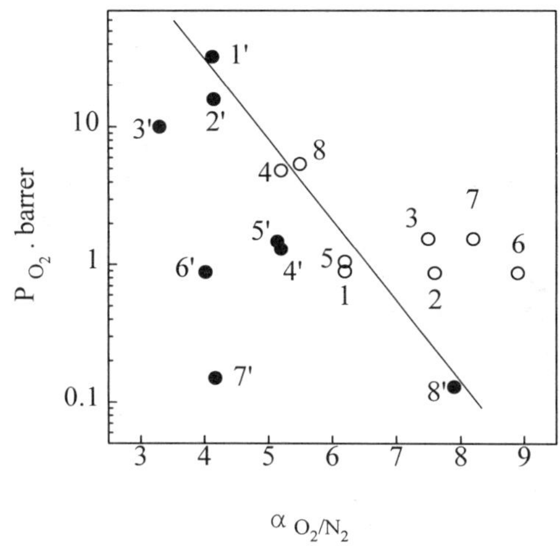


Figure 1. $P_{O_2}$ vs. $\alpha_{O_2/N_2}$ for modified PEK(S)-C polymers. 1: PES-C, 2: DMPES-C, 3: TMPES-C, 4: DIDMPES-C, 5: PEK-C, 6: DMPEK-C, 7: TMPEK-C, 8: DIDMPEK-C, 1': PMP, 2': PPO, 3': EC, 4': PSF, 5': PC, 6': PP, 7': Kapton, 8': Ube's PI.

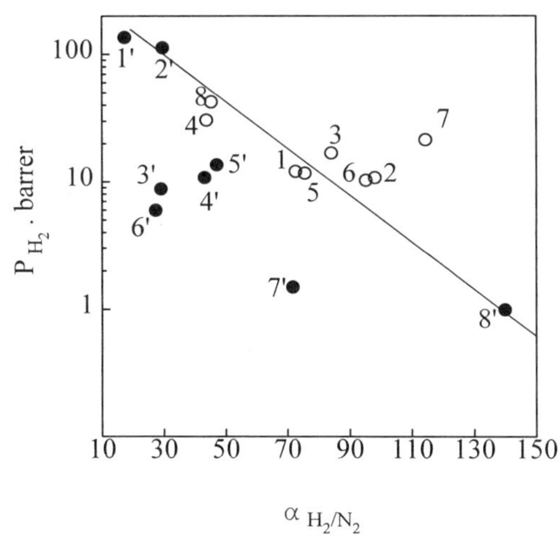


Figure 2. $P_{H_2}$ vs. $\alpha_{H_2/N_2}$ for modified PEK(S)-C polymers. 1: PES-C, 2: DMPES-C, 3: TMPES-C, 4: DIDMPES-C, 5: PEK-C, 6: DMPEK-C, 7: TMPEK-C, 8: DIDMPEK-C, 1': PMP, 2': PPO, 3': CA, 4': PSF, 5': PC, 6': PP, 7': Kapton, 8': Ube's PI.

### Literature Cited


1. Liu, K.J.; Zhang, H.C.; Chen, T.L. *Chin. Pat.* CN 85,101,721 **1987**.
2. Zhang, H.C.; Chen, T.L.; Yuan, Y.G. *Chin. Pat.* CN 85,108,751 **1987**.
3. Chen, T.L.; Yuan, Y.G.; Xu, J.P. *Chin. Pat.* CN 88,102,291 **1991**.
4. Chen, T.L.; Yuan, Y.G.; Xu, J.P. *Preprints of ICOM'87,* Tokyo, **1987,** p.249.
5. Liu, W.Y.; Wang, Z.G.; Chen, T.L.; Xu, J.P. *Proceedings of ICOM'90,* Chicago, **1990,** p.836.
6. Moe, M.B.; Koros, W.J.; Paul, D.R. *J. Polym. Sci.: Polym. Phys. Ed.* **1988,** *26,* 1931.
7. Wang, Z.G. *doctoral dissertation, CIAC CAS,* Changchun, 1994.
8. Wang, Z.G.; Chen, T.L.; Xu, J.P. *J. Appl. Polym. Sci.,* **1997,** *63,* 1127.
9. Wang, Z.G.; Chen, T.L.; Xu, J.P. *J. Appl. Polym. Sci.,* **1997,** *64,* 1725.
10. Sudgen, S. *J. Chem. Soc.,* **1927**, 1786.
11. Jeans, J. *An Introduction to the Kinetic Theory of Gases;* Cambridge Univ. Press: London, 1982, p.183.
12. Liu, Y.; Ding, M.X.; Xu, J.P. *Procedings of IMSTEC'92,* Sydney, Australia, Nov. 1992, p.240. (paper **E2-7**)
13. Robeson, L.M. *J. Membrane Sci.,* **1991,** *62,* 165.

## Chapter 20

# Fundamental and Practical Aspects of Mixed Matrix Gas Separation Membranes

**Rajiv Mahajan, Catherine M. Zimmerman, and William J. Koros**

**Department of Chemical Engineering, The University of Texas at Austin, Austin, TX 78712–1062**

Gas separation membranes combining the desirable gas transport properties of molecular sieving media and the attractive mechanical and low cost properties of polymers are considered. A fundamental analysis of predicted mixed matrix membrane performance based on intrinsic molecular sieve and polymer matrix gas transport properties is discussed. This assists in proper materials selection for the given gas separation. In addition, to explore the practical applications of this concept, this paper describes the experimental incorporation of 4A zeolites and carbon molecular sieves in a Matrimid matrix with subsequent characterization of the gas transport properties. There is a discrepancy between the predicted and the observed permeabilities of $O_2/N_2$ in the mixed matrix membranes. This discrepancy is analyzed. Some conclusions are drawn and directions for further investigations are given.

Despite rapid advances in polymeric gas separation membrane performance in the 1980's, recent efforts have yielded only small improvements. Six years ago, the "upper bound" tradeoff limit between $O_2$ permeability and $O_2/N_2$ selectivity was constructed (*1*), and it still defines the effective performance bounds for conventional soluble polymers. Consequently, an alternate approach to gas separation membrane construction is suggested to exceed current technology performance. Molecular sieves, such as zeolites and carbon molecular sieves (CMS), offer attractive transport properties but are difficult and expensive to process. A hybrid process exploiting the processability of polymers and the superior gas transport properties of molecular sieves may potentially provide enhanced gas separation properties.

## Theory.

Gas transport in zeolites, some molecular sieve carbons, and polymers is described by a sorption-diffusion mechanism. In these cases, the permeability coefficient, $\mathbf{P}_A$, of penetrant A is the product of a kinetic parameter, $\mathbf{D}_A$, the average diffusion coefficient and a thermodynamic parameter, $\mathbf{S}_A$, the solubility coefficient (*2*).

$$\mathbf{P}_A = \mathbf{D}_A \cdot \mathbf{S}_A \qquad (1)$$

The permselectivity, $\alpha_{A/B}$, describes the ideal ability of a membrane to separate gases A and B and may be written as the ratio of permeabilities of components A and B. Using equation 1, $\alpha_{A/B}$ can be written as the product of the diffusivity selectivity and solubility selectivity of the gas pair, viz.

$$\alpha_{A/B} = \frac{\mathbf{P}_A}{\mathbf{P}_B} = \frac{\mathbf{D}_A}{\mathbf{D}_B} \cdot \frac{\mathbf{S}_A}{\mathbf{S}_B} \qquad (2)$$

## Mixed Matrix Membranes.

Mixed matrix membranes are structures with molecular sieve entities embedded in a polymer matrix. To make them commercially attractive, we believe mixed matrix *composite* (MMC) membranes are preferable. These would be compatible with existing composite asymmetric membrane formation technology. Current asymmetric composite hollow fibers consist of an inexpensive porous polymeric support coated with a thin, higher performance polymer. Similar in construction, MMC membranes could replace the thin, higher performance polymeric layer with tightly packed (> 50 vol. %) molecular sieving media, such as zeolite or CMS, in a moderate performance polymeric scaffold. While supporting the molecular sieve phase, the polymeric matrix also connects the selective layer to the porous substructure which can conveniently be the same polymer to promote miscibility. Figure 1 illustrates the proposed MMC membrane formation.

In the past 25 years, relatively few attempts to increase gas separation membrane performance with dense film mixed matrices of zeolite and rubbery or glassy polymer have been reported. Table I summarizes practically all of the reported $O_2/N_2$ mixed matrix membranes. Permeabilities and permselectivities are specified as a range to encompass the various zeolite volume fractions studied. In general, an increase in permeability is observed with zeolite addition coupled with a slight increase in permselectivity. Despite the wide variety of combinations of zeolites with rubbery and glassy polymers, reported mixed matrix membranes fail to exhibit the desired $O_2/N_2$ performance increases. These failures have generally been attributed to defects between the matrix and molecular sieve domains. While this is certainly a possible practical source of failure, our work earlier (*8*) has addressed a more *fundamental source* caused by inattention to matching the transport properties of the molecular sieve and polymer matrix domains. This topic is discussed briefly prior to consideration of the practical defect issue noted above.

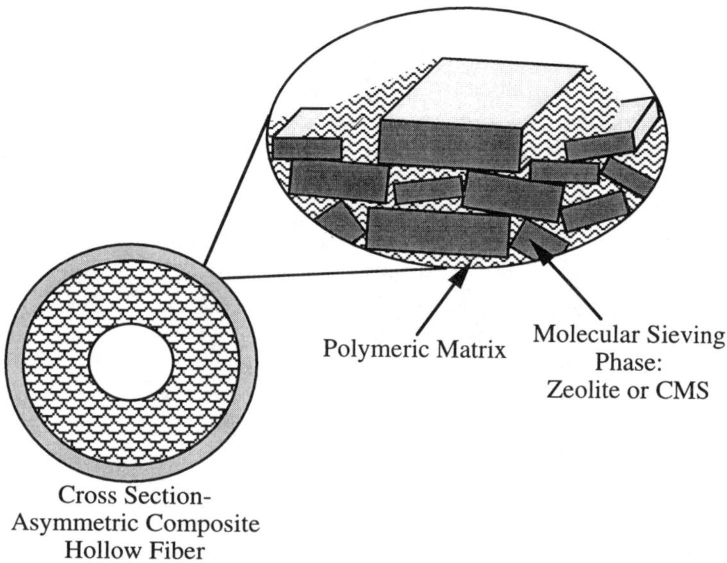


Figure 1. Proposed MMC membrane formation. (Reprinted from J. Membr. Sci., 137, C. M. Zimmerman, A. Singh and W. J Koros, Tailoring mixed matrix composite membranes for gas separations, 145 - 154, Copyright (1997), with permission from Elsevier Science)

Table I. Comparison of various polymer and mixed matrix membrane permeabilities and permselectivities.

| Molecular Sieve | Polymer Matrix[a] | Polymer $P_{O_2}$ (Barrers) | Mixed Matrix $P_{O_2}$ (Barrers) | Polymer $\alpha_{O_2/N_2}$ | Mixed Matrix $\alpha_{O_2/N_2}$ | Ref. |
|---|---|---|---|---|---|---|
| Zeolite 4A | PES | 0.5[b] | 0.4 - 1.1 | 3.7 | 3.9 - 4.4 | (*3*) |
| Silicalite | CA | - | - | 3.0 | 3.5 - 4.1 | (*4*) |
| Silicalite | SR | 600[c] | 560 - 730 | 2.1 | 2.5 - 2.9 | (*5*) |
| Silicalite | SR | 600[d] | 890 - 1370 | 2.1 | 2.5 - 2.7 | (*6*) |
| Silicalite | EPDM | 16[d] | 30 - 40 | 3.0 | 3.9 - 4.7 | (*6*) |
| Zeolite 13X | PSF | 1.4[e] | 1.4 - 1.6 | 5.6 | 5.0 - 5.6 | (*7*) |
| Zeolite 13X | PES | 0.5[b] | 0.3 - 0.5 | 3.7 | 3.8 - 4.3 | (*3*) |

[a] PES = polyethersulfone; CA = cellulose acetate; SR = silicone rubber, EPDM = ethylene-propylene rubber; PSF = polysulfone.

Measurements at (b) 25°C, (c) 30°C, (d) 24°C, (e) 35°C.

**Proper Materials Selection.**

Intrinsic material selection aspects can guide construction of mixed matrix membranes with desirable performance characteristics. The molecular sieving phase must accurately discriminate between the size and shape differences of spherocylindrical $O_2$ and $N_2$ molecules. Silicalite, commonly used in reported mixed matrix membrane studies, is a hydrophobic zeolite possessing channels with dimensions between 5.2 and 5.8 Å. Likewise, zeolite 13X possesses an aperture of 10 Å (*9*). Clearly, these materials are *not* molecular sieves for $O_2$ and $N_2$ molecules with lengths of 3.75 and 4.07 Å, respectively. Rather, an effective molecular sieve would possess an aperture size between the molecular dimensions of the two molecules. Zeolite 4A possesses an eight sided aperture with an effective aperture size of 3.8 Å and is appropriate for this application.

Polymer matrix selection determines minimum membrane performance while molecular sieve addition can only improve membrane selectivity in the absence of defects. Intrinsically, the matrix polymer selected must provide industrially acceptable performance. For example, a mixed matrix membrane using silicone rubber could exhibit properties similar to intrinsic silicone rubber properties, $O_2$ permeability of 933 Barrers and $O_2/N_2$ permselectivity of 2.1 (*8*). The resulting mixed matrix membrane properties would lie substantially below the upper bound trade-off curve for gas permeability and selectivity. In contrast, a polymer exhibiting economically acceptable permeability and selectivity is a likely candidate for a successful polymer matrix. A glassy polymer such as Matrimid polyimide (PI) is an example of such a material because it exhibits acceptable properties and current technology exists for formation of asymmetric hollow fibers for gas separation (*10*).

**Predicted MMC Membrane Performance.**

MMC membrane performance can be predicted using various theoretical expressions. Maxwell first analyzed steady-state dielectric properties in a conducting dilute suspension of identical spheres. Robeson *et al.* (*11*) applied a similar analysis to determine an effective permeability for a series of organic-organic mixed matrix membranes analogous to the desired defect-free molecular sieve-polymer membranes noted above where fundamental transport phenomena are similar

$$\mathbf{P}_{\mathrm{eff}} = \mathbf{P}_{\mathrm{c}}\left[\frac{\mathbf{P}_{\mathrm{d}} + 2\mathbf{P}_{\mathrm{c}} - 2\Phi_{\mathrm{d}}(\mathbf{P}_{\mathrm{c}} - \mathbf{P}_{\mathrm{d}})}{\mathbf{P}_{\mathrm{d}} + 2\mathbf{P}_{\mathrm{c}} + \Phi_{\mathrm{d}}(\mathbf{P}_{\mathrm{c}} - \mathbf{P}_{\mathrm{d}})}\right] \tag{3}$$

where $\mathbf{P}_{\mathrm{eff}}$ is the effective permeability, $\Phi$ is the volume fraction, and the subscripts d and c refer to the dispersed and continuous phases, respectively.

Guided by the criteria for proper materials selection discussed above, consider a zeolite 4A-PI membrane while recalling the intrinsic polyimide properties provide an acceptable minimum performance of $O_2$ permeability equal to 1.3 Barrers and $O_2/N_2$ permselectivity of 7.2. Since zeolites obey a sorption-diffusion mechanism, $O_2$ and $N_2$ gas permeability can be calculated from reported sorption and diffusivity data while accounting for competition between $O_2$ and $N_2$. Zeolite 4A Linde crystals have an $O_2$ permeability of approximately 0.77 Barrers and an $O_2/N_2$ permselectivity of approximately 37 at 35°C (*12-13*), markedly more selective than the polyimide.

Expected zeolite 4A-PI defect-free membrane performance can be estimated using equation 3 for a discontinuous zeolite phase. Conservative performance predictions assume gas molecules diffuse through both the high selectivity molecular sieve and the lower selectivity and higher permeability polymer matrix. Even higher performance can be anticipated if the mixed matrix morphology forms continuous pathways such that gas molecules can diffuse solely through the high selectivity molecular sieve phase. From membrane performance predictions shown in Figure 2, increased addition of zeolite 4A from 10 to 90 vol. % simultaneously decreases slightly the predicted MMC membrane $O_2$ permeability while leading to large increases in $O_2/N_2$ permselectivity.

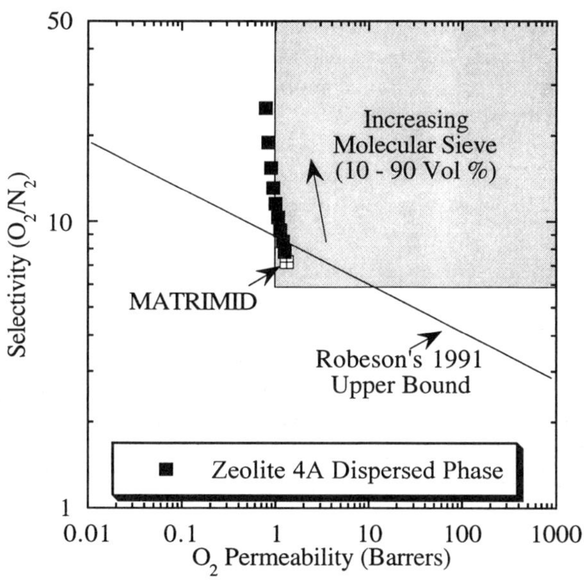


Figure 2. Predicted Zeolite 4A-PI MMC membrane performance from Maxwell's equation. The shaded boundary corresponds to the performance area typically thought to be commercially attractive based on current technology for forming asymmetric membranes with thin selective layers of 1000-2000 Å.

A similar analysis can also be made with carbon molecular sieves. Commercial carbon molecular sieves often have dead ended pores and are thus not suitable for steady state permeation. In this study a suitable (based on the above analysis) carbon molecular sieve developed within our group was used. The material has an $O_2$ permeability of approximately 20 Barrers and an $O_2/N_2$ permselectivity of approximately 12.5 at 35°C (*14*) and is thus markedly more selective and permeable than the polyimide. Again, use of equation 3 allows prediction of the performance of defect free CMS-PI membranes. The predictions suggest significant improvements in both permselectivity and permeability.

**Membrane Formation and Characterization.**

To explore the difficulties in practical implementation of the above concepts, mixed matrix membranes, with 20% molecular sieves (by volume), were prepared by solution deposition on top of a porous ceramic support. The ceramic supports used were Anodisc membrane filters which had 200 Å pores that open into 2000 Å pores and offer negligible resistance to gas flow. Initially the molecular sieve media, zeolites (4A crystals) or carbon molecular sieves, was dispersed in the solvent, dichloromethane, to remove entrapped air. After two hours, Matrimid was added to the mixture, and the solution was stirred for four hours. The solutions used varied in polymer content from 1-5 wt %. The solution was then deposited on top of the ceramic support, and the solvent was evaporated in a controlled manner. The membranes were then dried overnight at 90°C under vacuum. This was followed by a reactive intercalation post treatment technique (*15*) to eliminate defects. This technique involves imbibing a reactive monomer (e.g. diamine) from an inert solvent (e.g. heptane) into any micro defects. Next, a second reactive monomer (e.g. acid chloride) was introduced to reactively close defects by forming a low permeability polymer. The membranes were dried again to remove the inert solvent. Individual membrane thickness was determined by weight gain and varied from 5 to 25 μm.

Gas permeation measurements were made using equipment previously described (*16*). The membranes made were loaded into a Millipore cell, and the cell was sealed with a Viton O ring. The entire system was then evacuated upstream and downstream for 12 hours, after which gas was introduced to the upstream side. Upstream pressures used varied from 40-70 psia. On the permeant side, the changing pressure was measured with a Baratron pressure transducer (0-10 Torr) and plotted against time on a stripchart recorder. Steady state stripchart data were used to determine the downstream pressure increase with time.

**Results and Discussion.**

The results of the investigation are reported in Table II. The CMS-Matrimid and Zeolite 4A-Matrimid membranes give selectivities approaching those of the native polymer at best. Also, much higher permeabilities were observed than predicted by the model for both carbons molecular sieves and zeolites. These results suggest that there was improper contact between the two phases, probably due to dewetting of polymer chains from the sieve surface. In fact, a molecular sieve enclosed in a polymer cage with void space between the two phases appears to be the picture (shown in a cartoon representation in Figure 3) that best explains the observed data. Since the cage is 'closed', near intrinsic polymer selectivities result, yet the gap between the sieve and the polymer provides a less resistive route to gas diffusion, resulting in molecular sieve bypassing and higher apparent permeabilities. Based on the given picture of the molecular sieve-polymer interface, it was necessary to confirm that permeability increases were caused by voids rather than the more permeable carbons. Experiments were completed with an impermeable carbons commercially available as Black Pearls. This carbon was similar to the porous materials described above, except for the fact it was impermeable (*17*). As expected, incorporation of the impermeable carbons result in higher than anticipated permeabilities as well. This observation as well as SEM photographs appear to confirm the picture described above.

Gas Diffusion Pathway

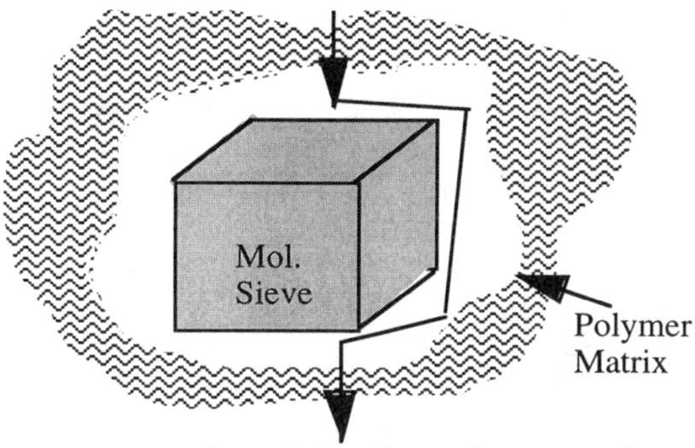


Figure 3. Cartoon representation of sieve in a polymer matrix

Table II. Mixed Matrix Membrane Performance, predicted vs observed.

| Membrane | $\alpha O_2/N_2$ | $PO_2$ (Barrers) |
|---|---|---|
| Matrimid | 7.2 | 1.32 |
| Carbon Mol. Sieve (20 vol%) in Matrimid (Predicted) | 7.6 | 2.1 |
| Carbon Mol. Sieve (20 vol%) in Matrimid (Observed) | 6.9 | 4.9 |
| Zeolite 4A (20 vol%) in Matrimid (Predicted) | 8.5 | 1.2 |
| Zeolite 4A (20 vol% ) in Matrimid (Observed) | 7.2 | 4.0 |
| Impermeable Carbons (20 vol%) in Matrimid (Predicted) | 7.2 | 0.96 |
| Impermeable Carbons (20 vol%) in Matrimid (Observed) | 7.0 | 4.2 |

All measurements at 25°C

The picture in Figure 3 can be quantified by extending the Maxwell model (discussed above) to the three phases involved here: the molecular sieving media, the polymer, and the void. It is simple to do so in this case because at all times the void phase lies between the molecular sieve and polymer phase. One can model the system as a pseudo two phase with the polymer being one phase and the combined molecular sieve and void being the other phase. Equation 3 can be used to obtain the permeability of the combined void and molecular sieve phase with the void phase as the continuous and molecular sieve as the disperse phase. The value of the combined permeability can be used along with the polymer permeability to explain membrane performance. Determination of gas permeability in the voids is necessary to determine the effective permeability of the 'void and molecular sieve phase'. Gas flow through the void can be assumed to be Knudsen for micron sized defects, thus the diffusivity of the gas through the void can be calculated using Eq. 4 (*18*)

$$\mathbf{D}_{A,K} = 9.7 \cdot 10^{-5} \cdot r \cdot \sqrt{\frac{T}{M_A}} \tag{4}$$

where $\mathbf{D}_{A,K}$ is the diffusion coefficient in $cm^2 / s$, r is the pore radius in Å, T is the absolute temperature, and $M_A$ is the gas molecular weight. In the diffusion coefficient calculation, the radius used in equation 4 was assumed to be half the distance between the polymer and sieve phase. This distance is a critical parameter in this modeling, and affects the behavior in three ways. Firstly as shown by equation 4, the radius is directly proportional to the diffusion coefficient in the void phase. Secondly, it affects the volume fraction of the void phase and sieve phase in the first application of the Maxwell Equation. Lastly, it affects the volume fraction of the combined phase in the polymer since only the volume fraction of molecular sieve phase can be controlled experimentally. The partition coefficient of the gas in the void, relative to the gas phase, can essentially be assumed to be one, i.e. a cubic centimeter of gas in the void is assumed to occupy the same volume as it would in the gas phase at the same temperature and pressure. Figure 4 indicates how $O_2$ permeability and $O_2/N_2$ permselectivity of a Zeolite 4A-Matrimid mixed matrix membrane with 20 vol. % zeolite at 25°C varies with defect size. The range of defect sizes between the sieve and polymer phase studied correspond roughly with those observed in SEM images. Selectivity and permeability calculations indicated that intrinsic polymer selectivity and a much higher permeability result. Using this model, the experimental Zeolite 4A-Matrimid oxygen permeability shown in Table II would be obtained if this distance was 2700 Å. SEM images such as that in Figure 5 indicate that this is around the right order of magnitude.

## Conclusions and Future Work.

Recent membrane materials development has failed to exceed the polymeric $O_2/N_2$ upper bound trade-off curve in the past six years. A MMC membrane processing scheme based on molecular sieving materials is attractive if a proper molecular sieve and polymer matrix is selected for the given separation. Using Maxwell's equation, estimates of MMC membrane performance indicate significant improvements should be achievable with elimination of non-selective defects. The preparation of actual

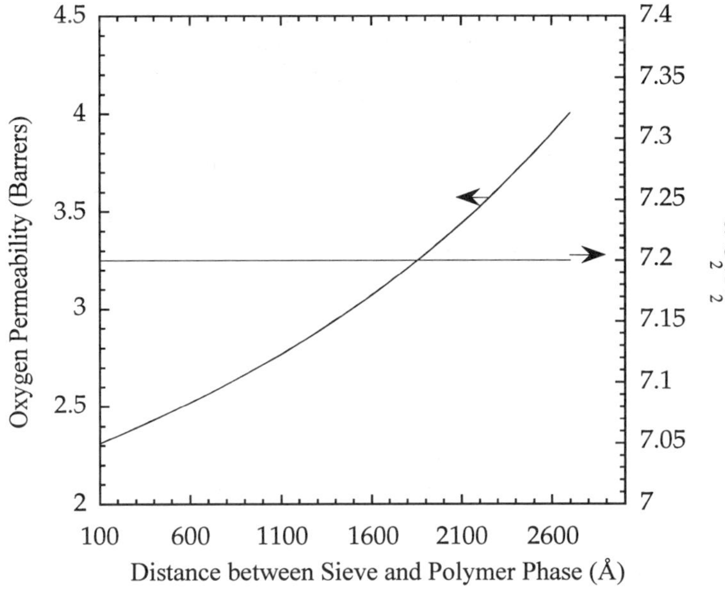


Figure 4. Predicted variation in $O_2$ permeability and $O_2/N_2$ selectivity as a function of distance between polymer and sieve.

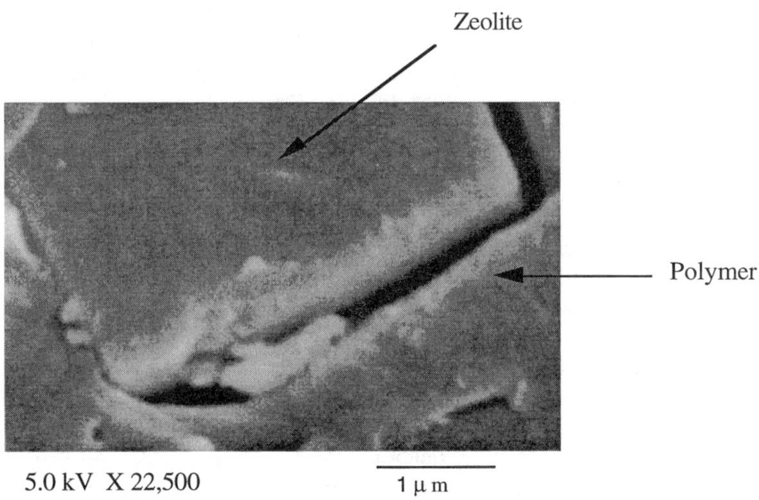


Figure 5. SEM showing cross section of a mixed matrix membrane.

mixed matrix membranes results in a three phase membrane: molecular sieve, polymer, and non selective defects. This is most likely due to dewetting of polymer polymer chains from the sieve surface. The 'sieve in a polymer cage' appears to be a picture that is consistent with observed data.

The most practical approach to the problem of dewetting on the sieve surface would be to encapsulate the sieve in a second polymeric phase that can either be grafted onto the sieve with coupling agents or adsorbed. At the same time, compatibility with the main polymeric phase is essential. The model developed above can easily be extended to these systems.

**Acknowledgments.**

The authors gratefully acknowledge the support of the National Science Foundation, Medal Inc. and the Texas Advanced Research Program.

**References**

1. Robeson, L. M. *J. Membrane Sci.* **1991**, 62, 165.
2. Singh, A.; Koros, W. J. *Ind. Eng. Chem. Res.* **1996**, 35, 1231.
3. Süer, M. G.; Baç, N.; Yilmaz, L. *J. Membrane Sci.* **1994**, 91, 77.
4. Kulprathipanja, S.; Neuzil, R. W.; Li, N. N. U.S. Patent 4,740,219 (**1988**)
5. Jia, M.; Peinemann, K.-V.; Behling, R.-D. *J. Membrane Sci.* **1991**, 57, 289.
6. Duval, J.-M.; Folkers, B.; Mulder, M.H.V.; Desgrandchamps, G.; Smolders, C.A. *J. Membrane Sci.* **1993**, 80, 189.
7. Gür, T. M. *J. Membrane Sci.* **1994**, 93, 283.
8. Zimmerman, C. M.; Singh, A.; Koros, W.J. *J. Membrane Sci.* **1997,** 137, 145.
9. Yang, R. T. *Gas separation by adsorption processes,* Butterworths Publications: Stoneham, MA, 1987; pp 23.
10. Ekiner, O. M.; Hayes, R. A. U.S. Patent 5,015,270, (**1991**)
11. Robeson, L. M.; Noshay, A.; Matzner, M.; Merriam, C. N. *Die. Ange. Makrom.Chem.* **1973**, 29, 47.
12. Ruthven D. M.; Derrah, R. I. *J. Chem. Soc., Faraday Trans.* **1975**, 71, 2031.
13. Kärger J.; Ruthven, D. M. *Diffusion in zeolites and other microporous solids*, Wiley-Interscience Publications, New York, NY, 1992; pp 579.
14. Singh, A. *Membrane materials with enhanced selectivity: an entropic interpretation,* Ph.D. Dissertation, The University of Texas at Austin (1997)
15. Ekiner, O. M.; Hayes R. A.; Manos, P. U.S. Patent 5,091,216, (**1992**)
16. O'Brien, K. C.; Koros, W. J.; Barbari, T. A. *J. Membrane Sci.* **1984**, 29, 229.
17. *Cabot Carbon Blacks,* product information folder.
18. Hines A. L.; Maddox, R. N. *Mass transfer fundamentals and applications,* Prentice-Hall Publications, Englewood Cliffs, NJ, 1985; pp 45-48.

Chapter 21

# $CO_2$ Sorption Properties of Annealed Syndiotactic Polystyrene

**K. Tsutsui, Y. Tsujita, H. Yoshimizu, and T. Kinoshita**

**Department of Materials Science and Engineering, Nagoya Institute of Technology, Gokiso-cho, Showa-ku, Nagoya 466–8555, Japan**

Thermal and spectral properties of annealed cast syndiotactic polystyrene (SPS) films were investigated. The sample annealed at lower temperature exhibited an ordered conformation but a lack of crystal regularity. This result indicated the presence of a mesophase. $CO_2$ sorption properties of these films were also examined and dual-mode sorption parameters were obtained. In particular, the hole saturation constant for Langmuir sorption ($C_H$') reflected the fine structure of the annealed SPS and exhibited a different annealing time dependence when annealed at different temperatures. From these results, we propose that the samples annealed at lower temperature contain many nanovoids, which form as a result of solvent desorption.

Highly syndiotactic polystyrene (SPS) crystallizes in two different conformations (TT and TTGG). One of the crystalline forms, the $\delta$ form (TTGG conformation), is prepared by solution-casting*(1,2)*. The $\delta$ form includes solvent molecules and is transformed into the $\gamma$ form by annealing, which is accompanied by solvent desorption. This transition from the $\delta$ to the $\gamma$ form has been investigated by various methods*(3,4)*. In this study, we investigated $CO_2$ sorption properties of SPS films, which were prepared by solution-casting and annealed at different temperatures. $CO_2$ sorption properties probe the structural reorganization that occurs during the transition from the $\delta$ to the $\gamma$ form because sorption of a penetrant molecule is sensitive to physical properties of the polymer matrix such as degree of crystallinity and fractional free volume*(5)*.

## Experimental

Highly syndiotactic polystyrene (SPS) ($M_w$ = 211,000, $M_w$ / $M_n$ = 3.45) was kindly supplied by Idemitsu Petrochemical Co.Ltd. Film samples cast from toluene solution were annealed at 100, 120, 140, and 170℃ for various times in vacuum. The annealed SPS films have been labeled according to the following scheme : XAY, where X is the annealing time (hours) and Y is the annealing temperature (℃). Differential scanning calorimetry (DSC) thermograms in the range 25-300℃ were performed using a Perkin Elmer DSC 7 at a heating rate of 20℃/min. under a nitrogen gas purge. Thermogravimetric analysis (TGA) data in the range 25-450℃ were obtained using a Seiko TG/DTA220 at a heating rate of 20℃/min under a nitrogen gas purge. Wide-angle X-ray diffraction patterns in the range 5-25° $2\theta$ were recorded with nickel-filtered Cu K $\alpha$ radiation by a Rigaku RAD-RC diffractometer. The scan rate was 1°/min. and the collection number was 5. Infrared spectra were obtained using a Jasco IR-700 type IR spectrometer operating at $1 cm^{-1}$ resolution in the range 4000-400 $cm^{-1}$. The number of accumulation cycles was 1. $CO_2$ sorption isotherms at 25℃ were obtained by the Quartz Crystal Microbalance (QCM) method. Thin films were coated on the electrode of the QCM by spin-casting and annealed under appropriate conditions. The $CO_2$ used in this sorption study was at least 99.9% pure and was used without further purification.

## Characterization of the annealed cast SPS

Films prepared by solution-casting contain the $\delta$ form of crystallites. When they are annealed below 190℃, they are first reorganized into the $\gamma$ form and, subsequently, from the $\gamma$ to the $\alpha$ form at 190-200℃. This transition from the $\gamma$ to the $\alpha$ form is accompanied by a conformational change from TTGG to TT. We examined IR spectra to detect the conformational transition accompanying the transition from the $\gamma$ to the $\alpha$ form. IR absorbances at 933 $cm^{-1}$ and 1222 $cm^{-1}$ are assigned to TTGG and TT conformations, respectively. Figure 1 shows the annealing time dependence of the relative absorbance of TT and TTGG conformations of the cast SPS annealed at 120℃, 140℃ and 170℃, respectively. The absorbance values ($R_1$ and $R_2$) at 1222 $cm^{-1}$ and 933 $cm^{-1}$ are normalized by the absorbance at 1183 $cm^{-1}$ as an internal standard. For samples annealed at 120℃, $R_1$ remains zero even if the annealing time is more than 100 h. Annealing the cast film at 140℃ or 170℃ makes $R_1$ increase with annealing time. The higher the annealing temperature, the more $R_1$ changes. The value of $R_2$ of the cast film is 1.24 and when films are annealed at 120℃, 140℃, and 170℃, the values decrease rapidly and become sensibly constant with increasing annealing time. These results may indicate that the transition from TTGG to TT conformation occurs below the $\gamma$ to $\alpha$ transition temperature (Tc = 190℃), which corresponds to the maximum conformation transition temperature for a large crystallite. In other words, it is possible that the conformational transition from TTGG to TT for a small crystallite occurs at temperatures lower than Tc. Upon annealing in the vicinity of Tc, both conformations exist. However, when samples are annealed below 120℃, i.e. at

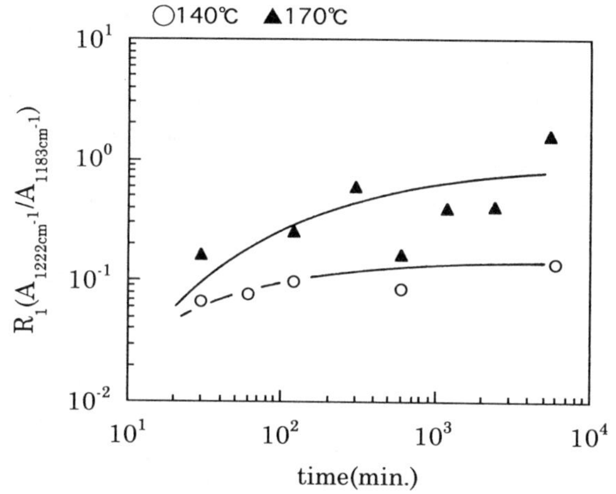


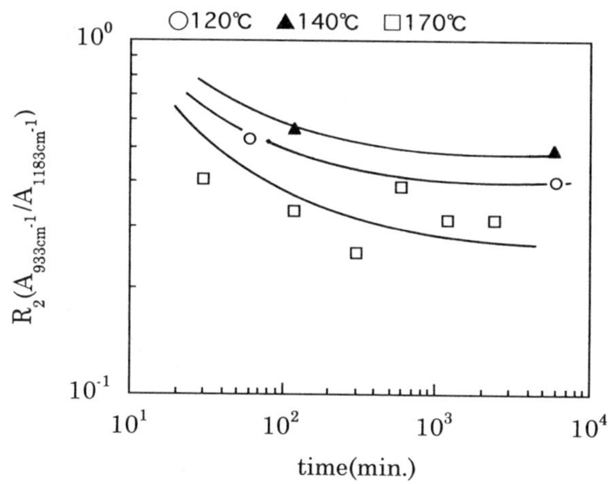


**Figure 1** Effect of annealing time and temperature on chain conformation in annealing time and temperature. $R_1$ and $R_2$ characterize the relative amounts of polymer in the TT and TTGG conformations, respectively.

much lower temperatures than Tc, the TTGG conformation decreases slightly, and the conformational transition to TT never occurs by annealing.

Figure 2 shows DSC thermograms of cast films, 1A120, 1A140, and 1A170. The endothermic peaks at 270℃ in all samples correspond to the melting of the $\alpha$ form. Tgs are in the range of 105 ~ 110℃. A large endothermic peak for the as-cast film and 1A120 appears at 150℃. This peak is due to desorption of complexed solvent as demonstrated from thermogravimetric analysis (TGA) (Figure 3). It is probable that the SPS amorphous chains reorganize, judging from the observation of a broad exothermic peak appearing at 150~200℃. On the other hand, the thermograms of samples annealed near Tc indicate that there is no endothermic peak at 150℃. The other endothermic peaks, which are sharper as the annealing temperature approaches Tc, involve the transition from the $\gamma$ to the $\alpha$ form of crystallites. The crystallite size of the $\gamma$ form might be small and there might be many crystal defects in the $\gamma$ form which is transformed from the $\delta$ form by low temperature annealing. Therefore, the $\gamma$ form may transform into the $\alpha$ form during DSC scans below Tc. In the case of annealing near Tc, the transition from the $\delta$ to the $\gamma$ form is likely to occur and the transformed $\gamma$ form may consist of large crystallites with few crystal defects. An endothermic peak resulting from the transition from the $\delta$ to the $\gamma$ form is observed at 190℃ because there exists a stable $\gamma$ form due to annealing near Tc. The DSC thermograms of the samples annealed for 100h at 100, 120, 140, and 170℃ are shown in Figure 4. Solvent is still present in 100A100, but no solvent is present in the other samples. The absence of endothermic peaks around Tc indicates the presence of small and imperfect $\gamma$ form crystallites for 100A100. The important feature of these results is that this sample exhibits TTGG conformation based on the IR spectra but contains small and imperfect $\gamma$ form crystals according to the DSC study. That is, a mesophase, which has an ordered TTGG conformation but lacks crystalline regularity, exists in 100A120. This is also confirmed by X-ray profiles shown in Figure 5. The X-ray profiles of (A), (B), (C) and (D) are typical of the samples consisting of $\delta$ form crystals, mesophase, $\gamma$ form and $\alpha$ form crystals, respectively. These results suggest the existence of a relatively regular microvoid structure in the mesophase which is different from so called Langmuir microvoids in glassy polymers. Hereafter, we call this structure a nanopore.

### $CO_2$ sorption properties of the annealed SPS samples

The sorption isotherms of the SPS samples annealed at 120℃ and 170℃ are shown in Figure 6. The sorption data in these figures are not corrected for the crystalline content of the samples. Atactic polystyrene (APS), which is annealed at 120℃ for 1h(1A120(APS)), is also shown in this figure. The sorption isotherms of all of the samples are typical of those of glassy polymers. The $CO_2$ sorption level in the

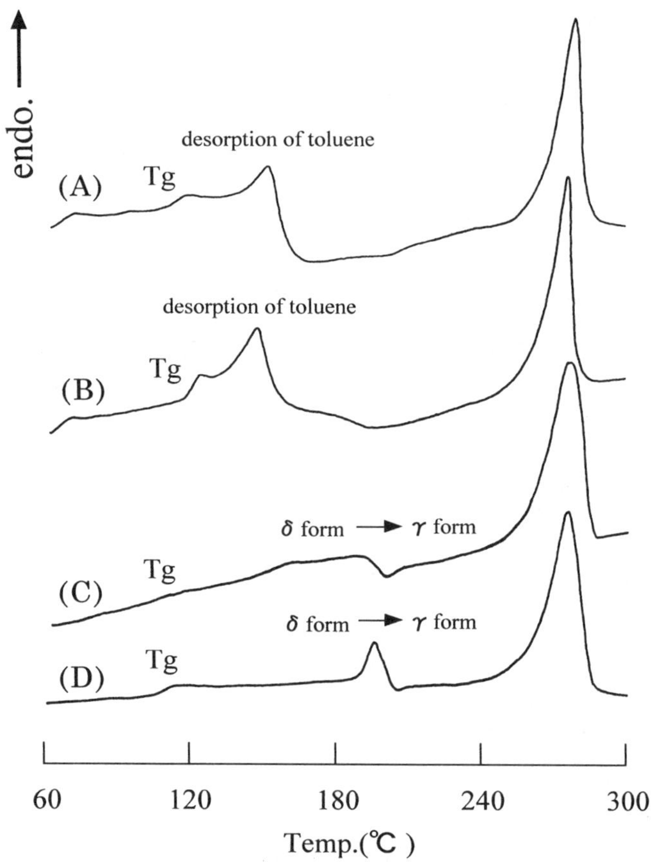


**Figure 2** DSC thermograms of (A)as-cast, (B)1A120, (C)1A140, and (D)1A170.

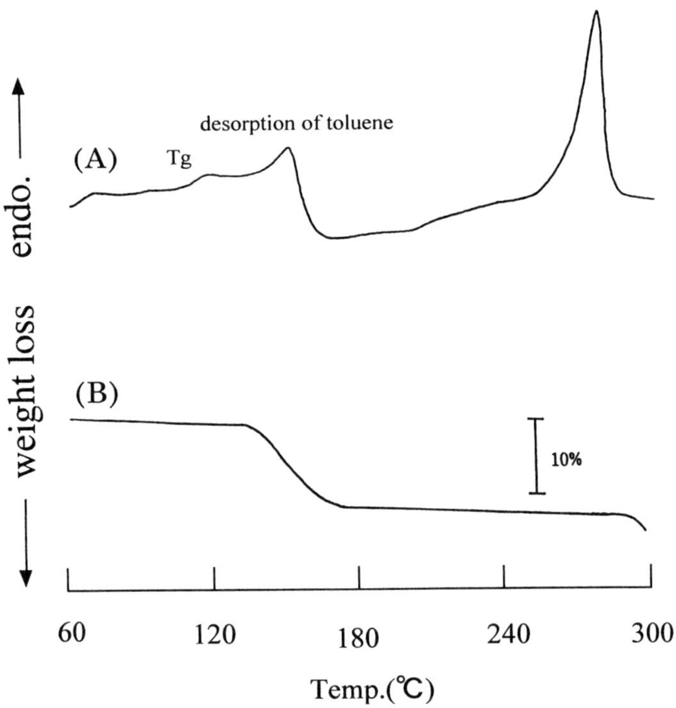


**Figure 3** Thermal analysis of as-cast SPS ; (A)DSC thermogram and (B)TG curve.

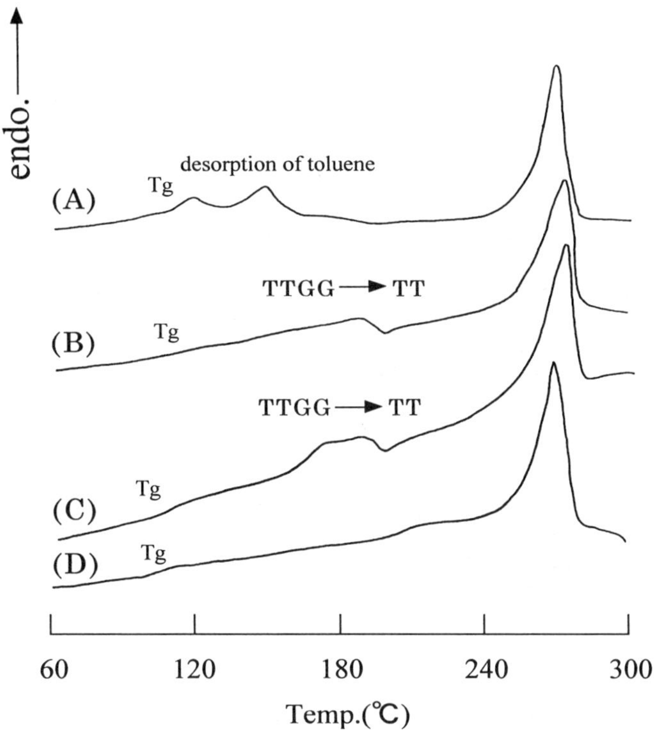


**Figure 4** DSC thermograms of (A)100A100, (B)100A120, (C)100A140, and (D)100A170.

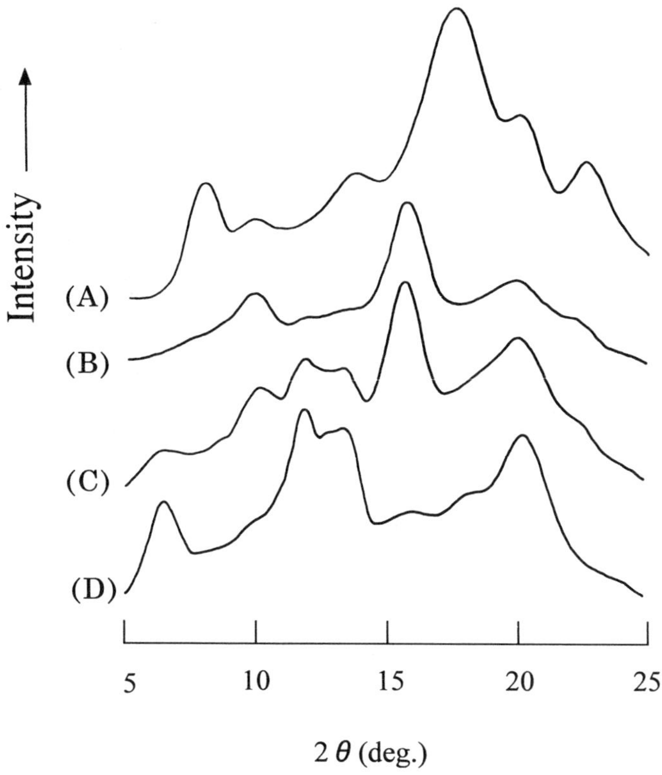


**Figure 5** X-ray diffraction profiles of (A)as-cast, (B)100A120, (C)5A170, and (D)100A170.

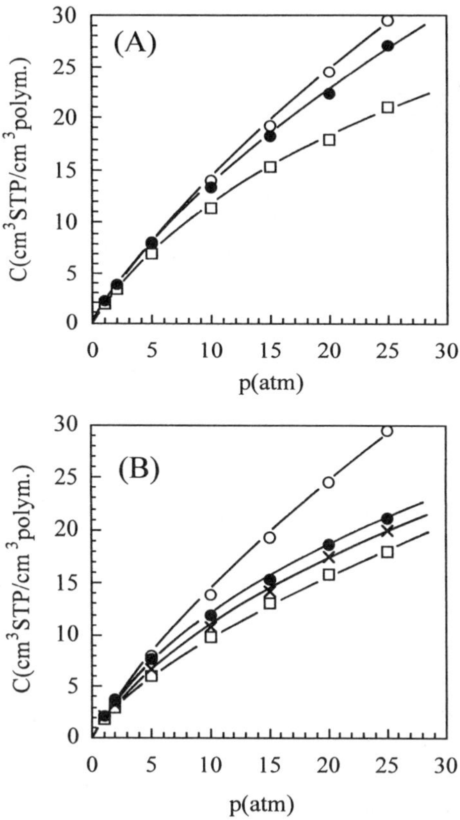


**Figure 6** (A)$CO_2$ sorption isotherms at 25℃ of 1A120(APS) (○), 1A120(●),100A120(□);(B)$CO_2$ sorptions at 25℃ of 1A120(APS) (○), 1A170(●),1A170(slowly-cooled)(□), and 5A170(×).

annealed SPS is less than that in 1A120(APS) because of the crystallinity in the annealed SPS. The $CO_2$ sorption level of the annealed SPS is larger than that of 1A120(APS) when the sorption data are reported on an amorphous basis. Because of a lack of crystalline regularity, sorption into the nanopores between the polymer chains of the mesophase occurs. One can assume that sorption into the nanopores is Langmuir type sorption. The plasticization by $CO_2$ can be ignored at pressures up to 25 atm since both sorption and desorption isotherms are almost identical. To evaluate the $CO_2$ sorption behaviour more fully, the dual-mode sorption model has been applied to the experimental sorption isotherms. The resulting dual-mode sorption parameters are presented in Table I. The parameters $k_d$ and b of the SPS samples annealed at 120℃ decrease as annealing time increases. On the other hand, $C_H$' increases with annealing time. $C_H$' reflects the amount of Langmuir microvoids per unit polymer volume. Interestingly, $C_H$' of the annealed SPS is larger than that of the amorphous APS, irrespective of the presence of the crystalline phase. Upon annealing at 170℃, $k_d$ does not change markedly as annealing time increases, but $C_H$' decreases remarkably with increasing annealing time. The difference in the annealing time dependence of $C_H$' between 120℃ and 170℃ is very important for understanding the reorganization of the crystalline material from the $\delta$ to the $\gamma$ form. Chatani et al. (3) reported that SPS samples complexed with solvents exhibited a monoclinic crystal lattice ($P_21/a$) with a = 17.48Å, b =13.27Å, c = 7.71Å, and $\gamma$ = 122°. The unit lattice contains eight monomers $(TTGG)_2$ and two solvent molecules which are sandwiched between the two chain segments in the unit lattice. From these results, we propose that the solvent first desorbs from the unit lattice during the annealing process and, subsequently, the two chains pack in the direction of the b axis depending on annealing temperature as shown in Figure 7. The DSC and IR measurements mentioned above showed that annealing at 120℃ induced full desorption of the solvent but resulted in imperfect $\gamma$ form crystals. This may cause the increase of $C_H$' since the solvent desorption site still remains. That is, there are many nanopores between the polymer chains in the mesophase. On the other hand, reorganization from the $\delta$ to the $\gamma$ form may occur perfectly by annealing at 170℃, resulting in a decrease in the number of nanopores. This process may cause the decrease of $C_H$' with annealing time.

From $CO_2$ sorption measurements, annealing at low temperature induces the formation of a mesophase containing many nanopores. The reorganization of the crystalline form and mesophase is represented schematically in Figure 8. One can see many more nanopores in the SPS samples annealed for long times at low temperature.

## Literature Cited


1. Roels, T.; Deberdt, F.; Berghmans, H. *Macromolecules* **1994**, *27*, 6216.
2. Zimba, C. G.; Rabolt, J. F.; English, A. D. *Macromolecules* **1989**, *22*, 2867.

**Table I** Dual-mode sorption parameters of various annealed samples

| sample | $C_H'$ | $k_D$ x$10^2$ | b x $10^3$ |
|---|---|---|---|
| 1A120(APS) | 5.86 | 1.29 | 3.10 |
| 1A120 | 7.19 | 1.10 | 2.97 |
| 100A120 | 13.27 | 0.58 | 1.53 |
| 1A170 | 13.55 | 0.57 | 1.71 |
| 1A170 (slowly-cooled) | 10.40 | 0.53 | 1.74 |
| 5A170 | 8.53 | 0.68 | 2.54 |

$C_H'$ :Langmuir sorption capacity; $cm^3$ STP/$cm^3$ polym.

$k_D$ :Henry's law solubility coefficient; $cm^3$ STP/ ($cm^3$ polym. cmHg).

b :affinity constant of Langmuir site; $cmHg^{-1}$

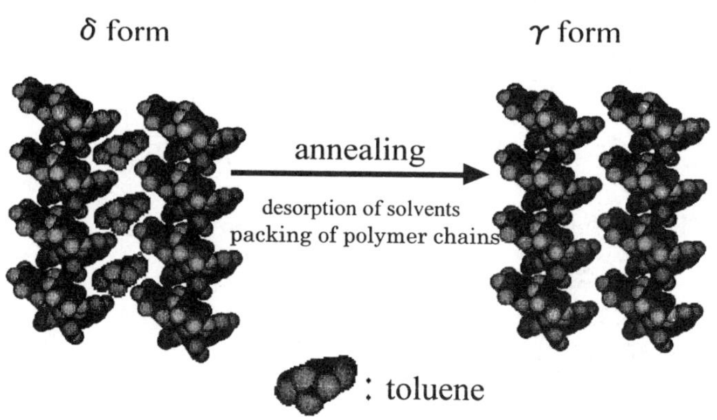


**Figure 7** Schematic representation of reorganization from the $\delta$ to the $\gamma$ form in the of SPS-toluene system.

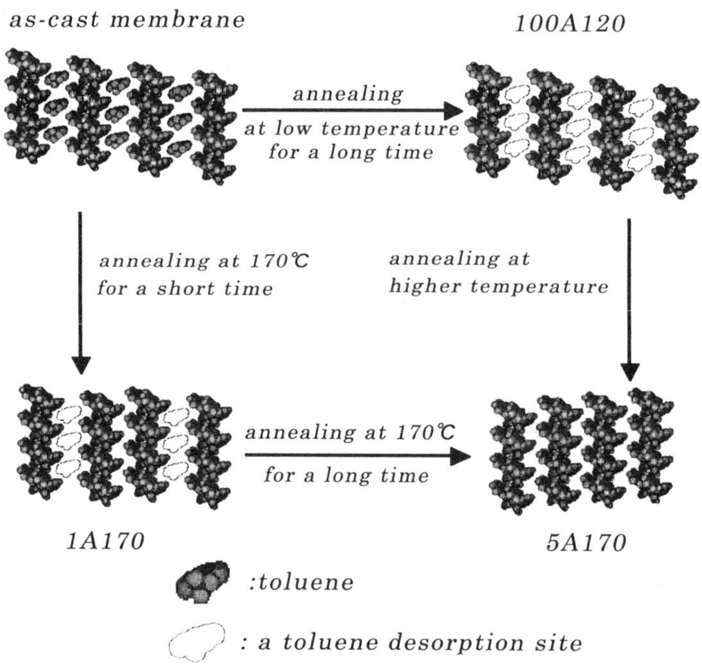


**Figure 8** Schematic representation of the annealed as-cast SPS films.

3. Chatani, Y.; Shimane, Y.; Inoue, Y.; Inagaki, T.; Ishioka, T. *Polymer* **1992**, *33*, 488.
4. Wang, Y. K.; Savage, J. D.; Yang, D.; Hsu, S. L. *Macromolecules* **1992**, *25*, 3659.
5. Crank, J.; Park, G. S. *Diffusion in polymers*; Academic Press: London, 1968.

## Chapter 22

# Measurement of Oxygen Permeation in Mixed Gas Systems: An Alternative to Gas Chromatography

**N. S. M. Stevens and M. E. Rezac**

**School of Chemical Engineering, Georgia Institute of Technology, 778 Atlantic Drive, Atlanta, GA 30332–0100**

An inexpensive, electrochemical probe, sensitive to oxygen partial pressure is used to measure the oxygen composition in a gas permeate stream following gas separation. Separation of air by a barrier material, poly(ethylene carbonate), is shown as an example of probe operation. Separation and calibration techniques demonstrate probe accuracy and reproducibility.

In food packaging, the permeability of oxygen through a barrier material is of utmost importance. Oxygen increases the decomposition rate of the food, and therefore, a packaging material's ability to restrict oxygen transport must be known. In the production of purified nitrogen from air, polymeric membranes are often utilized. Knowledge of both oxygen and nitrogen transport rates and separation selectivity is required for system design. While independent measurement of pure oxygen and pure nitrogen transport rates allow for a prediction of separation efficiency, measurement of a mixture of the two gases may exhibit significant differences. Lack of agreement between the two methods is most apparent when there is plasticization of the polymer matrix by one of the penetrants (*1*) or when the material exhibits time-dependent properties, as in this study. In these cases, simultaneous measurement of the transport rates of both gases is needed.

Measurement of the simultaneous transport rates of two of more gases through a solid is typically achieved by applying a mixture of known composition and pressure to one side of a material and measuring the rate of gas build-up and its composition on the other side. This technique is detailed by Pye and co-workers (*2*). The technical challenge in such a system is accurate measurement of the gas composition at very low pressures (*2,3*). The method most frequently employed to analyze the gas composition of product streams is gas chromatography. Unfortunately, results for air separation are sometimes difficult to interpret due to the inability of typical gas chromatography columns to effectively separate these materials. Further, permeate gas is usually collected under vacuum (*2,3*). Transport of gases which are below atmospheric pressure to the gas chromatograph is challenging (*2,3*). Small leaks in these systems may be difficult to identify, and because the leak gas will consist of air ($O_2$ and $N_2$), a leak will artificially alter results.

An alternative to gas chromatography is an electrochemical oxygen sensor. These inexpensive probes are commonly used in laboratories for atmospheric determination of oxygen. Connection of this probe to a sensitive multimeter allows for probe utilization under vacuum, eliminating the need for transport of gases to a separate analytical system.

## Background and Theory

**Gas Transport.** Transport through a dense polymer film can be characterized by a film's permeability, *P*:

$$P = \frac{\text{flux} * \text{thickness}}{\text{driving force}}. \tag{1}$$

For a mixed gas, the permeability of component A is described by

$$P_A = \frac{J_A * l}{\Delta P_A} \tag{2}$$

where $J_A$ is the flux of component A through the material; $l$ is the thickness of the material; and $\Delta P_A$ is the pressure differential across the material.

The efficiency of the separation, or the ability for one gas to be selectively permeated relative to another, is described by a material's selectivity, $\alpha$. For mixed gases, this component is defined as

$$\alpha_{i/j} = \frac{y_i / y_j}{x_i / x_j} \tag{3}$$

where $y_i$ and $y_j$ are the mole fractions of components i and j in the permeate stream and $x_i$ and $x_j$ are the mole fractions of components i and j in the feed stream.

**Electrochemical Potential.** Electrochemical potential of a half-reaction can be described by the Nernst equation:

$$E = E^\circ - \frac{RT}{nF} \ln \frac{a_r}{a_o} \tag{4}$$

where E is the reduction potential; E° is the standard potential; n is the number of electrons; F is the Faraday constant; T is the temperature; R is the gas constant; and $a_r$ and $a_o$ are the activities of the reducing species and oxidizing species, respectively. The general half-reaction of oxygen is described by

$$O_2 + 4e^- \leftrightarrow 2O^{2-}. \tag{5}$$

The Nernst equation for this half-reaction can be simplified so that

$$E \approx E^\circ - \frac{RT}{nF} \ln P_{O2} \tag{6}$$

where $P_{O2}$ is the partial pressure of oxygen applied to the probe. As the partial pressure of oxygen increases, the reduction potential also increases. For low pressures, the natural log of $P_{O2}$ is almost equal to the pressure, and therefore, potential is linear with respect to the partial pressure of oxygen.

**Current Methods for Composition Analysis.** The most frequently used method for determination of permeate gas composition is gas chromatography *(2-6)*. In addition to the gas chromatograph, other analytical techniques, such as a lone thermal conductivity detector, mass spectrometry, and confinement of the more condensable gas in a series of cold traps, can be employed *(2,7-12)*.

A thermal conductivity detector, commonly utilized on a gas chromatograph, can be used by itself for measurement of both the transport rate and diffusion of one component in a permeate stream *(2,7-9)*. In this technique, a reference gas at a controlled temperature is used to zero the detector *(8,9)*. The detector provides a signal based on changes in the thermal conductivity of the gas stream, physically measured as the heat of dissipation, and at low concentrations, this signal is linear with respect to the partial pressure of the sample gas *(8,9)*. A thermal conductivity detector can be employed in-situ with or without application of a reference sweep gas on the permeate side. The reference gas and sample gas must have large differences in their thermal conductivity for this detection method to be effective. Thus, hydrogen and helium are most commonly used as reference gases *(9)*. Because of this limitation, a thermal conductivity meter is most often used with a gas chromatograph for mixed gas measurements *(2, 8)*.

Because the majority of permeate samples are collected below atmospheric pressure, transport to the gas chromatograph or alternative analytical devices can be a challenge. This transfer is traditionally achieved by use of either an inert carrier gas or by the vacuum trap method. Transport by an inert carrier gas involves a 'sweep' gas which is allowed to flow through the permeate chamber after sufficient sample has been collected. The product gas is 'carried' to the gas chromatograph. The flow rate of the inert gas must be accurately known for calculation of gas permeability coefficients *(2,3)*. Further, the gas chromatograph must be purged with the inert gas prior to sample injection *(3)*.

In the vacuum trap method, the analytical device is separated from the permeate chamber by an isolation valve *(3)*. The analytical chamber containing the gas chromatograph is evacuated as a test is run. After ample product gas is collected, the vacuum source is isolated from the analytical chamber, and the valve separating the analytical and permeate chambers is opened. Product gas flows to the analytical device as the pressure in the two chambers equilibrates. The vacuum-trap method can be very effective. The accuracy of this method is limited by the presence of any leaks in the system.

Although current systems are accurate, they are often complex. The method presented in this study provides a simple and inexpensive solution.

## Experimental Method

**Materials.** Poly(ethylene carbonate), was kindly supplied by PAC Polymers, Incorporated. Reagent grade methylene chloride ($T_{bp}$ = 40°C) was obtained from J. T. Baker (purity 99.8%) and used without further modification. Breathing quality air (measured to be 20.7% oxygen) was purchased from Air Products and was used for gas permeation measurements as received.

Poly(ethylene carbonate), of the trade name QPAC-25, is a copolymer of ethylene oxide and carbon dioxide. This unique polymer has been employed as both a barrier material and a binder. The structure of QPAC-25 is shown in Figure 1.

The glass transition temperature of this material ranges from 10 to 28°C, and the density of this material is 1.42g/cm$^3$ (*13*).

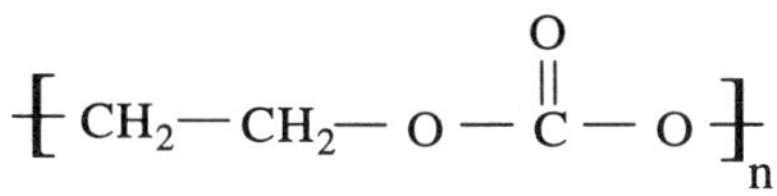

Figure 1. Structure of poly(ethylene carbonate)

**Films.** A solution of the poly(ethylene carbonate) and methylene chloride, 2.2% polymer by weight, was mixed and filtered. Dense films of approximately 30μm thickness were cast from the QPAC-25 solution into rings onto glass plates in a methylene chloride environment. The films remained in a solvent rich environment for twelve hours. The films were then removed from the glass plate and placed under vacuum for an additional twelve hours to remove excess solvent. After this time, a test sample was placed in the permeation test system for analysis.

**Equipment.** The method for measurement of gas permeation in a constant volume-variable pressure apparatus has been previously described (*1*). For this study, a similar system was modified to include an in-system oxygen detector (see Figure 2).

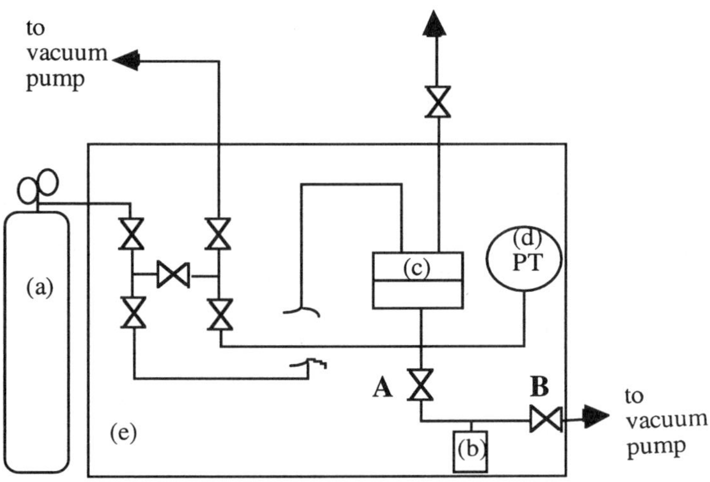


Figure 2. Experimental test system modified with the addition of an oxygen probe: (a) air cylinder, (b) oxygen electrode, (c) test cell holder, (d) pressure transducer, and (e) constant temperature (35°C) box.

An oxygen electrode (VWR OX630) was purchased from VWR and connected via a valve to the permeate reservoir of a permeation test system. The probe has a 95% response in 30 seconds at pressures ranging from 0 to 25 psi. Further, the manufacturer claims that this sensor is not poisoned when in contact with either $H_2S$ or $CO_2$. A Fluke Multimeter, model 45, with resolution of 0.001 mV was employed to report probe potential.

The probe is connected to the permeate chamber of the system. Due to a slow leak rate through the probe itself, the probe was placed in a separate chamber from the permeate. Vacuum was applied to the probe at all times, except during measurement of permeate gas analysis.

**Calibration Procedure.** The probe potential exhibits a linear relationship with the partial pressure of the oxygen for the pressure range explored in this study. Thus, calibration at varying pressures and varying compositions was required for system operation. Additionally, the probe baseline tends to drift with time, as shown in Figure 3. Therefore, the probe must be recalibrated occasionally to ensure accuracy.

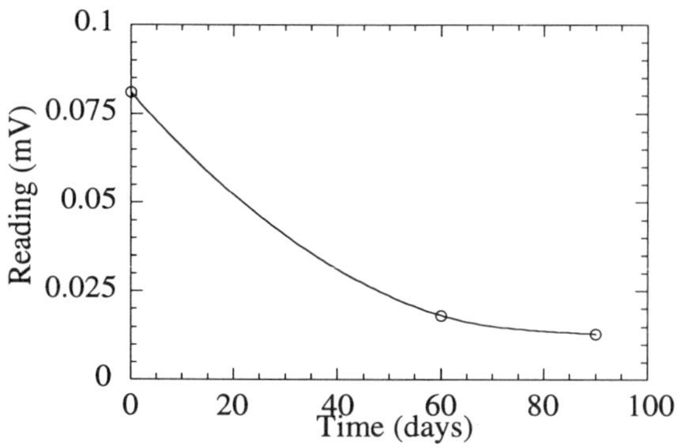


Figure 3. Drift of probe with time for 100% nitrogen at a pressure of 10 torr.

Oxygen/nitrogen calibrations were performed for total pressures of 2 torr, 5 torr, and 10 torr and mixtures ranging from 0% to 100% oxygen. The system was held at a constant temperature of 35.0°C, and the probe was maintained under vacuum. The permeate reservoir was filled with a mixture of nitrogen and oxygen of known composition and pressure. Valve B, as shown in Figure 2, was then closed, to isolate the probe chamber from the vacuum source. Valve A was opened allowing gas to flow into the empty probe chamber. A time period of eight minutes was allowed for both the probe voltage and chamber pressure to equilibrate. The voltage reading on the multimeter and the initial pressure in the permeate system (prior to release into the probe chamber) were recorded. The time period was chosen so that the maximum time for probe and pressure equilibration was allowed but so that the probe leak rate (0.005 torr/min) did not affect the probe or pressure reading. The final system pressure could also be used for calibration, but the initial system pressure was chosen for use in this study for simplicity.

**Experimental Procedure.** Mixed gas transport through a 30μm QPAC-25 film was monitored for a total of seven days using a pressure gradient of 64.7 psia and a temperature of 35.0°C. As the permeate volume was collected, total system permeability was measured by recording permeate pressure increase with time. After 2 to 10 torr of gas was collected as permeate, the feed pressure was released. Valve B, as shown in Figure 2, was closed, and valve A was opened, allowing transport of the permeate gas to the probe chamber. After 8 minutes, the oxygen composition was recorded. Once oxygen analysis was complete, the feed pressure was reapplied, and the permeate and probe chambers were evacuated for preparation for the next run. Because the permeability of gas through the sample changed during each experimental run, the time weighted average of the permeability coefficient was taken.

The technique employed provides an accurate measure of the oxygen composition in a stream. In the experiements completed here, the feed stream was a binary mixture of oxygen and nitrogen. Thus, it was possible to determine the composition of the nitrogen in the permeate stream by difference. If more complex mixtures are evaluated, additional analysis may be necessary.

## Results and Discussion

**System Calibration.** Calibration lines are shown in Figures 4 and 5. A linear relationship exists between the detector reading and the partial pressure of oxygen in the system. Because the meter reading is proportional to the system pressure and the precision of the multimeter is a constant ± 0.001 mV, the accuracy of the probe increases with an increasing system pressure, as detailed in Table 1. Therefore, the larger the pressure in the permeate volume, the more accurate the result, as predicted by the Nernst equation. A linear relationship exists between the detector reading and the partial pressure of oxygen in the system for the low pressures examined in this study.

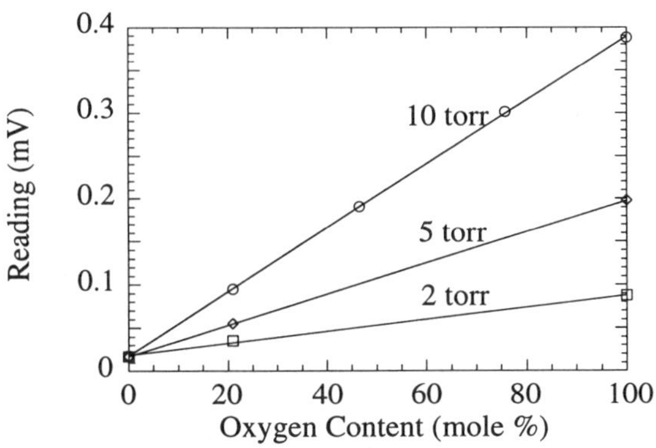


Figure 4: Calibration plot for the oxygen sensor at varying pressures.

Table 1. Increased accuracy with increased permeate pressure.

| Permeate Pressure | Oxygen Accuracy |
|---|---|
| 2 torr | ± 1.4% |
| 5 torr | ± 0.6% |
| 10 torr | ± 0.3% |

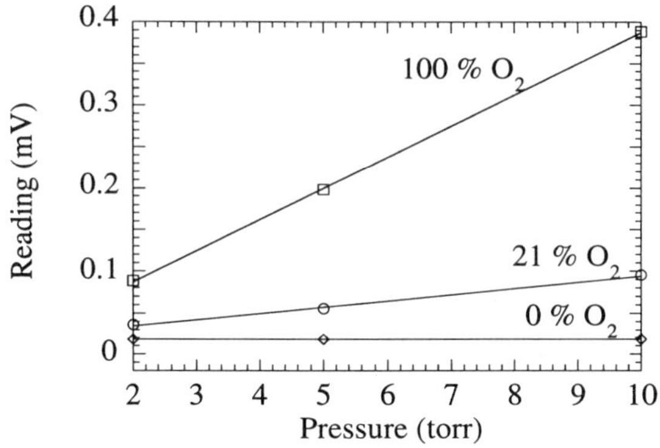


Figure 5. Calibration plot for the oxygen sensor at varying concentrations.

**Transport Through Poly(ethylene carbonate).** The permeability coefficients of oxygen and nitrogen through QPAC-25 are shown in Figures 6. Figure 7 presents the actual selectivity of the polymer under the conditions investigated. Transport varies with time, most possibly due to the slow loss of residual casting solvent remaining in the sample when testing begins. Data were collected at approximately 24 hour intervals. During the first 80 hours, variations in the permeability coefficients are the most significant, introducing the most experimental error. As transport approaches steady-state, experimental error is substantially reduced.

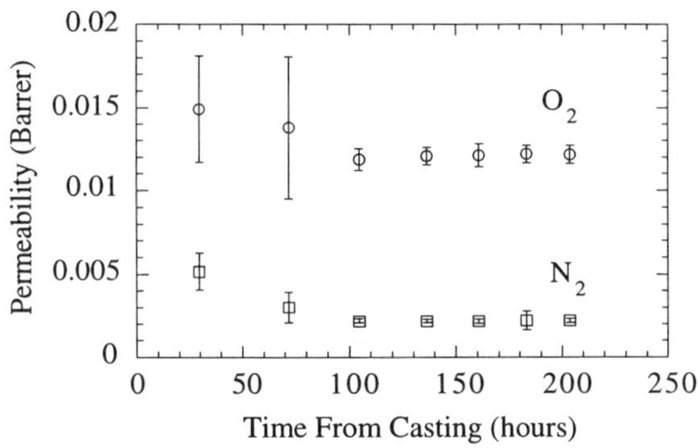


Figure 6. Oxygen and Nitrogen permeability at 35 °C through the poly(ethylene carbonate) film as a function of time.

Accurate and simultaneous oxygen and nitrogen transport measurements are essential for systems such as the QPAC-25 evaluated here. The steady-state transport rate of oxygen and nitrogen through the QPAC-25 (measured after 200 hours drying time) is very slow (0.012 Barrer and 0.002 Barrer, respectively). Further the time-dependent behavior monitored is only apparent during the first 100 hours of evaluation. If pure gas measurements were to be attempted, it may be possible to get 3 to 4 data points during the initial 100 hours. Examination of Figures 6 and 7 clearly indicate that accurate determination of instantaneous selectivity would not be possible from the pure gas data. Thus, in-situ composition measurement is crucial for systems that exhibit time-dependent behavior.

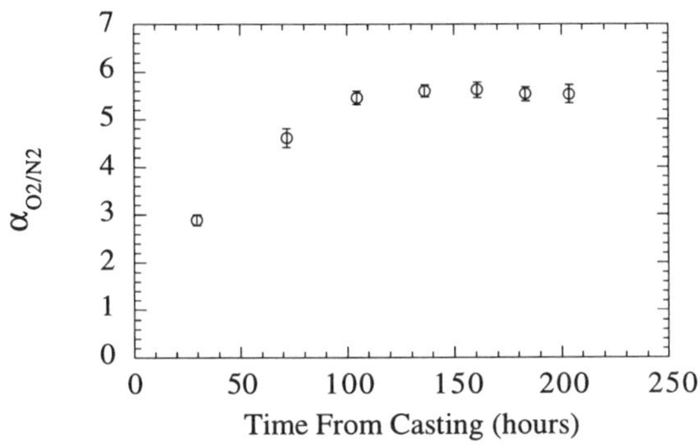


Figure 7. Mixed gas selectivity as a function of time.

After steady state was reached, the pure gas permeabilities of these films were measured. Independent measurement of nitrogen and oxygen resulted in values that were identical to those reported above. Further, these values are consistent with those reported in the product literature for this polymer (*13*).

**Probe Considerations.** Experimental results have shown that although this is a simple technique, several considerations must be made for accurate use. The probe baseline drift with time must be corrected by frequent calibrations, especially during initial sensor service. The probe also loses this baseline when it is removed from vacuum for an extended period of time. This is due to the long time required for the solid probe to return to chemical equilibrium at the low pressure. Return to the original baseline may take up to two weeks, depending on the length of time which the probe was exposed to the atmosphere. In this study, the probe was maintained under vacuum to prevent baseline drift.

The leak rate of the probe chamber prevented the determination of oxygen transport in-situ. If this leak rate could be minimized, instantaneous oxygen measurements would lend themselves to the study of systems as steady-state is approached. The time-lag method for estimation of the diffusion coefficient could be extended to mixed gas systems.

While the electrochemical sensor used in this study is sensitive to oxygen, a different probe would be required for use with other gases. Ternary gas systems require the use of several electrochemical sensors as opposed to one.

## Conclusions

This study has shown the use of an inexpensive oxygen sensor as an alternative to gas chromatography in the analysis of mixed gas streams. This system is simple and provides reproducible data. However, when using an electrochemical sensor, one must be cognizant of probe drift with time and time required for the probe potential to equilibrate. Reduction of the leak rate through the probe system may enable in-situ measurements of oxygen concentration with time.

The sensor utilized in this study is sensitive to oxygen. For expansion of this detection technique a different gas, a probe designed specifically for its detection is required.

## Acknowledgments

The authors would like to acknowledge the donors of the Petroleum Research Fund, administered by the American Chemical Society, for partial support of this research and PAC Polymers, Inc., Greenville, DE, for kindly donating the poly(ethylene carbonate). N. S. M. Stevens acknowledges the Texaco Foundation for financial support through an educational fellowship.

## References

1. Jordon, S.M.; Koros, W. J. *J. Polym. Sci. Polym. Phys. Ed.* **1990,** *28*, 795.
2. Pye, D. G.; Hoehn, H. H.; Panar, M. *J. Memb. Sci.* **1976,** *20,* 287.
3. O'Brien, K. C.; Koros, W. J.; Barbari, T. A.; Sanders, E.S. *J. Appl. Polym. Sci.* **1986,** *20,* 287.
4. Agrawal, J. P.; Sourirajan, S. *J. Appl. Polym. Sci.* **1970,** *14,* 1303.
5. Caskey, T. L. *Mod. Plast.* **1967,** *45,* 447.
6. Chern, R. T.; Koros, W. J.; Yui, B.; Hopfenberg, H. B.; Stannett, V. T. *J. Polym. Sci. Polym. Phys. Ed.* **1984,** *22*, 1061.
7. Pasternak, R. A.; Schimscheimer, J. F.; Heller, J. *J. Polym. Sci.: Part A-2* **1970,** *8,* 467.
8. Yasuda, H.; Rosengren, K. J. *J. Appl. Polym. Sci.* **1970,** *14,* 2839.
9. Ziegl, K. D.; Frensdorff, H. K.; Blair, D. E. *J. Polym. Sci., Part A-2* **1969,** *7,* 809.
10. Adam, S. J.; David, C. E. *23rd International SAMPE Technical Conference* **1991,** 23, 428.
11. Meyer, J. A.; Rogers, C.; Stannett, V.; Szwarc, M. *Tappi* **1957,** *40,* 142.
12. Rogers, C.; Meyer, J. A.; Stannett, V.; Szwarc, M. *Tappi* **1956,** *39,* 737.
13. PAC Polymers, Inc. Properties Sheet **1992**.

# Indexes

# Author Index

# Subject Index

## D

G

## H



## I



## K



## L



## M

N



O



P

**T**



**V**



**W**



**X**



**Y**



**Z**